D0482359

Simplified Design of Structural Steel

Simplified Design of Structural Steel

The Late Harry Parker, M.S.
Formerly Professor of Architectural Construction
University of Pennsylvania

FIFTH EDITION

prepared by

JAMES AMBROSE, M.S.
Professor of Architecture
University of Southern California

A Wiley-Interscience Publication

JOHN WILEY & SONS

New York • Chichester • Brisbane • Toronto • Singapore

Library of Congress Cataloging in Publication Data:

Parker, Harry, 1887–
Simplified design of structural steel.

"A Wiley-Interscience publication."
Includes bibliographical references and index.
1. Building, Iron and steel. 2.Steel, Structural.

I. Ambrose, James E. II. Title.
TA684.P33 1983 624.1′821 83-1180
ISBN 0-471-89766-3

Printed in the United States of America

10 9 8 7 6 5 4 3 2

Preface to the Fifth Edition

The publication of this edition presents the opportunity for yet another generation to use this enduringly popular work by Professor Parker. The purpose of the book remains essentially unchanged from that described in the preface to the first edition. If anything, the need for a simplified approach to the subject has become even more pressing as the complexity of codes and practices increases. In the preparation of this edition the effort was made to adhere as closely as possible to the spirit of the original work while bringing the content into reasonable agreement with present needs and practices. The task of preparing this edition was made immeasurably easier by the excellent work done by Harold D. Hauf on the preceding edition.

Two major modifications have been undertaken in this edition. The first consists of the revisions made necessary by the changes presented in the 1978 edition of the standard specification of the American Institute of Steel Construction and the 1980 edition of the *Manual of Steel Construction*. The second modification consists of the incorporation of computations and data in metric (SI) units. Although the main body of the work remains in traditional English (now called US) units, equivalent values in SI units have been given whenever practicable.

I am indebted to the American Institute of Steel Construction (AISC) for their permission to reprint and adapt an extensive amount of material from their publication *Manual of Steel Construction*, 8th ed. I am also grateful to the International Conference of Building Officials, publishers of the 1982 edition of the *Uniform Building Code*, for permission to quote from that publication and to the Steel Deck Institute and the Steel Joist Institute for permission to use their publications.

Without a doubt the quality of this publication is due in large part to the excellent work of the editors and production staff at John Wiley and Sons, whose support and consistency I tend to take for granted.

Finally, I must acknowledge the patience of my colleagues, the support of my family, and the valuable assistance given in the preparation of the manuscript by my wife Peggy and my daughter Julie.

JAMES AMBROSE

Westlake Village, California
March 1983

Preface to the First Edition

"Simplified Design of Structural Steel" is the fourth of a series of elementary books dealing with the design of structural members used in the construction of buildings. The first volume, "Simplified Engineering for Architects and Builders," discussed rather briefly the design of structural members of timber, steel, and reinforced concrete. The primary objective of the first volume was to present the basic principles of design for younger men having no preliminary training in engineering. The book being elementary in character and the scope limited, many important subjects were necessarily omitted.

The present volume treats of the design of the most common structural steel members that occur in building construction. The solution of many structural problems is difficult and involved but it is surprising, on investigation, how readily many of the seemingly difficult problems may be solved. The author has endeavored to show how the application of the basic principles of mechanics simplifies the problems and leads directly to a solution. Using tables and formulas blindly is a dangerous procedure; they can only be used with safety when there is a clear understanding of the underlying principles upon which the tables or formulas are based. This book deals principally in the practical application of engineering principles and formulas and in the design of structural members. The derivations of the most commonly used formulas are given in order that the reader may comprehend fully why certain formulas are appropriate in the solution of specific problems.

In preparing material for this book the author has assumed that the reader is unfamiliar with the subject. Consequently the discussions advance by easy stages, beginning with problems relating to simple direct stresses and continuing to the more involved examples. Most of the fundamental principles of mechanics are reviewed and, in general, the only preparation needed is a knowledge of arithmetic and high school algebra.

The text has been arranged for use in the classroom as well as for home study. The tables essential to structural engineering have been included so that no additional reference books are required.

In addition to discussions and explanations of design procedure, it has been found that the solution of practical examples adds greatly to the value of a book of this character. Consequently, a great portion of the text consists of the solution of illustrative examples. The examples are followed by problems to be solved by the student.

The author is deeply grateful to the American Institute of Steel Construction for its kindness in granting permission to reproduce tables and data from its manual "Steel Construction." In general, the American Institute of Steel Construction specifications have been followed in the preparation of this book. Thanks are extended also to the American Welding Society for permission to reproduce data from its "Code for Arc and Gas Welding in Building Construction," and to The Lincoln Electric Company of Cleveland, Ohio, publishers of "Procedure Handbook of Arc Welding Design and Practice." This company has graciously permitted the reproduction of the data and design procedure from its excellent and comprehensive volume. The cooperation of the societies and company mentioned above is greatly appreciated; without such cooperation a book of this character would not be possible.

The author proposes no new methods of design nor short cuts of questionable value. Instead, he has endeavored to present concise and clear explanations of the present-day design methods with the hope that the reader may obtain a foundation of sound principles of structural engineering.

HARRY PARKER

High Hollow, Southampton, Pa.
March 1945

Contents

1

Introduction

1-1. Sources of Design Information

The most widely used source of information for the design of steel structures is the *Manual of Steel Construction,* published by the American Institute of Steel Construction (Ref. 1). This book is commonly referred to as the AISC Manual and we refer to it as such in this book. The current (8th) edition, copyrighted in 1980, contains the 1978 *Specification for the Design, Fabrication, and Erection of Structural Steel for Buildings*. This specification, in its several successive editions, has been generally adopted for reference by code-enforcing agencies in the United States. When referring to this document we call it the AISC Specification.

In addition to the specification and an accompanying extensive commentary, the AISC Manual contains essential data that pertain to the standard structural steel products used for the construction of building structures. There is also considerable general information and numerous design aids to assist the working structural designer. Although we have reprinted or adapted all of the information from the AISC Manual that is required for the work in this book, it is recommended that the serious reader acquire a personal copy of this highly useful reference.

Although the AISC is the principal service organization in the area of steel construction, many other industrial and professional organizations provide material for the designer. The American Society for Testing and Materials (ASTM) establishes widely used standard specifications for types of steel, for welding and connector materials, and for various production and fabrication processes. Standard grades of steel and other materials are commonly referred to by short versions of their ASTM designation codes. Other organizations of note are the Steel Deck Institute (SDI), the Steel Joist Institute (SJI), and the American Iron and Steel Institute (AISI). Some of the materials distributed by these organizations are listed in the references at the back of this book.

Readers who wish to pursue topics in the area of design of steel structures beyond the level of development presented in this book are advised to seek the latest editions of one of the books used as a text on the topic in schools of civil engineering. For a general reference an excellent book is *Steel Structures: Design and Behavior* by Salmon and Johnson (Ref. 3).

1-2. Nomenclature

The standard symbols (called the general nomenclature) used in steel design work are those established in the AISC Specification. Although an attempt is being made to standardize symbols for structural work, some specialized nomenclature remains for each of the areas of design: soils, wood, steel, concrete, and masonry. The following is a list of symbols used in this book; it is an abridged version of a more extensive list in the AISC Manual:

A Cross-sectional area (sq in.)
Gross area of an axially loaded compression member (sq in.)

A_e Effective net area of an axially loaded tension member (sq in.)

A_n Net area of an axially loaded tension member (sq in.)

C_c Column slenderness ratio separating elastic and inelastic buckling

E Modulus of elasticity of steel (29,000 ksi)

F_a Axial compressive stress permitted in a prismatic member in the absence of bending moment (ksi)

F_b Bending stress permitted in a prismatic member in the absence of axial force (ksi)

F'_e Euler stress for a prismatic member divided by factor of safety (ksi)

F_p Allowable bearing stress (ksi)

F_t Allowable axial tensile stress (ksi)

F_u Specified minimum tensile strength of the type of steel or fastener being used (ksi)

F_v Allowable shear stress (ksi)

F_y Specified minimum yield stress of the type of steel being used (ksi). As used in this Manual, "yield stress" denotes either the specified minimum yield point (for those steels that have a yield point) or specified minimum yield strength (for those steels that have no yield point)

F'_y The theoretical maximum yield stress (ksi) based on the width–thickness ratio of one-half the unstiffened compression flange, beyond which a particular shape is not "compact." See AISC Specification Sect. 1.5.1.4.1.2.

$$= \left[\frac{65}{b_f/2t_f}\right]^2$$

F'''_y The theoretical maximum yield stress (ksi) based on the depth–thickness ratio of the web below which a particular shape may be considered "compact" for any condition of combined bending and axial stresses. See AISC Specification Sect. 1.5.1.4.1.4.

$$= \left[\frac{257}{d/t_w}\right]^2$$

I Moment of inertia of a section (in.^{4})

I_x Moment of inertia of a section about the X–X axis (in.^{4})

I_y Moment of inertia of a section about the Y–Y axis (in.^{4})

J Torsional constant of a cross section (in.^{4})

K Effective length factor for a prismatic member

L Span length (ft)
Length of connection angles (in.)

L_c Maximum unbraced length of the compression flange at which the allowable bending stress may be taken at $0.66F_y$ or as determined by AISC Specification Formula (1.5-5a) or Formula (1.5-5b), when applicable (ft)
Unsupported length of a column section (ft)

L_u Maximum unbraced length of the compression flange at which the allowable bending stress may be taken at $0.6F_y$ (ft)

M Moment (k-ft)

M_D Moment produced by dead load

M_L Moment produced by live load

M_p Plastic moment (k-ft)

M_R Beam resisting moment (k-ft)

N Length of base plate (in.)
Length of bearing of applied load (in.)

N_e Length at end bearing to develop maximum web shear (in.)

P Applied load (kips)
Force transmitted by a fastener (k)

P_e Euler buckling load (k)

Q_s Axial stress reduction factor where width–thickness ratio of unstiffened elements exceeds limiting value given in Specification Sect. 1.9.1.2, Appendix C

R Maximum end reaction for $3\frac{1}{2}$ in. of bearing (kips)
Reaction or concentrated load applied to beam or girder (kips)
Radius (in.)

R_i Increase in reaction (R) in kips for each additional inch of bearing

S Elastic section modules (in.^{3})

S_x Elastic section modulus about the X–X (major) axis (in.^{3})
S_y Elastic section modulus about the Y–Y (minor) axis (in.^{3})
V Maximum permissible web shear (k)
Statical shear on beam (k)
Z Plastic section modulus (in.^{3})
Z_x Plastic section modulus with respect to the major (X–X) axis (in.^{3})
Z_y Plastic section modulus with respect to the minor (Y–Y) axis (in.^{3})
b_f Flange width of rolled beam or plate girder (in.)
d Depth of column, beam, or girder (in.)
Nominal diameter of a fastener (in.)
e_o Distance from outside face of web to the shear center of a channel section (in.)
f_a Computed axial stress (ksi)
f_b Computed bending stress (ksi)
f'_c Specified compression strength of concrete at 28 days (ksi)
f_p Actual bearing pressure on support (ksi)
f_t Computed tensile stress (ksi)
f_v Computed shear stress (ksi)
g Transverse spacing locating fastener gage lines (in.)
k Distance from outer face of flange to web toe of fillet of rolled shape or equivalent distance on welded section (in.)
l For beams, distance between cross sections braced against twist or lateral displacement of the compression flange (in.)
For columns, actual unbraced length of member (in.)
Length of weld (in.)
m Cantilever dimension of base plate (in.)
n Number of fasteners in one vertical row
Cantilever dimension of base plate (in.)
r Governing radius of gyration (in.)
r_x Radius of gyration with respect to the X–X axis (in.)
r_y Radius of gyration with respect to the Y–Y axis (in.)

s Longitudinal center-to-center spacing (pitch) of any two consecutive holes (in.)

t Girder, beam, or column web thickness (in.)
Thickness of a connected part (in.)
Wall thickness of a tubular member (in.)
Angle thickness (in.)

t_f Flange thickness (in.)

t_w Web thickness (in.)

x Subscript relating symbol to strong axis bending

y Subscript relating symbol to weak axis bending

Δ Beam deflection (in.).

1-3. Units of Measurement

At the time of preparation of this edition the building industry in the United States is still in a state of confused transition from the use of English units (feet, pounds, etc.) and the new metric-based system referred to as the SI units (for Systeme Internationalle). Although a complete phase-over to SI units seems inevitable, at the time of this writing the construction-materials and products suppliers in the United States are still resisting it. Consequently the AISC Manual and most building codes and other widely used references are still in the old units. (The old system is now more appropriately called the United States system, because England no longer uses it!) Although it results in some degree of clumsiness in the work, we have chosen to give the data and computations in this book in both units as much as is practicable. The technique is generally to perform the work in US units and immediately follow it with the equivalent work in SI units enclosed in brackets [thus] for separation and identity.

Table 1-1 lists the standard units of measurement in the US system with the abbreviations used in this work and a description of the type of use in structural work. In similar form Table 1-2 gives the corresponding units in the SI system. For more ready access the conversion units used in shifting from one system to the other appear on the inside back cover of this book.

TABLE 1-1. Units of Measurement: US System

Name of unit	Abbreviation	Use
Length		
Foot	ft	large dimensions, building plans, beam spans
Inch	in.	small dimensions, size of member cross sections
Area		
Square feet	ft^2	large areas
Square inches	$in.^2$	small areas, properties of cross sections
Volume		
Cubic feet	ft^3	large volumes, quantities of materials
Cubic inches	in^3	small volumes
Force, mass		
Pound	lb	specific weight, force, load
Kip	k	1000 pounds
Pounds per foot	lb/ft	linear load (as on a beam)
Kips per foot	k/ft	linear load (as on a beam)
Pounds per square foot	lb/ft^2, psf	distributed load on a surface
Kips per square foot	k/ft^2, ksf	distributed load on a surface
Pounds per cubic foot	lb/ft^3, pcf	relative density, weight
Moment		
foot-pounds	ft-lb	rotational or bending moment
inch-pounds	in-lb	rotational or bending moment
kip-feet	k-ft	rotational or bending moment
kip-inches	k-in	rotational or bending moment
Stress		
Pounds per square foot	lb/ft^2, psf	soil pressure
Pounds per square inch	$lb/in.^2$, psi	stresses in structures
Kips per square foot	k/ft^2, ksf	soil pressure
Kips per square inch	$k/in.^2$, ksi	stresses in structures
Temperature		
Degree Fahrenheit	°F	temperature

TABLE 1-2. Units of Measurement: SI System

Name of unit	Abbreviation	Use
Length		
Meter	m	large dimensions, building plans, beam spans
Millimeter	mm	small dimensions, size of member cross sections
Area		
Square meters	m^2	large areas
Square millimeters	mm^2	small areas, properties of cross sections
Volume		
Cubic meters	m^3	large volumes
Cubic millimeters	mm^3	small volumes
Mass		
Kilogram	kg	mass of materials (equivalent to weight in US system)
Kilograms per cubic meter	kg/m^3	density
Force: (load on structures)		
Newton	N	force or load
Kilonewton	kN	1000 Newtons
Stress		
Pascal	Pa	stress or pressure (one pascal = one N/m^2)
Kilopascal	kPa	1000 Pascals
Megapascal	MPa	1,000,000 Pascals
Gigapascal	GPa	1,000,000,000 Pascals
Temperature		
Degree Celcius	°C	temperature

1-4. Computations

In most professional design firms structural computations are now commonly done with computers, particularly when the work is complex or repetitive. Anyone aspiring to participation in professional design work is advised to acquire the background and experience necessary to the application of computer-aided techniques. The computational work in this book is simple and can be performed easily with a pocket calculator. The reader who has not already done so is advised to obtain one. The "pocket slide rule" type with 8-digit capacity is quite sufficient.

Structural computations can for the most part be rounded off. Accuracy beyond the third place is seldom significant, and this is the level used in this work. In some examples more accuracy is carried in early stages of the computation to ensure the desired degree in the final answer. All the work in this book, however, was performed on an eight-digit pocket calculator.

2

Structural Elements of Steel

A wide variety of the elements used in building construction are made from steel. The work in this book deals only with the ordinary components of steel-structure production: rolled structural shapes, fabricated structural shapes, steel connectors, cold-formed elements, and prefabricated structural joists.

2-1. Rolled Structural Shapes

The products of the steel rolling mills used as beams, columns, and other structural members are known as *sections* or *shapes* and their designations are related to the profiles of their cross sections. American Standard I-beams (Fig. 2-1*a*) were the first beam sections rolled in the United States and are currently produced in sizes of 3 to 24 in. in depth. The wide flange shapes (Fig. 2-1*b*) are a modification of the I cross section and are characterized by parallel flange surfaces as contrasted with the tapered inside flange surfaces of Standard I-beams; they are available in

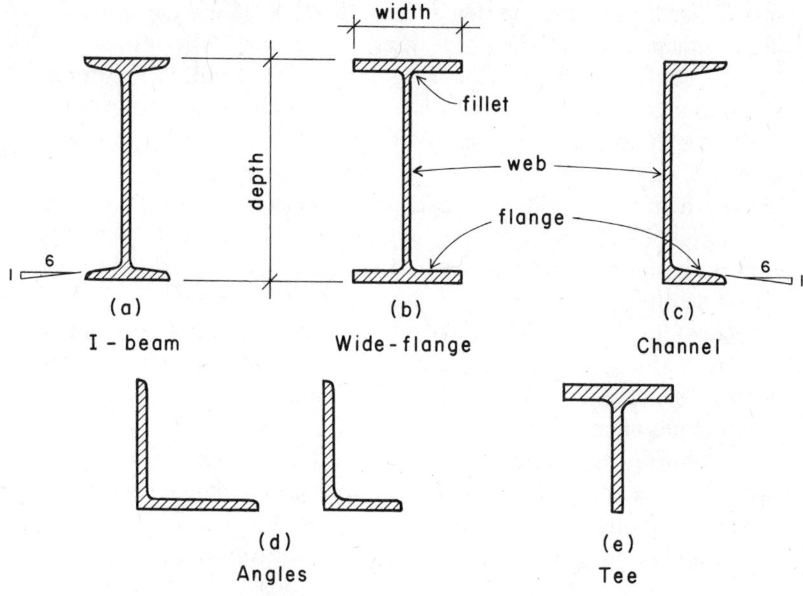

FIGURE 2-1. Rolled structural shapes.

depths of 4 to 36 in. In addition to the Standard I and wide flange sections, the structural steel shapes most commonly used in building construction are channels, angles, tees, plates, and bars. The tables in Appendix A list the dimensions and weights of some of these shapes with other properties that are identified and discussed in Chapters 3 through 9. Complete tables of structural shapes are given in the AISC *Manual of Steel Construction* (Ref. 1).

2-2. Wide Flange Shapes

In general, wide flange shapes have greater flange widths and relatively thinner webs than Standard I-beams; and, as noted above, the inner faces of the flanges are parallel to the outer faces. These sections are identified by the alphabetical symbol W, followed by the *nominal* depth in inches and the weight in pounds per linear foot. Thus the designation W 12 × 26 indicates a wide flange shape of nominal 12-in. depth, weighing 26 lb per lin ft.

The actual depths of wide flange shapes vary within the nominal depth groupings. By reference to Appendix A, it is found that a W 12 × 26 has an actual depth of 12.22 in., whereas the depth of a W 12 × 35 is 12.50 in. This is a result of the rolling process during manufacture in which the cross-sectional areas of wide flange shapes are increased by spreading the rolls both vertically and horizontally. The additional material is thereby added to the cross section by increasing flange and web thickness as well as flange width (Fig. 2-1*b*). The resulting higher percentage of material in the flanges makes wide flange shapes more efficient structurally than Standard I-beams. A wide variety of weights is available within each nominal depth group.

In addition to shapes with profiles similar to the W 12 × 26, which has a flange width of 6.490 in., many wide flange shapes are rolled with flange widths approximately equal to their depths. The resulting H configurations of these cross sections are much more suitable for use as columns than the I profiles. By reference to Appendix A it is found that the following shapes, among others, fall into this category: W 14 × 90, W 12 × 65, W 10 × 60, and W 8 × 40. It is recommended that the reader compare these shapes with others listed in their respective nominal depth groups in order to become familiar with the variety of geometrical relationships.

2-3. Standard I-Beams

American Standard I-beams are identified by the alphabetical symbol S, the designation S 12 × 35 indicating a Standard shape 12 in. deep weighing 35 lb per lin ft. In Appendix A it is shown that this section has an *actual* depth of 12 in., a flange width of 5.078 in., and a cross-sectional area of 10.3 sq in. Unlike wide flange sections, Standard I-beams in a given depth group have uniform depths, and shapes of greater cross-sectional area are made by spreading the rolls in one direction only. Thus the depth remains constant, whereas the width of flange and thickness of web are increased. A comparison of sections S 12 × 35 and S 12 × 50 in Appendix A will clarify these relationships. Since a bar of steel 1 sq in. in cross section and 1 ft long weighs approximately 3.4 lb, the weight per linear foot of any structural shape is 3.4 times the cross-sectional area.

All Standard I-beams have a slope on the inside faces of the flanges of $16\frac{2}{3}\%$, or 1 in 6. In general, Standard I-beams are not so efficient structurally as wide flange sections and consequently are not so widely used. Also, the variety available is not nearly so large as that for wide flange shapes. Characteristics that may favor the use of American Standard I-beams in any particular situation are constant depth, narrow flanges, and thicker webs.

2-4. Standard Channels

The profile of an American Standard channel is shown in Fig. 2-1*c*. These shapes are identified by the alphabetical symbol C. The designation C 10 × 20 indicates a Standard channel 10 in. deep and weighing 20 lb per lin ft. Appendix A shows that this section has an area of 5.88 sq in., a flange width of 2.739 in., and a web thickness of 0.379 in. Like the Standard I-beams, the depth of a particular group remains constant and the cross-sectional area is increased by spreading the rolls to increase flange width and web thickness. Because of their tendency to buckle when used independently as beams or columns, channels require lateral support or bracing. They are generally used as elements of built-up sections such as columns and lintels. However, the absence of a flange on one side makes channels particularly suitable for framing around floor openings.

2-5. Angles

Structural angles are rolled sections in the shape of the letter L. Appendix A gives dimensions, weights, and other properties of equal and unequal leg angles. Both legs of an angle have the same thickness.

Angles are designated by the alphabetical symbol L, followed by the dimensions of the legs and their thickness. Thus the designation L 4 × 4 × $\frac{1}{2}$ indicates an equal leg angle with 4-in. legs, $\frac{1}{2}$ in. thick. By reference to Appendix A it is found that this section weighs 12.8 lb per lin ft and has an area of 3.75 sq in. Similarly, the designation L 5 × 3$\frac{1}{2}$ × $\frac{1}{2}$ indicates an unequal leg angle with one 5-in. and one 3$\frac{1}{2}$-in. leg, both $\frac{1}{2}$ in. thick. Appendix A shows that this angle weighs 13.6 lb per lin ft and has an area of 4 sq in. To change the weight and area of an angle of a given leg length the thickness of each leg is increased, as shown in Fig. 1-3*a*. Thus, if the leg thickness of the L 5 × 3$\frac{1}{2}$ × $\frac{1}{2}$ is increased to $\frac{5}{8}$ in., Appendix A shows that the resulting L 5 × 3$\frac{1}{2}$ × $\frac{5}{8}$ has a weight of 16.8 lb per lin ft and an area of 4.92 sq in. It should be noted that this method of spreading the rolls changes the leg lengths slightly.

Single angles are often used as lintels and pairs of angles, as members of light steel trusses. Angles were formerly used as elements of built-up sections such as plate girders and heavy columns, but the advent of the heavier wide-flange shapes has largely eliminated their usefulness for this purpose. Short lengths of angles are common connecting members for beams and columns.

2-6. Structural Tees

A structural tee is made by splitting the web of a wide flange shape (Fig. 2-1*f*) or a Standard I-beam. The cut, normally made along the center of the web, produces tees with a stem depth equal to half the depth of the original section. Structural tees cut from wide flange shapes are identified by the symbol WT; those cut from Standard I shapes, by ST. The designation WT 6 × 53 indicates a structural tee with a 6-in. depth and a weight of 53 lb per lin ft. This shape is produced by splitting a W 12 × 106 shape. Similarly, ST 9 × 35 designates a structural tee 9 in. deep, weighing 35 lb per lin ft, and cut from a S 18 × 70. Tables of properties of these shapes appear in Appendix A. Structural tees are used for the chord members of welded steel trusses and for the flanges in certain types of plate girder.

2-7. Plates and Bars

Plates and bars are made in many different sizes and are available in all the structural steel specifications listed in Table 4-1.

Flat steel for structural use is generally classified as follows:

Bars: 6 in. or less in width, 0.203 in. and more in thickness
6 to 8 in. in width, 0.230 in. and more in thickness

Plates: More than 8 in. in width, 0.230 in. and more in thickness
More than 48 in. in width, 0.180 in. and more in thickness

Bars are available in varying widths and in virtually any required thickness and length. The usual practice is to specify bars in increments of $\frac{1}{4}$ in. for widths and $\frac{1}{8}$ in. in thickness.

For plates the preferred increments for width and thickness are the following:

Widths: Vary by even inches, although smaller increments are obtainable

Thickness: $\frac{1}{32}$ in. increments up to $\frac{1}{2}$ in.
$\frac{1}{16}$ in. increments of more than $\frac{1}{2}$ in. to 2 in.
$\frac{1}{8}$ in. increments of more than 2 in. to 6 in.
$\frac{1}{4}$ in. increments of more than 6 in.

The standard dimensional sequence when describing steel plate is

$$\text{thickness} \times \text{width} \times \text{length}$$

All dimensions are given in inches, fractions of an inch, or decimals of an inch.

Column base plates and beam bearing plates may be obtained in the widths and thicknesses noted. For the design of column base plates and beam bearing plates see Arts. 11-13 and 9-7, respectively.

2-8. Designations for Structural Steel Elements

As noted in Arts. 2-2 and 2-3, wide flange shapes are identified by the symbol W, American Standard beam shapes, by S. It was also

TABLE 2-1. Standard Designations for Structural Steel Elements

Type of element	Designation
Wide flange shapes	W 12 × 27
American Standard Beams	S 12 × 35
Miscellaneous shapes	M 8 × 18.5
American Standard channels	C 10 × 20
Miscellaneous channels	MC 12 × 45
Angles—equal legs	L 4 × 4 × $\frac{1}{2}$
Angles—unequal legs	L 5 × $3\frac{1}{2}$ × $\frac{1}{2}$
Structural tees—cut from wide flange shapes	WT 6 × 53
Structural tees—cut from American Standard beams	ST 9 × 35
Structural tees—cut from miscellaneous shapes	MT 4 × 9.25
Plate	PL $\frac{1}{2}$ × 12
Structural tubing: square	TS 4 × 4 × 0.375
Pipe	Pipe 4 Std.

pointed out that W shapes have essentially parallel flange surfaces, whereas S shapes have a slope of approximately $16\frac{2}{3}\%$ on the inner flange faces. A third designation, M shapes, covers miscellaneous shapes that cannot be classified as W or S; these shapes have various slopes on their inner flange surfaces and many of them are of only limited availability. Similarly, some rolled channels cannot be classified as C shapes. These are designated by the symbol MC.

Table 2-1 lists the standard designations used for rolled shapes, formed rectangular tubing, and round pipe.

2-9. Cold-Formed Steel Products

Many structural elements are formed from sheet steel. Elements formed by the rolling process must be heat-softened, whereas those produced from sheet steel are ordinarily made without heating the steel; thus the common description for these elements is *cold-formed*. Because they are typically formed from very thin sheets, they are also referred to as *light-gage* steel products.

Figure 2-2 illustrates the cross sections of some of these products. Large corrugated or fluted sheets are in wide use for wall paneling and for structural decks for roofs and floors. Use of these elements for floor decking is discussed in Chapter 10. These products are made by a number of manufacturers and information regarding their structural properties may be obtained directly from the m. General information on structural decks may also be obtained from the Steel Deck Institute (see Refs. 8 and 9).

Cold-formed shapes range from the simple L, C, U, etc., to the special forms produced for various construction systems. Structures for some buildings may be almost entirely comprised of cold-formed products. Several manufacturers produce patented kits of these components for the formation of predesigned, packaged building structures. The design of cold-formed elements is described in the *Cold-Formed Steel Design Manual*, published by the American Iron and Steel Institute (Ref. 7).

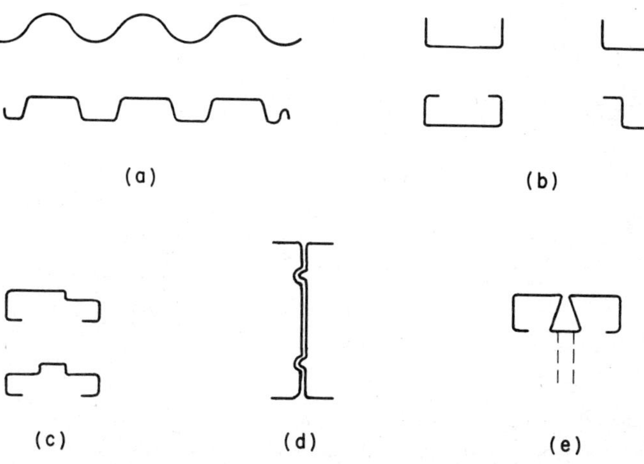

FIGURE 2-2. Cold-formed structural shapes.

2-10. Fabricated Structural Joists

A number of special products are formed of both hot-rolled and cold-formed elements for use as structural members in buildings. Open-web steel joists consist of prefabricated, light steel trusses. For short spans and light loads a common design is that shown in Fig. 2-3*a* in which the web consists of a single, continuous bent steel rod and the chords of steel rods or cold-formed elements. For larger spans or heavier loads the forms more closely resemble those for ordinary light steel trusses; single angles, double angles, and structural tees constitute the truss members. Open-web joints for floor framing are discussed in Art. 10-9.

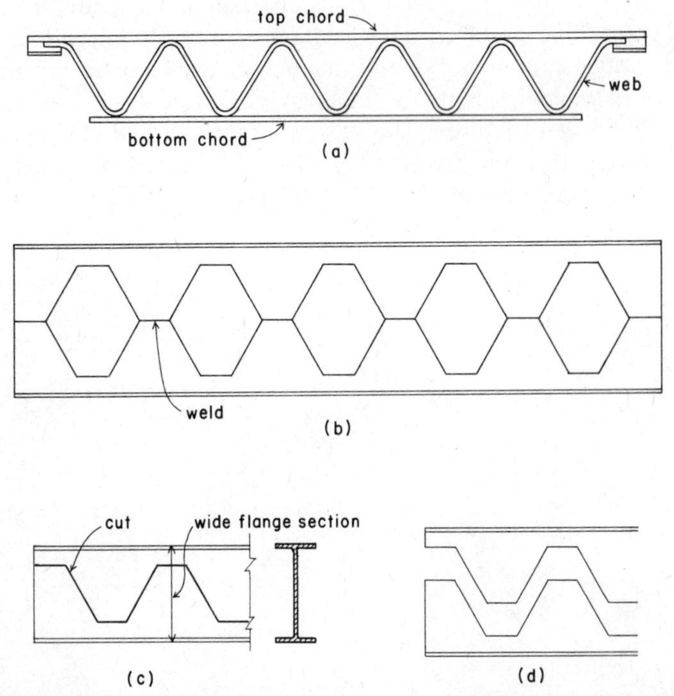

FIGURE 2-3. Fabricated structural joists: (*a*) open-web steel joist; (*b*) castelated steel joist; (*c*) and (*d*) cutting the parts for a castelated joist from a wide-flange shape.

Another type of fabricated joist is shown in Fig. 2-3*b*. This member is formed from standard rolled shapes by cutting the web in a zigzag fashion. The resulting product has a greatly reduced weight-to-depth ratio when compared with the lightest of the rolled shapes.

2-11. Steel Connectors

Connection of structural steel members that consist of rolled elements is typically achieved by direct welding or by steel rivets or bolts. Riveting for building structures has become generally obsolete in favor of high-strength bolts. The design of bolted connections is discussed in Chapter 12; simple welded connections are discussed in Chapter 13. In general, welding is preferred for shop fabrication and bolting for field connections.

Thin elements of cold-formed steel may be attached by welding, bolting, or by sheet metal screws. Thin deck and wall paneling elements are sometimes attached to one another by simple interlocking at their abutting edges; the interlocked parts are sometimes folded or crimped to give further security to the connection.

3

Stresses and Strains

As a prelude to discussing the strength and behavior of steel elements under load, it would be well to establish a clear understanding of the concept of *unit stress*. Throughout this book many terms are used with which the reader may or may not be familiar, depending on the extent of experience in structural work. In any event, it is important that these terms and the concept to which they apply be understood precisely.

3-1. Direct Stress

Stresses are the means by which a body develops internal resistance to external forces. The hanger bar shown in Fig. 3-1*a* supports a suspended load P that acts along the vertical axis of the bar. The load constitutes an external force that tends to stretch the bar and the bar resists the tendency to elongate by developing an internal tensile force equal in magnitude to the external force. This internal force is developed by stresses in the material of the bar. Under this condition of axial loading the tensile stress produced is called *direct stress*.

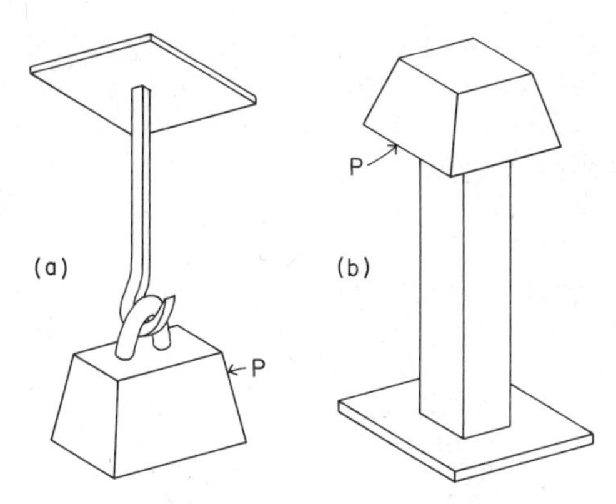

FIGURE 3-1. Direct stress.

A characteristic of direct stress is that the internal force may be assumed to be evenly distributed over the cross sectional area of the body under stress. Thus, if the load P in Fig. 3-1a is 30 kips [133 kN] and the cross-sectional area of the hanger bar is 2 square inches (written in.^{2}) [1290 mm^2], each unit of area is stressed to $30 \div 2 = 15$ kips per square inch (written ksi) or 15,000 pounds per square inch (psi) [103 MPa]. This tensile stress per unit area is called a *unit stress* to distinguish it from the total internal force of 30 kips. By calling the load or external force P, the area of the bar's cross-section A, and the unit stress f the fundamental relationship governing direct stress may be stated as

$$f = \frac{P}{A} \quad \text{or} \quad P = f \times A \quad \text{or} \quad A = \frac{P}{f}$$

When using this relationship remember the assumptions on which it is based: the loading is axial and the stresses are evenly (uniformly) distributed over the cross section. Note also that if any two of the quantities are known the third may be found.

The first form of the direct stress relationship gives the computed stress when P and A are known. The second form gives the allowable load when the allowable stress and the bar's area are known. The third form gives the required cross-sectional area when the load to be carried and the allowable stress are known.

The load P on the short block B exerts an axial force that tends to shorten the block (see Fig. 3-1*b*). This external force is resisted by an internal compressive force equal to P. Again the unit compressive stress is expressed by the direct stress formula $f = P/A$.

We may define a unit stress as an internal resistance per unit area that results from an external force. To be exact we should also consider the weight of the member. In the case of the hanger the unit tensile stress immediately above the hook would be slightly higher than that at a point near the ceiling because of the weight of the bar. Similarly, the unit compressive stress at the base of the block would be slightly greater than that at a cross section just under the load. The relative significance of this variation depends on the ratio of the magnitude of the load to the weight of the structural member. Because of the relatively high strength of steel, load-carrying capacities of steel structures usually greatly exceed the weight of the structural members.

3-2. Kinds of Stress

The three basic kinds of stress are *tension, compression,* and *shear*. Tension tends to stretch structural members, compression tends to shorten them, and shear tends to make parts of a structure slide past one another. Direct tensile and compressive stresses act at right angles to the cross sections of the members considered but shearing stress acts parallel to the cross section. This is illustrated in Fig. 3-2, in which two steel bars are connected by a bolt. Under the action of the forces P there is a tendency for the bolt to fail by shearing at the plane of contact between the two bars (Fig. 3-2*c*). If P is 8500 lb [38 kN] and the bolt has a diameter of 7/8 in. [22 mm], with a cross-sectional area of 0.6013 in.^{2} [388 mm^2], the unit shearing stress in the bolt is found as follows:

$$f = \frac{P}{A} = \frac{8500}{0.6013} = 14{,}136 \text{ psi}$$

$$\left[f = \frac{38 \times 10^3}{388} = 98 \text{ MPa} \right]$$

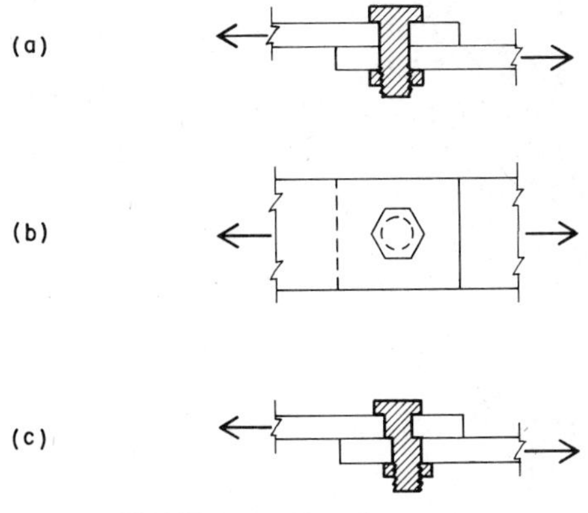

FIGURE 3-2. Shearing stress.

It will be noted that this formula is similar to the one used for direct tension and compression. There is also an analogous assumption that the shearing stress is uniformly distributed over the cross section. However, it must be understood clearly that *the physical situations represented by the two cases are quite different.*

Another situation in which shearing stress can be important is illustrated in Fig. 3-3*a*, which shows a beam that supports a uniformly distributed load over its length. It is evident from the sketch that the loaded beam might fail by simply dropping between the walls, as indicated in Fig. 3-3*b*. Although the nature of beam action is such that failure would probably occur in some other manner, the tendency to drop between the walls would nevertheless be present. This tendency for one part of a beam to

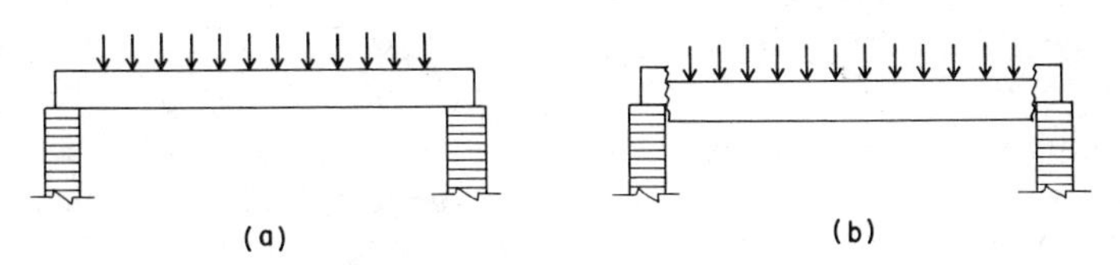

FIGURE 3-3. Vertical shear in beams.

move vertically with respect to an adjacent part is called the *vertical shear*. It is designated by the symbol V and the determination of its magnitude is explained in Chapter 5. It should be pointed out here, however, that the resistance to shear in this case is not uniformly distributed over the cross section of the beam; consequently the unit shearing stress cannot be found simply by dividing the value of the vertical shear by the cross-sectional area of the beam. This is considered further in Art. 3-4.

Problem 3.3.A.* A steel bar 1.25-in. [32-mm] square supports a tensile load of 19,000 lb [85 kN]. Compute the unit tensile stress in the bar.

Problem 3.3.B. A steel rod 1.5 in. [38 mm] in diameter supports a tensile load of 21,000 lb [93 kN]. Compute the unit tensile stress.

Problem 3.3.C.* A steel bar 0.75-in. [19-mm] thick supports a tensile load of 32,000 lb [142 kN]. What should its width be if the allowable unit tensile stress is 22,000 psi [150 MPa]?

Problem 3.3.D. An L $4 \times 3 \times \frac{5}{16}$ is used as a hanger to support a load of 50,000 lb [222 kN]. Compute the unit stress. (See Appendix Table A.7.)

Problem 3.3.E. What should the diameter of a steel bolt be to resist a shearing force of 13,000 lb [58 kN] if the allowable shearing stress is 22,000 psi [150 MPa]?

3-3. Bending Stresses

The tensile and compressive stresses that accompany beam action are not simple direct stresses and cannot be computed by the formula $f = P/A$. Figure 3-4*a* illustrates a rectangular beam with supports at each end. Figure 3-4*b* is an exaggerated drawing of the shape the beam tends to assume when loaded. The material in the upper portion of the beam is in compression and that in the lower portion, in tension. Figure 3-4*c* is an enlarged detail that represents a segment of the beam cut at a section near the middle of the span. The line marked *neutral surface* locates the plane between the upper and lower surfaces above which the stresses are compressive and below which they are tensile. The line in which the neutral surface cuts the cross section of the beam is called the *neutral axis* (Fig. 3-4*e*). In beams of rectangular or other symmetrical cross section (such as I-beams and wide flange

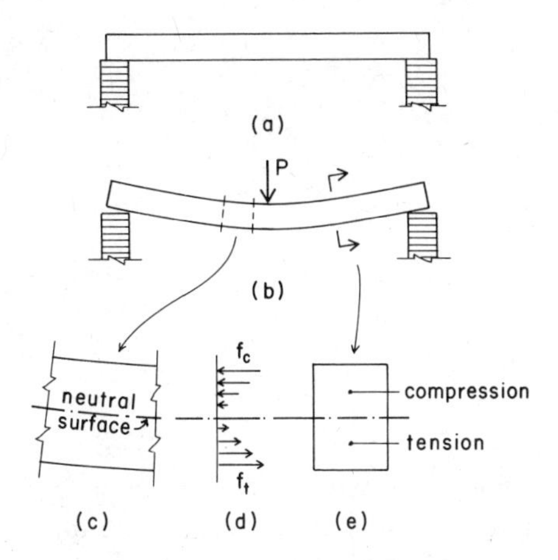

FIGURE 3-4. Bending stress.

shapes) the neutral surface occurs midway between the top and bottom surfaces.

Figure 3-4*d* shows the distribution of the stresses at the cut section. The maximum tensile and compressive stresses that occur at the upper and lower surfaces of the beam are called *extreme fiber stresses* or simply *bending stresses* (f_b). The stresses decrease in magnitude toward the neutral surface, where they become zero. If the magnitude of the bending stress at the extreme fibers does not exceed certain limits, the stresses over the cross section are directly proportional to their distances from the neutral surface. In the design of beams the problem is to select a structural shape that will sustain the given loading without exceeding the allowable bending stress for the type of structural steel. The expression used to compute the value of the bending stress in tension or compression is known as the *beam formula* or the *flexure formula* and is developed in Chapter 6.

3-4. Horizontal Shear

In addition to the bending stresses developed when a beam deflects, as indicated in Fig. 3-4*b*, there is a horizontal shearing

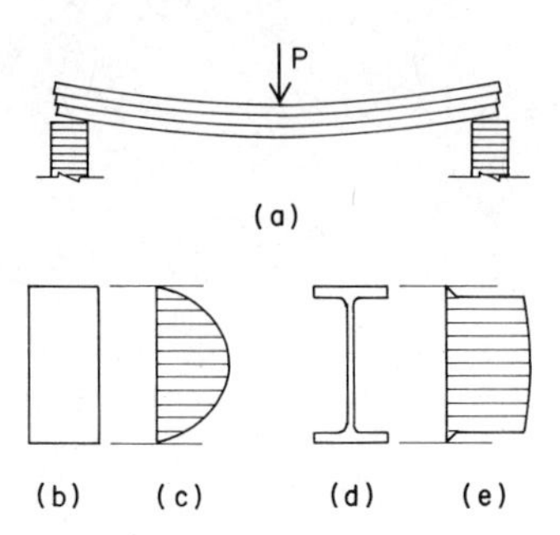

FIGURE 3-5. Horizontal shear in beams.

stress related to the vertical shear previously discussed. Figure 3-5*a* represents a beam of the same size and span length as the one in Fig. 3-4*b* but instead is made up of three independent strips. If subjected to the same loading, its deflection would be greater than that of the solid beam and slipping would occur along the surfaces of contact between the independent strips (Fig. 3-5*a*). This same tendency is present in the solid beam but the action is restrained by its resistance to *horizontal shear*.

It can be shown that at any point in a beam the intensity of the horizontal shear is equal to the intensity of the vertical shear. The shearing stresses, however, are not distributed uniformly over the cross-sectional area. For rectangular beam cross sections (Fig. 3-5*b*), the stresses vary as shown by the parabola (Fig. 3-5*c*); the length of a horizontal line represents the magnitude of the stress at that point in the depth of the beam. From the figure it is apparent that the maximum unit shearing stress occurs at the neutral surface. Its magnitude is 1.5 times the average unit stress and may be computed by the equation

$$f_v = \frac{3}{2} \times \frac{V}{A}$$

where f_v = the maximum unit shearing stress, horizontal or vertical,

V = the maximum vertical shear,

and

A = the area of the rectangular section.

This equation applies only to rectangular beams and consequently must be modified when wide flange and I-beam sections in which most of the material lies in the flanges are investigated. Figure 3-5*e* indicates the distribution of the shearing stresses in these structural shapes. Because the maximum shearing stress occurs at the neutral surface and the value at the extreme fibers is zero, the flanges have little influence on resistance to shear. It is therefore customary to ignore the material in the flanges and to consider only the web as resisting the shear. Based on this assumption, the following approximate expression for unit shearing stress is customarily used:

$$f_v = \frac{V}{A_w}$$

where f_v = the unit shearing stress,
V = the maximum vertical shear,

and

A_w = the area of web (actual depth of section times the web thickness).

Here, of course, the value of f_v is really the *average* unit shearing stress over the area of the web. The fact that the maximum unit stress is somewhat greater than the average value (Fig. 3-5*e*) is handled in practice by assigning a value for the *allowable* shearing stress low enough to compensate for this difference.

Example. A W 12 × 35 is used as a beam subjected to a vertical shear at the supports (Fig. 3-3) of 50,000 lb [222 kN]. Find the value of the unit shearing stress.
Solution: Referring to Appendix Table A-1, we find that the depth of this section is 12.50 in. [318 mm] and the web thickness is 0.300 in. [7.62 mm]. Then the area of the web is 12.50 × 0.300 = 3.75 in^2 [2423 mm^2]. Therefore

$$f_v = \frac{V}{A_w} = \frac{50{,}000}{3.75} = 13{,}333 \text{ psi}$$

$$\left[f_v = \frac{222 \times 10^3}{2423} = 91.6 \text{ MPa}\right]$$

Problem 3.5.A. If the maximum vertical shear force on an S 18 × 70 is 100,000 lb [445 kN], determine the value of the unit shearing stress.

Problem 3.5.B.* If the maximum vertical shearing force on a W 18 × 71 is 100,000 lb [445 kN], determine the value of the unit shearing stress.

3-5. Elastic Limit, Yield Point, and Ultimate Strength

When a member is subjected to a load there is always an accompanying change in length or shape of the member. Such a change is called a *deformation* or *strain*. For compression and tension the deformations are, respectively, a shortening and lengthening. The deformation accompanying bending is called deflection. Shear produces an angular distortion.

In order to investigate the relation between the magnitudes of deformation and the accompanying stresses, it is convenient to record them on a sheet of graph paper. Figure 3-6 illustrates such a chart. As an example, a structural steel bar was placed in a testing machine and subjected to a tensile test. To record the

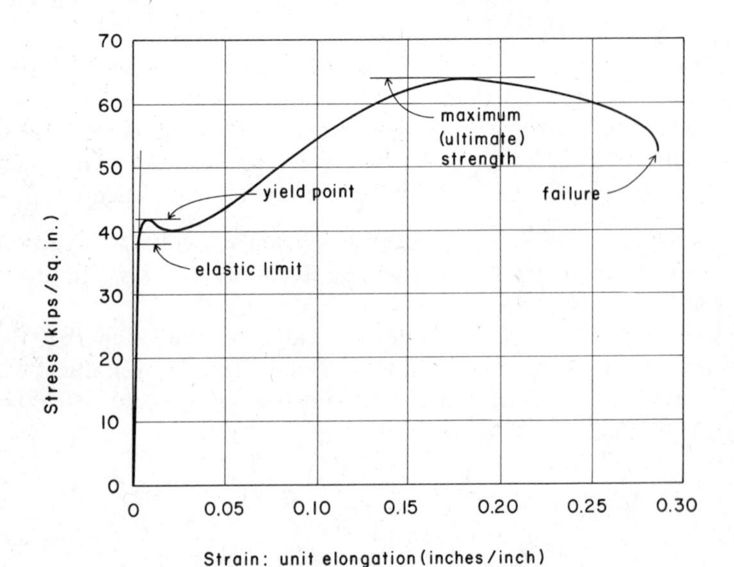

FIGURE 3-6. Stress-strain behavior of ductile steel.

stresses and deformations a vertical scale, representing the unit stresses in pounds per square inch, was laid off and a horizontal scale, shown at the bottom of the chart, was made for the deformations. The latter scale indicates the deformations of the specimen in inches per inch of length. At various stresses points were plotted on the chart to indicate the accompanying deformations. When the bar was finally stressed to rupture the different points were connected; the curve, represented by the heavy black line, presents graphically the results of the test.

Note that up to a unit stress of about 38,000 psi [260 MPa] the curve is a straight line. This is important, for it shows that the deformations up to this stress are directly proportional to the applied loads, a verification of Hooke's law that stress is proportional to deformation. It is seen that beyond this stress of about 38,000 psi the curve bends to the right, showing that the deformations are increasing more rapidly than the applied loads. This unit stress below which the deformations are directly proportional to the unit stresses is called the *elastic limit* or sometimes the *proportional limit*. If, during the test, the loads had been removed before the elastic limit had been reached, the bar would have returned to its original length. If the unit stress exceeds the elastic limit and the load is removed, the bar will not return to its original length. There is a permanent lengthening, and this increase in length is called a *permanent set*. Hence it is seen that the unit stresses in structural members should always be well within the elastic limit.

Just beyond the elastic limit it is seen that for a short distance the curve is almost horizontal, indicating a *slight amount of deformation without an increase in stress*. This unit stress is called the *yield point*. In testing steel specimens the yield point may be determined more accurately than the elastic limit; the two stresses, however, are quite close together. With respect to structural steel, the yield point is a particularly important unit stress. The AISC Specification, in giving the various allowable unit stresses, designates these stresses as fractions of the yield point.

Beyond the yield point the curve begins to flatten out until the greatest stress is reached. This is the *ultimate strength*. It occurs at or immediately before rupture. It should be noted that the deformations increase rapidly beyond the yield point. The specimen not only elongates but there is also a reduction in the cross-sectional area. Failure begins at the ultimate strength, the greatest unit stress reached, and rupture occurs at the point marked *breaking strength.*

3-6. Modulus of Elasticity

The *modulus of elasticity* of a material is a number that indicates its degree of *stiffness*. A material is said to be stiff if its deformation is relatively small when the unit stresses are high. As an example, a steel rod 1 in.^{2} [645 mm^2] in cross-sectional area and 10 ft [3.05 m] in length will elongate about 0.008 in. [0.2 mm] under a tensile load of 2000 lb [9 kN]. A piece of wood of the same dimensions and with the same tensile load will stretch about 0.24 in. [6 mm]. We say that the steel is stiffer than the wood because for the same unit stress the deformation is not so great.

It is quite a simple matter to compute the deformations for tensile and compressive loads. It is important to remember, however, that for such computations we must be careful to see that the unit stress does not exceed the elastic limit of the material. In Fig. 3-6 it was noted that the deformation curve up to the elastic limit was a straight line because the deformations and stresses were directly proportional. This proportion may be expressed as *the unit stress divided by the unit deformation.* The name for this ratio is the *modulus of elasticity*. It is represented by the letter E and has the same value for compression and tension for most structural materials.

E = the modulus of elasticity (psi or ksi)
P = the applied force in pounds or kips
l = the length of the member in inches
A = the cross-sectional area of the member in square inches
e = the total deformation of the member in inches
s = the unit deformation in inches per inch
f = the unit stress in pounds per square inch

Because, by definition, the modulus of elasticity is the unit stress divided by the unit deformation,

$$E = \frac{f}{s} \quad \text{or} \quad E = \frac{P/A}{e/l} \quad \text{or} \quad E = \frac{Pl}{Ae} \quad \text{or} \quad e = \frac{Pl}{AE}$$

There are five terms in this equation and if any four are known the fifth term may be found. In computing the deformation by the use of this formula, we must, of course, know the modulus of elasticity of the material. For structural steel E = 29,000,000 psi [200 GPa]. Depending on the species, E for wood varies from about 1,000,000 to 2,000,000 psi [7 to 14 GPa]. Bear in mind that the foregoing formula is valid only when the unit stress lies within the elastic limit of the material. Let us try an example.

Example. A bar 1 × 2 in. [25 × 50 mm] in cross section and 20 ft [6.1 m] long is made from structural steel with a yield point of 42,000 psi [290 MPa]. Compute its total elongation under a tensile load of 36,000 lb [160 kN].
Solution: The bar has an area of 1 × 2, or 2 in.^{2} [1250 mm^2]; the unit stress is therefore 36,000/2 = 18,000 psi. Because this is considerably lower than the yield point, the elastic deformation formula is applicable. Then

$$e = \frac{Pl}{AE} \quad \text{or} \quad e = \frac{36{,}000 \times 20 \times 12}{2 \times 29{,}000{,}000} = 0.149 \text{ in.}$$

$$\left[e = \frac{160 \times 6.1 \times 10^3}{1250 \times 200} = 3.90 \text{ mm}\right]$$

is the total elongation. Note the use of the factor of 12 in the American unit formula to convert the 20-ft length to inches.

Problem 3.6.A. In Art. 3-6 it was stated that a steel rod of certain dimensions and a unit stress of 2000 psi would elongate about 0.008 in. Verify this by computation.

Problem 3.6.B.* A steel rod 1 in. [25 mm] in diameter and 8 in. [200 mm] long is placed in a testing machine and tested for tension. Under a load of 15,000 lb [67 kN] the deformation is 0.0052 in. [0.13 mm]. Compute the modulus of elasticity.

3-7. Factor of Safety

The strength of a structural member must be greater than the force it is required to resist. The uncertainties with regard to determination of actual loading and uniformity in quality of materials require that some reserve strength be built into the design.

This degree of reserve strength is the *factor of safety*. Although there is no general agreement on an exact definition of this term, the following discussion will serve to fix the concept in mind.

Consider a structural steel that has an ultimate tensile strength of 70,000 psi, a yield point stress of 36,000 psi, and an allowable unit tensile stress of 22,000 psi. If the factor of safety is defined as the ratio between the ultimate strength and the allowable stress, its value is 70,000/22,000, or 3.18. On the other hand, if it is defined as the ratio of the yield point stress to the allowable stress, its value is 36,000/22,000, or 1.64. This is a considerable variation, and because failure of a structural member begins when it is stressed beyond the elastic limit the higher value may be misleading. It must be borne in mind that stresses beyond the elastic limit produce deformations that are *permanent* and thus change the shape of a member, even though there may not be danger of collapse.

3-8. Allowable Stresses

In our discussion of unit stresses we made a distinction between the computed value of a particular kind of stress and the allowable value. It is the latter that is controlled by the requirements of local building codes. As noted earlier, there is some variation among these codes and the allowable unit stresses in some are more conservative than in others. The user of this book should obtain a copy of the building requirements that govern local construction and be familiar with those items that pertain to structural steel.

A table of allowable unit stresses (Table 4-3) appears in the following chapter. When referring to the table note that stress values are expressed in kips per square inch (ksi) rather than in pounds per square inch (psi). This practice follows that used in the AISC Manual (Ref. 1).

Review Problems

Problem 3.8.A.* If the allowable unit tensile stress of steel is 22,000 psi [150 MPa], will a 1.5-in. [38-mm] diameter rod support a suspended load of 50,000 lb [222 kN]?

Problem 3.8.B. Assume that the diameter of the rivet shown in Fig. 3-2 is 0.75 in. [19 mm]. If the allowable unit shearing stress of the rivet steel is 15,000 psi [103 MPa], compute the maximum value of P.

Problem 3.8.C. Compute the maximum vertical shear that a W 10 × 22 can resist if the allowable unit shearing stress is 14,500 psi [100 MPa]. (See Table A.1 in the Appendix for properties of W shapes.)

Problem 3.8.D. Compute the maximum vertical shear force that an S 10 × 25 can resist if the allowable unit shearing stress is 14,500 psi [100 MPa]. (See Appendix Table A.2.)

Problem 3.8.E.* A steel rod 1 in. [25 mm] in diameter and 18 ft 6 in. [5.64 m] long is subjected to a tensile load of 15,000 lb [66.7 kN]. Compute the total elongation.

Problem 3.8.F. A steel rod 0.5 in. [13 mm] in diameter and 9 in. [229 mm] long is subjected to a tensile load of 4000 lb [18 kN]. Its total elongation under this loading is 0.007 in. [0.18 mm]. Find the unit stress in the rod under these conditions.

4

Types of Steel

4-1 Grades of Structural Steel

Steel that meets the requirements of the American Society for Testing and Materials (abbreviation: ASTM) Specification A36 is the grade of structural steel commonly used to produce rolled steel elements for building construction. It must have an ultimate tensile strength of 58 to 80 ksi and a minimum yield point of 36 ksi. It may be used for bolted, riveted, or welded fabrication. The current edition of the AISC Manual (Ref. 1) lists the following steels as those available for the production of rolled structural shapes, plates, and bars:

Structural Steel, ASTM A36

Structural Steel with 42,000 psi Minimum Yield Point, ASTM A529

High-Strength Low-Alloy Structural Manganese Vanadium Steel, ASTM A441

High-Strength Low-Alloy Columbium-Vanadium Steel of Structural Quality, ASTM A572

High-Strength Low-Alloy Structural Steel, ASTM A242
High-Strength Low-Alloy Structural Steel with 50,000 Minimum Yield Point to 4-in. Thick, ASTM A588
Quenched and Tempered Alloy Steel, A514 (for plates only)

Table 4-1 lists the critical stress properties of these steels and the various rolled products that are produced from them. Table 4-2 gives the groupings of rolled shapes referred to in Table 4-1. Of primary concern is the minimum yield stress, designated F_y, on which most allowable design stresses are based. The other limiting stress is the ultimate tensile stress, designated F_u (see Fig. 3-6). A few design stresses, mostly those relating to connection design, are based on the ultimate tensile stress. For some grades the ultimate stress is given as a range rather than a single value, in which case it is advisable to use the lower value for design unless a higher value can be verified by a specific supplier for a particular rolled product.

4-2. Structural Carbon Steels

Prior to 1963 a steel designated ASTM A7 was the basic product for structural purposes. It had a yield point of 33 ksi and was used primarily for riveted fabrication. With the increasing demand for bolted and welded construction A7 steel became less useful, and in a short time A36 steel was the material of choice for the majority of structural products. When a slightly higher strength is desired A529 steel is sometimes used, although its availability is limited to lighter weight rolled shapes and plates and bars up to 0.5 in. thick.

4-3. High-Strength Steels

These steels have yield points higher than those of the carbon steels used for the same products; thus the design allowable stresses will be higher and increased strength is possible for elements of the same size. A441 is a high-strength steel intended for use primarily in welded construction. A572 is suitable for bolting,

TABLE 4-1. ASTM Structural Steel Grades for Rolled Products[a]

Steel Type	ASTM Designation	F_y Minimum Yield Stress (ksi)	F_u Tensile Stress[c] (ksi)	Shapes[b] Group per ASTM A6					Plates and Bars										
				[d]1	2	3	4	5	To 1/2″ Incl.	Over 1/2″ to 3/4″ Incl.	Over 3/4″ to 1 1/4″ Incl.	Over 1 1/4″ to 1 1/2″ Incl.	Over 1 1/2″ to 2″ Incl.	Over 2″ to 2 1/2″ Incl.	Over 2 1/2″ to 4″ Incl.	Over 4″ to 5″ Incl.	Over 5″ to 6″ Incl.	Over 6″ to 8″ Incl.	Over 8″
Carbon	A36	32	58-80																■
	A36	36	58-80[e]	■	■	■	■	■	■	■	■	■	■	■	■	■	■	■	
	A529	42	60-85	■					■										
High-Strength Low-Alloy	A441	40	60													■	■	■	
	A441	42	63				■	■					■	■	■				
	A441	46	67			■					■	■							
	A441	50	70	■	■				■	■									
	A572—Grade 42	42	60	■	■	■	■	■		■	■	■	■	■	■	■	■		
	A572—Grade 50	50	65	■	■	■	■	■	■	■	■	■	■						
	A572—Grade 60	60	75	■	■				■	■	■								
	A572—Grade 65	65	80	■					■	■	■								
Corrosion-Resistant High-Strength Low-Alloy	A242	42	63				■	■					■	■	■				
	A242	46	67			■					■	■							
	A242	50	70	■	■				■	■									
	A588	42	63														■	■	
	A588	46	67													■			
	A588	50	70	■	■	■	■	■	■	■	■	■	■	■	■				
Quenched & Tempered Alloy	A514[f]	90	100-130												■	■	■		
	A514[f]	100	110-130						■	■	■	■	■	■					

[a] Shaded portion in table indicates availability of products for each grade.
[b] See Table 4-2.
[c] Minimum unless a range is shown.
[d] Includes bar-size shapes.
[e] For shapes over 426 lb/ft minimum of 58 ksi only applies.
[f] Plates only.

TABLE 4-2. Structural Shape Size Groupings for Tensile Property Classification

Structural Shape	Group 1	Group 2	Group 3	Group 4	Group 5
W Shapes	W 24x55, 62	W 36x135 to 210 incl	W 36x230 to 300 incl	W 14x233 to 550 incl	W 14x605 to 730 incl
	W 21x44 to 57 incl	W 33x118 to 152 incl	W 33x201 to 241 incl	W 12x210 to 336 incl	
	W 18x35 to 71 incl	W 30x99 to 211 incl	W 14x145 to 211 incl		
	W 16x26 to 57 incl	W 27x84 to 178 incl	W 12x120 to 190 incl		
	W 14x22 to 53 incl	W 24x68 to 162 incl			
	W 12x14 to 58 incl	W 21x62 to 147 incl			
	W 10x12 to 45 incl	W 18x76 to 119 incl			
	W 8x10 to 48 incl	W 16x67 to 100 incl			
	W 6x9 to 25 incl	W 14x61 to 132 incl			
	W 5x16, 19	W 12x65 to 106 incl			
	W 4x13	W 10x49 to 112 incl			
		W 8x58, 67			
M Shapes	to 20 lb/ft incl				
S Shapes	to 35 lb/ft incl	over 35 lb/ft			
HP Shapes		to 102 lb/ft incl	over 102 lb/ft		
American Standard Channels (C)	to 20.7 lb/ft incl	over 20.7 lb/ft			
Miscellaneous Channels (MC)	to 28.5 lb/ft incl	over 28.5 lb/ft			
Angles (L), Structural & Bar-Size	to ½ in. incl	over ½ to ¾ in. incl	over ¾ in.		

Notes: Structural tees from W, M and S shapes fall in the same group as the structural shape from which they are cut.

Group 4 and Group 5 shapes are generally contemplated for application as compression members. When used in other applications or when subject to welding or thermal cutting, the material specification should be reviewed to determine if it adequately covers the properties and quality appropriate for the particular application. Where warranted, the use of killed steel or special metallurgical requirements should be considered.

TABLE 4-3. Allowable Unit Stresses for Structural Steel: ASTM A36[a]

Type of stress and conditions	See discussion in this book in	Stress designation	AISC specification	Allowable Stress (ksi)	Allowable Stress (MPa)
Tension					
1. On the gross (unreduced) area	Art. 12-4	F_t	0.60 F_y	22	150
2. On the effective net area, except at pinholes			0.50 F_u	29	125
3. Threaded rods on net area at thread			0.33 F_u	19	80
Compression	Chapter 11	F_a	See discussion		
Shear					
1. Except at reduced sections	Art. 9-5	F_v	0.40 F_y	14.5	100
2. At reduced sections	Art. 12-7		0.30 F_u	17.4	120
Bending					
1. Tension and compression on extreme fibers of compact members braced laterally, symmetrical about and loaded in the plane of their minor axis	Chapter 9	F_b	0.66 F_y	24	165

2. Tension and compression on extreme fibers of other rolled shapes braced laterally			$0.60\ F_y$	22	150
3. Tension and compression on extreme fibers of solid round and square bars, on solid rectangular sections bent on their weak axis, on qualified doubly symmetrical I & H shapes bent about their minor axis			$0.75\ F_y$	27	188
Bearing					
1. On contact area of milled surfaces		F_p	$0.90\ F_y$	32.4	225
2. On projected area of bolts and rivets in shear connections	Chapter 12		$1.50\ F_u$	87	600

[a] $F_y = 36$ ksi; assume that $F_u = 58$ ksi; some table values are rounded off as permitted in the AISC Manual (Ref. 1). For SI units $F_y = 250$ MPa, $F_u = 400$ MPa.

riveting, or welding. A514 is used exclusively for plates in built-up sections for large structures: girders, piers, arches, and so on.

4-4. Corrosion-Resistant Steels

Because of their chemical composition, these steels exhibit higher degrees of resistance to atmospheric corrosion. For A242 this resistance is four to six times that of structural carbon steel; for A588 the resistance is about eight times that of A441. Several brands of A588 steel are available; each represents the proprietary product of a different manufacturer. These steels are used in the bare (uncoated) condition for exposed construction. Exposure to normal atmosphere causes formation of an oxide on the surface which adheres tightly and protects the steel from further oxidation.

4-5. Allowable Stresses for Structural Steel

For structural steel the AISC Specification expresses the allowable unit stresses in terms of some percent of the yield stress F_y or the ultimate stress F_u. Selected allowable unit stresses used in design are listed in Table 4-3. Specific values are given for ASTM A36 steel, with values of 36 ksi for F_y and 58 ksi for F_u. This is not a complete list, but it generally includes the stresses used in the examples in this book. Reference is made to the more complete descriptions in the AISC Specification which is included in the AISC Manual (Ref. 1). There are in many cases a number of qualifying conditions, some of which are discussed in other portions of this book. Table 4-3 gives the location of some of these discussions.

4-6. Cold-Formed Products

The grades of steel discussed so far are those used only for hot-rolled products: shapes, plates, and bars. Thin sheets of steel are formed in a wide range of thicknesses and a great variety of steels. Structural products formed by cold processes of stamping, rolling, or bending are generally made from steels with stress properties approximately in the range of the steels used for hot-rolled products. Although the behavior and design of these elements is not discussed in this book, the use of some of them is covered in Chapter 10. The design of these products is contained in *Cold-Formed Steel Design Manual* (Ref. 7), prepared by the American Iron and Steel Institute.

4-7. Grades of Steel for Connectors

The AISC Specification designates the type of material permitted in bolted and welded connections. High-strength bolts must meet one of the following specifications: ASTM A325, ASTM A449, or ASTM A490. Bolts other than high strength must conform to ASTM A307. Allowable stresses for bolts are discussed in Chapter 12 under design procedures for bolted connections. Filler material for welding is required to meet the specifications of the American Welding Society for the type of welding process and type of steel. Allowable stresses in welds are discussed in Chapter 13.

5

Beam Actions

5-1. General

A beam is a structural member that resists transverse loads. Sometimes it is defined as a structural member on which the applied loads are perpendicular to its longitudinal axis. The loads tend to bend the beam and are said to induce *flexure* or *bending*. The forces developed by the beam's supports are called *reactions*. In particular situations of use beams are often given other names. *Joists* are beams used in closely spaced sets. Large beams, or those that support a number of other beams, are called *girders*.

5-2. Types of Beams

There are several types of beam that may be identified by the number, kind, and position of the supports. In Figure 5-1 a number of common beams are shown with the exaggerated shape that each beam assumes when loaded.

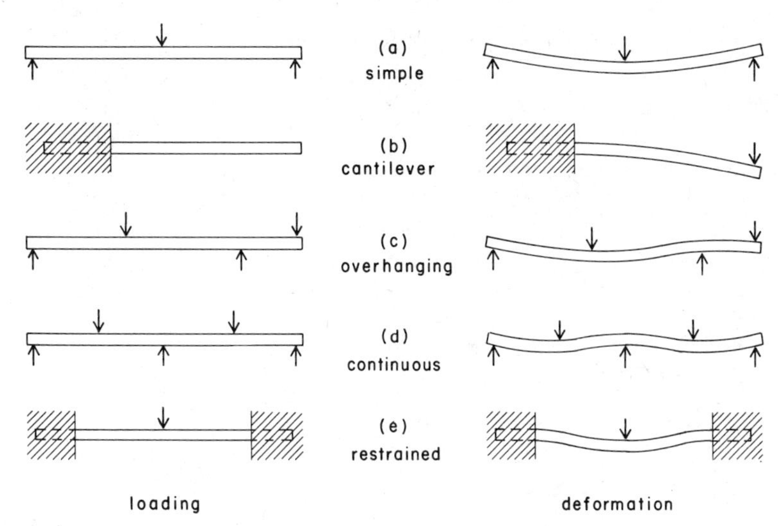

FIGURE 5-1. Types of beam.

A *simple beam* rests on a support at each end, the beam ends being free to rotate. A large percentage of the beams in steel-frame buildings are designed as simple beams (Fig. 5-1*a*). When there is no restraint against rotation at the supports, we say that the beam is *simply supported.*

A *cantilever beam* projects beyond its support. A beam embedded in a wall and extending beyond the face of the wall is a typical example. This is illustrated in Fig. 5-1*b*. This beam is said to be *fixed* or *restrained* at the support.

An *overhanging beam* is a beam whose end or ends project beyond its supports. Figure 5-1*c* indicates an overhanging beam. The projecting end is a cantilever in which the stresses are similar to those in the beam shown in Fig. 5-1*b*. It will be seen later, however, that the stresses in the portion of the beam between the supports are different from the stresses in the simple beam shown in Fig. 5-1*a*.

A *continuous beam* is supported on more than two supports (Fig. 5-1*d*). Continuous beams are frequently used in welded steel floor framing.

A *restrained beam* has one or both ends restrained or *fixed* against rotation (Fig. 5-1*e*). Fixed-end conditions occur when the end of a beam is rigidly connected to its supporting member.

When designing beams it is advantageous to make a diagram to show the loading conditions and to sketch directly below it the shape the beam tends to assume when it bends. The ability to visualize this deformation curve is important because, as we shall see later, the curve indicates changes in the character of the bending stresses along the beam's length.

5-3. Loading

The loads supported by beams are classified as *concentrated* or *distributed*. A concentrated load is one that extends over so small a portion of the beam length that it may be assumed to act at a point, as indicated in the diagrams of Fig. 5-1. A girder in a building takes concentrated loads at the points at which the floor beams frame into it (Fig. 5-2). Also, the load exerted by a column that rests on a beam or girder is a concentrated load. Actually this load extends over a short length of the beam, represented by the width of the column base. For practical purposes, however, the load acts at the axis of the column. Similarly, a beam resting on a masonry wall produces a downward force that is resisted by the upward reaction. The reaction is distributed over the length of the beam (or its bearing plate) that rests on the wall, but the reaction may be considered to act at the midpoint of the beam's bearing area on the wall.

A distributed load is one that extends over a substantial portion or the entire length of a beam. Most distributed loads are *uniformly distributed*; that is, they have a uniform magnitude for each unit of length, such as pounds per linear foot, kips per linear foot, or kilonewtons per linear meter.

Figure 5-2*a* shows a portion of a floor framing plan. The diagonal crosshatching represents the area supported by one of the beams. This area is 8 ft [2.44 m] (the sum of one-half the distances to the next beam on each side), multiplied by the span length of 20 ft [6.10 m]. The beam is supported at each end by girders that

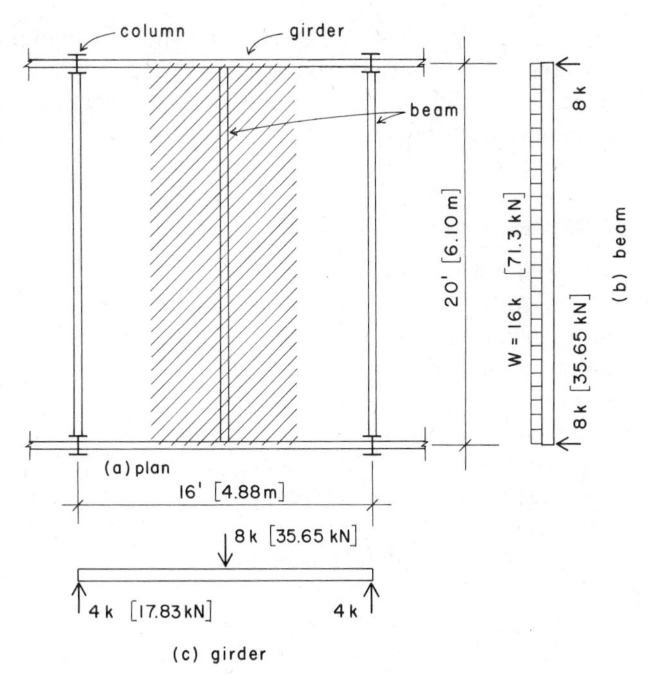

FIGURE 5-2

span between the columns. Imagine that the floor load is 100 lb per sq ft [4.79 kN/m²], including the weight of the construction. The total load on the beam is then 8 × 20 × 100 = 16,000 lb, or 16 kips [2.44 × 6.10 × 4.79 = 71.3 kN]. It is common to designate this total load as *W*. Another way to determine the load is to say that it is 8 × 100 = 800 lb per ft [2.44 × 4.79 = 11.69 kN/m], in which case it would be designated as *w* to indicate that it is a *unit load* rather than the total load on the beam. Most distributed loads are uniform and are constant over the entire length of the beam, as in this example. With different framing arrangements, however, the distributed loads may sometimes not be uniform or may extend over a limited portion of the beam length.

Because the beam in Fig. 5-2 is symmetrically loaded, both end reactions are the same and are equal to one-half the total load on the beam. From this single beam each girder thus receives a concentrated load of one-half the beam load at the center of the girder span. Figures 5-2*b* and 5-2*c* are the conventional representations of the beam and girder load diagrams, respectively.

5-4. Moments

The *moment* of a force is its tendency to cause rotation about a given point or axis. The *magnitude* of a moment is the product of the force times the perpendicular distance from its line of action to the point about which the moment is taken. The *sense* of a moment is the indication of its direction of rotation: clockwise or counterclockwise, for example. Moments are expressed in compound units such as foot-pounds, inch-pounds, kip-feet, or kip-inches, or in SI units: newton-meters or kilonewton-meters. It is important to remember that the moment of a force must be determined with respect to a particular point or axis.

Figure 5-3*a* shows a cantilever beam with a concentrated load of 100 lb [445 N] placed 4 ft [1.22 m] from the face of the supporting wall. In this position the moment of the force about point *A* is $4 \times 100 = 400$ ft-lb [$445 \times 1.22 = 542.9$ N-m]. If the load is moved 2 ft [0.61 m] farther to the right, the moment of the force about point *A* is 600 ft-lb [814.4 N-m]. When the load is moved to the end of the beam the moment about the same point *A* is 800 ft-lb [1085.8 N-m].

Figure 5-3*b* shows a cantilever beam 10 ft [3.048 m] in length with a uniformly distributed load of 200 lb per ft [2.92 kN/m] extending over a length of 6 ft [1.83 m] at the position indicated.

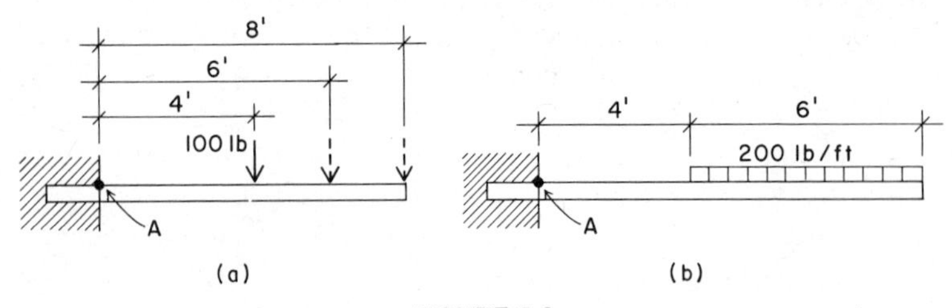

FIGURE 5-3

Because w = 200 lb per ft, W (the total distributed load) = 200 × 6 = 1200 lb [2.92 × 1.83 = 5.34 kN]. In computing the moment of a distributed load about some specific point, we consider that the load is acting at its center of gravity. For the distributed load in Fig. 5-3*b* the center of gravity is 3 ft [0.914 m] from the right end of the beam. Suppose we are asked to find the moment of this load about point A at the face of the wall. With the total load acting at its center of gravity the moment is 1200 × 7 = 8400 ft-lb [5.34 × 2.134 = 11.40 kN-m]. The distance 7 ft [2.134 m] is called the *lever arm* or *moment arm*. Point A is sometimes referred to as the *center of moments*. The important thing to remember is that in computing a moment we must have in mind the particular point about which the moment is taken.

Soon we shall write an *equation of moments* in which the moments of several forces will occur. In writing such an equation, the moment of each force must be taken *about the same point*. Careful attention to identifying the center of moments and the lengths of the moment arms of the forces involved will lead to a ready understanding of the concept of moments, which is fundamental to much of structural analysis and design. It is also essential to a precise understanding of moments that their values be expressed in the correct units; if one does not know whether a moment magnitude is foot-pounds or inch-pounds, one does not know the value of the moment.

5-5. Laws of Equilibrium

When a body is acted on by a number of forces each force tends to move the body. If the forces are of such magnitude and position that their combined effect produces no motion of the body, the forces are said to be in *equilibrium*.

The three fundamental laws of equilibrium follow:

1. The algebraic sum of all the horizontal forces equals zero.
2. The algebraic sum of all the vertical forces equals zero.
3. The algebraic sum of the moments of all the forces about any point equals zero.

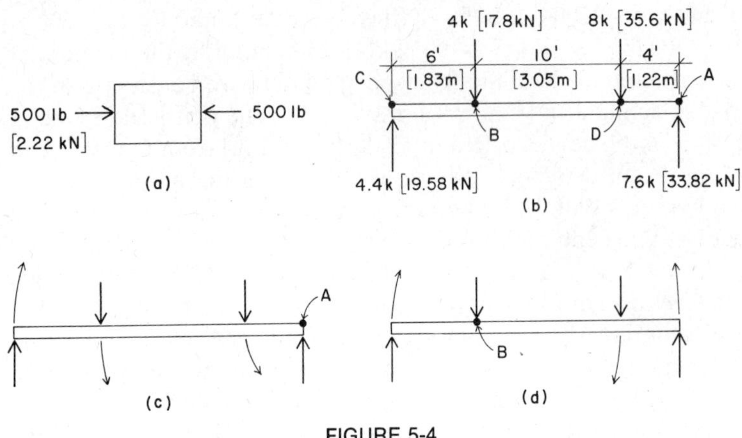

FIGURE 5-4

Figure 5-4*a* represents a body subjected to two equal horizontal forces with the same line of action. Obviously the forces are in equilibrium, one balancing the other. Call the force acting toward the right positive and the force acting toward the left negative. Then, in accordance with the first law of equilibrium,

$$500 - 500 = 0 \quad \text{or} \quad 500 \text{ lb} = 500 \text{ lb}$$

Figure 5-4*b* is a simple beam. Four vertical forces act on this beam and are in equilibrium. The two downward forces, or loads, are 4000 and 8000 lb [17.8 and 35.6 kN]. The two upward forces, or reactions, are 4400 and 7600 lb [19.57 and 33.80 kN]. If we call the upward forces positive and the downward forces negative, by the second law of equilibrium

$$4400 + 7600 - 4000 - 8000 = 0$$

$$[17.8 + 35.6 - 19.58 - 33.82 = 0]$$

Accordingly, if the forces are in equilibrium, we may say that the sum of the downward forces equals the sum of the upward forces or that the sum of the loads equals the sum of the reactions. If, for instance, we had known the magnitudes of the loads and also the magnitude of the left reaction, we could have written

$$4000 + 8000 = 4400 + R_2$$

$$[17.8 + 35.6 = 19.58 + R_2]$$

which yields $R_2 = 7600$ lb [33.82 kN].

Now consider the third law of equilibrium. Take the point marked A on the line of action of the right support and write an equation of moments. (See Fig. 5-4*c*.) Consider moments that tend to cause clockwise rotation as positive and those that tend to cause counterclockwise rotation as negative. The directions of the rotations are indicated by the curved arrow lines. Then

$$(4400 \times 20) - (4000 \times 14) - (8000 \times 4) = 0$$

or

$$(4400 \times 20) = (4000 \times 14) + (8000 \times 4)$$

or

$$88{,}000 \text{ ft-lb} = 88{,}000 \text{ ft-lb}$$

$$[(19.58 \times 6.10) = (17.8 \times 4.27) + (35.6 \times 1.22)]$$

$$[119.4 \text{ kN-m} = 119.4 \text{ kN-m}]$$

It is now apparent that in accordance with the third law of equilibrium the sum of the moments that tend to cause clockwise rotation equals the sum of the moments that tend to cause counterclockwise rotation.

Note carefully that in writing the equation of moments the moment of each force is taken with respect to the same point, point A in this case. Attention is also called to the moment of the 7600 lb [33.82 kN] force about point A. Because the line of action of this force passes through point A, it can cause no rotation about the point. The lever arm of the moment is zero, or $7600 \times 0 = 0$. We may say that the moment of a force about a point in its line of action is zero. When writing an equation of moments, then, it will be unnecessary to consider the moment of a force if the center of moments is on the line of action of the force.

According to the third law of equilibrium, we may consider moments about *any* point. Let us take point B in Fig. 5-4d and see if the law holds. Then

$$(4400 \times 6) + (8000 \times 10) = (7600 \times 14)$$
$$26{,}400 + 80{,}000 = 106{,}400$$
$$106{,}400 \text{ ft-lb} = 106{,}400 \text{ ft-lb}$$
$$[(19.58 \times 1.83) + (35.6 \times 3.05) = (33.82 \times 4.27)]$$
$$[35.83 + 108.58 = 144.41]$$
$$[144.41 \text{ kN-m} = 144.41 \text{ kN-m}]$$

Here again we have omitted writing the moment of the 4000-lb [17.8 kN] force about point B, for we know that its lever arm is zero.

Another example that will serve to fix in mind the principle of moments is given in Fig. 5-5. The beam is 9 ft [2.74 m] long with a single pedestal-type support located 6 ft [1.83 m] from the left end. What should the magnitude of the load be at the right end of the beam to produce equilibrium? Call this unknown force x and consider the point of support as the center of moments. Then, because the moment of the force that tends to cause clockwise rotation must equal the moment of the force that tends to cause counterclockwise rotation, we may write

$$3 \times x = 800 \times 6 \quad \text{or} \quad 3x = 4800 \quad \text{and} \quad x = 1600 \text{ lb}$$

$$[0.915 \times x = 3.56 \times 1.83 \quad \text{or} \quad 0.915x = 6.515 \quad \text{and} \quad x = 7.12 \text{ kN}]$$

The reaction supplied by the single support is 800 + 1600 or 2400 lb [3.56 + 7.12 or 10.68 kN].

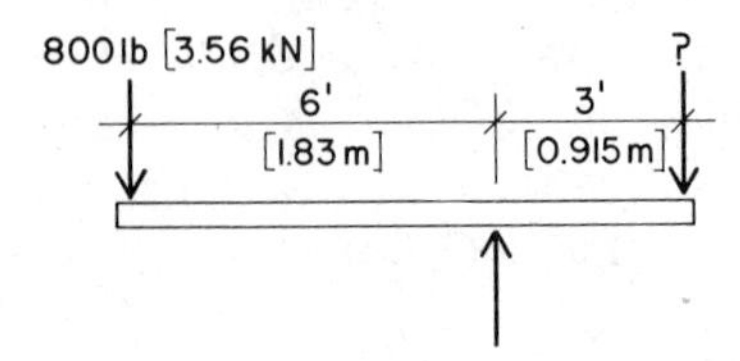

FIGURE 5-5

Problem 5.5.A. Write the two equations of moments for the four forces shown in Fig. 5-4*b*, taking points *C* and *D* as the centers of moments.

5-6. Determination of Reactions

As noted earlier, *reactions* are the forces at the supports of beams that hold the loads in equilibrium. We have seen that the sum of the loads is equal to the sum of the reactions. It is the third law of equilibrium that enables us to compute the magnitudes of the reactions. Throughout this book the left reaction is denoted R_1 and the right reaction is denoted R_2. Another designation frequently used is R_L and R_R.

A simple beam 12 ft [3.66 m] in length has a concentrated load of 1800 lb [8.00 kN] at a point 9 ft [2.74 m] from the left support, as shown in Fig. 5-6. Let us compute the reactions. To do this we simply write the sum of moments that tend to cause clockwise rotation and equate it to the sum of the moments that tend to cause counterclockwise rotation. Then, taking R_2 (the right reaction) as the center of moments,

$$12 \times R_1 = 1800 \times 3 \quad \text{or} \quad 12R_1 = 5400 \quad \text{and} \quad R_1 = 450 \text{ lb}$$

$$[3.66 \times R_1 = 8.00 \times 0.915 \quad \text{or} \quad 3.66R_1 = 7.32 \quad \text{and} \quad R_1 = 2.0 \text{ kN}]$$

To find R_2 we know that $R_1 + R_2$ equals the load on the beam. Then

$$R_1 + R_2 = 1800 \quad \text{or} \quad 450 + R_2 = 1800$$

and

$$R_2 = 1800 - 450 = 1350 \text{ lb}$$

$$[R_2 = 8.0 - 2.0 = 6.0 \text{ kN}]$$

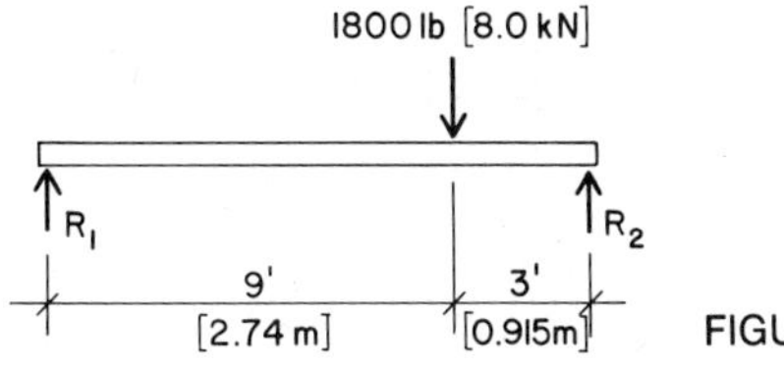

FIGURE 5-6

To see whether we have made an error we can check the magnitude of R_2 by writing an equation of moments about R_1 as the center of moments. Then

$$12 \times R_2 = 1800 \times 9 \quad \text{or} \quad 12R_2 = 16{,}200 \quad \text{and} \quad R_2 = 1350 \text{ lb}$$

Note: For the SI computation we use distance values with more digits of accuracy to verify the reactions.

$$[3.6576 \times R_2 = 8.0 \times 2.7432 \quad \text{or} \quad 3.6576R_2 = 21.9456]$$

$$\left[R_2 = \frac{21.9456}{3.6576} = 6.0 \text{ kN}\right]$$

Example 1. The simple beam in Fig. 5-7 is 15 ft [4.5 m] long and has three concentrated loads, as indicated. Compute the reactions.

Solution: Taking R_2 as the center of moments,

$$15R_1 = (400 \times 12) + (1000 \times 10) + (600 \times 4)$$

$$15R_1 = 17{,}200 \quad \text{and} \quad R_1 = 1146.7 \text{ lb}$$

$$[4.5R_1 = (1.78 \times 3.6) + (4.45 \times 3.0) + (2.67 \times 1.2)]$$

$$[4.5R_1 = 22.962 \quad \text{and} \quad R_1 = 5.10 \text{ kN}]$$

To compute R_2 take R_1 as the center of moments. Then

$$15R_2 = (400 \times 3) + (1000 \times 5) + (600 \times 11)$$

$$15R_2 = 12{,}800 \quad \text{and} \quad R_2 = 853.3 \text{ lb}$$

$$[4.5R_2 = (1.78 \times 0.9) + (4.45 \times 1.5) + (2.67 \times 3.3)]$$

$$[4.5R_2 = 17.088 \quad \text{and} \quad R_2 = 3.80 \text{ kN}]$$

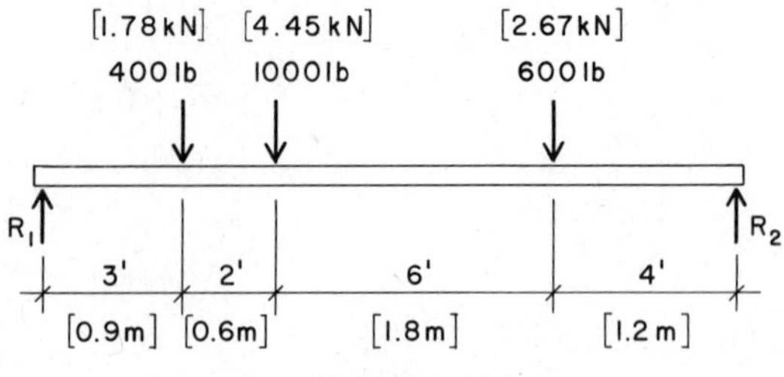

FIGURE 5-7

To check the results

$$400 + 1000 + 600 = 1146.7 + 853.3$$

$$2000 \text{ lb} = 2000 \text{ lb}$$

$$[1.78 + 4.45 + 2.67 = 5.10 + 3.80]$$

$$[8.90 \text{ kN} = 8.90 \text{ kN}]$$

Example 2. Compute the reactions for the overhanging beam shown in Fig. 5-8.

Solution: Taking R_2 as the center of moments, we equate the clockwise moments to the counterclockwise moments:

$$18R_1 + (600 \times 2) = (200 \times 22) + (1000 \times 10) + (800 \times 4)$$

$$18R_1 = 16{,}400 \quad \text{and} \quad R_1 = 911.1 \text{ lb}$$

$$[5.4R_1 + (2.67 \times 0.6) = (0.89 \times 6.6) + (4.45 \times 3.0) + (3.56 \times 1.2)]$$

$$[5.4R_1 = 21.894 \quad \text{and} \quad R_1 = 4.05 \text{ kN}]$$

Taking R_1 as the center of moments,

$$18R_2 + (200 \times 4) = (1000 \times 8) + (800 \times 14) + (600 \times 20)$$

$$18R_2 = 30{,}400 \quad \text{and} \quad R_2 = 1688.9 \text{ lb}$$

$$[5.4R_2 + (0.89 \times 1.2) = (4.45 \times 2.4) + (3.56 \times 4.2) + (2.67 \times 6.0)]$$

$$[5.4R_2 = 40.584 \quad \text{and} \quad R_2 = 7.52 \text{ kN}]$$

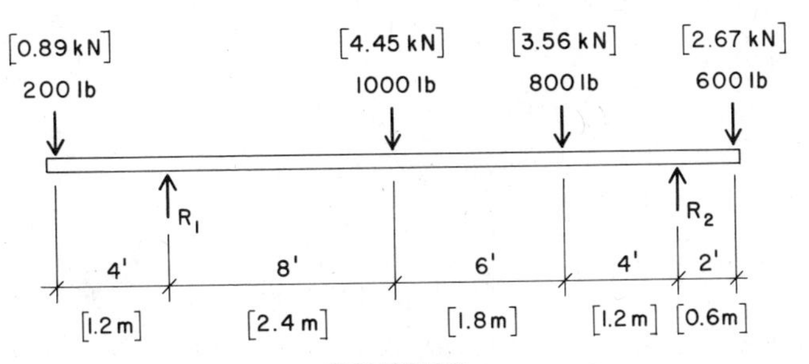

FIGURE 5-8

Check

$$200 + 1000 + 800 + 600 = 911.1 + 1688.9$$

$$2600 \text{ lb} = 2600 \text{ lb}$$

$$[0.89 + 4.45 + 3.56 + 2.67 = 4.05 + 7.52]$$

$$[11.57 \text{ kN} = 11.57 \text{ kN}]$$

Example 3. The simple beam shown in Fig. 5-9 has one concentrated load and a uniformly distributed load over part of its length. Compute the reactions.

Solution: The total uniformly distributed load is $200 \times 8 = 1600$ lb [$2.92 \times 2.4 = 7.0$ kN]. The center of gravity of this load is at the center of the 8 ft [2.4 m] distance; therefore, insofar as the determination of the reactions is concerned, this loading is the same as that shown in Fig. 5-9*b*. The equation of moments with R_2 as the center of moments is

$$20R_1 = (2200 \times 14) + (200 \times 8 \times 4)$$

$$20R_1 = 37{,}200 \quad \text{and} \quad R_1 = 1860 \text{ lb}$$

$$\begin{bmatrix} 6.0R_1 = (9.79 \times 4.2) + (2.92 \times 2.4 \times 1.2) \\ 6.0R_1 = 49.5276 \quad \text{and} \quad R_1 = 8.25 \text{ kN} \end{bmatrix}$$

Taking R_1 as the center of moments,

$$20R_2 = (2200 \times 6) + (200 \times 8 \times 16)$$

$$20R_2 = 38{,}800 \quad \text{and} \quad R_2 = 1940 \text{ lb}$$

$$\begin{bmatrix} 6.0R_2 = (9.79 \times 1.8) + (2.92 \times 2.4 \times 4.8) \\ 6.0R_2 = 51.2604 \quad \text{and} \quad R_2 = 8.54 \text{ kN} \end{bmatrix}$$

Check

$$2200 + 1600 = 1860 + 1940$$

$$3800 \text{ lb} = 3800 \text{ lb}$$

$$\begin{bmatrix} 9.79 + 7.0 = 8.25 + 8.54 \\ 16.79 \text{ kN} = 16.79 \text{ kN} \end{bmatrix}$$

Problems 5.6.A, B, C,* D, E.* Compute the reactions for the beams shown in Fig. 5-10*a*, *b*, *c*, *d*, and *e*.

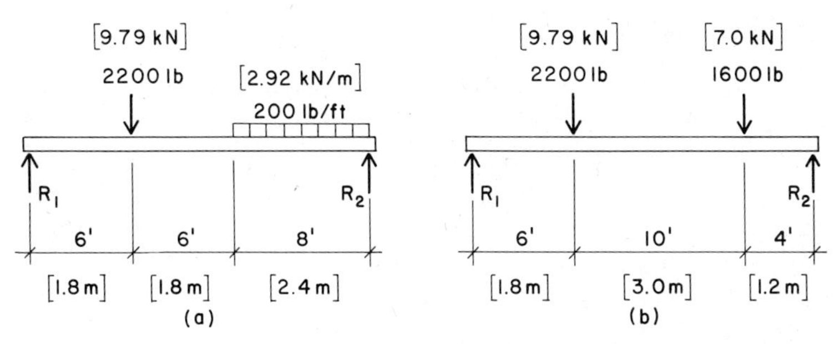

FIGURE 5-9

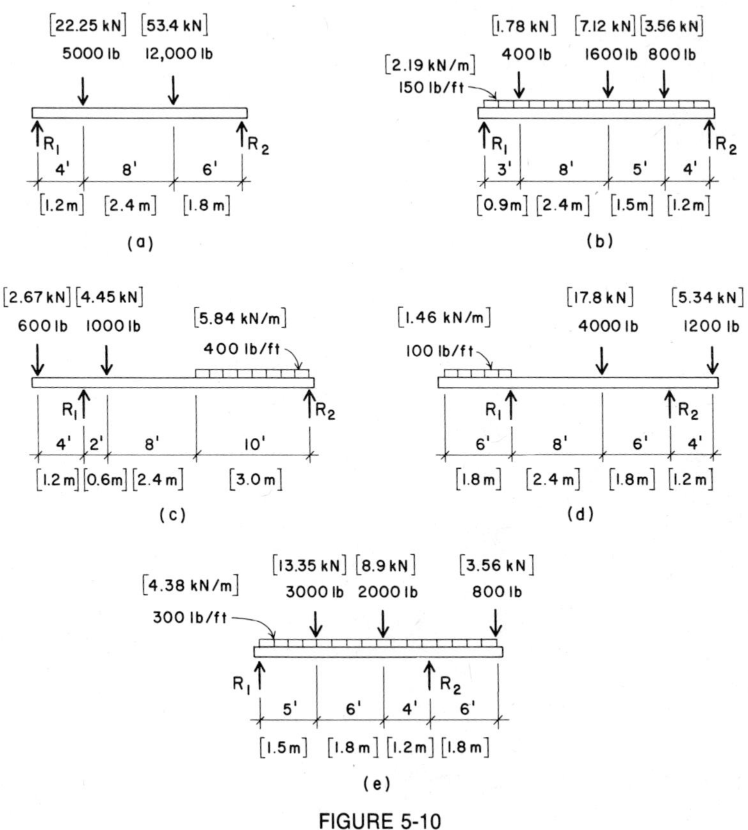

FIGURE 5-10

5-7. Vertical Shear

As explained in Art. 3-3, the *vertical shear* (V) is the tendency for one part of a member to move vertically with respect to an adjacent part. Referring to Fig. 3-3, if the total uniformly distributed load on the beam is W, the loading is symmetrical and each reaction is $W/2$ or 7200 lb. If we imagine a vertical plane cutting a section through the beam flush with the face of the left supporting wall, the reaction of the wall on the beam to the left of the section is upward. Just to the right of the section the load on the beam acts downward, and the magnitude of the tendency for the left and right portions to slide past each other (the vertical shear) is equal to the value of the reaction $W/2$ lb. Thus we say that at a section through the beam infinitely close to the support $V = W/2$ lb.

At any other vertical section taken along the beam span this same tendency exists for the portion to the left of the section to slip past the portion to the right. Therefore we may formulate the following definition:

The magnitude of the vertical shear at any section of a beam is equal to the algebraic sum of all the vertical forces on one side of the section.

To understand how readily the value of the vertical shear may be computed for any section of a beam let us call the upward forces (the reactions) positive and downward forces (the loads) negative. For the present let us consider only the forces to the *left* of the section. Then, in accordance with the foregoing definition, we can say that *the magnitude of the vertical shear at any section of a beam is equal to the sum of the reactions minus the sum of the loads to the left of the section.* To find the value of the shear at a particular section we simply note the forces to the left and then write shear = reactions − loads. Of course, we could consider the forces to the right of the section and find that the magnitude of the shear is the same, but in the following examples we consider the forces on the left. This procedure will avoid confusion with respect to algebraic sign, which is discussed later.

Let us try this rule on the beam shown in Fig. 5-6. We have found that the left reaction is 450 lb [2.0 kN] and that the right reaction is 1350 lb [6.0 kN]. When writing an equation for computing the value of the vertical shear it is convenient to identify the section at which the shear is taken by using a subscript: $V_{(x=4)}$. For this illustration this terminology indicates that the value of the vertical shear is taken at 4 ft [1.2 m] from the left end of the beam.

Now referring to Fig. 5-6 and remembering the values computed for the reactions, let us write the value of the shear at 2 ft [0.6 m] from R_1. Repeat the foregoing rule and observe that the only force to the left of the section is the left reaction. Then

$$\text{shear} = \text{reaction} - \text{loads}$$

or

$$V_{(x=2)} = 450 - 0$$

Thus

$$V_{(x=2)} = 450 \text{ lb}$$

$$[V_{(x=0.6)} = 2 \text{ kN}]$$

Because there are no loads up to the load of 1800 lb [8 kN], the value of the shear in the beam is the same magnitude at any section in this portion of the beam. (From $x = 0$ to $x = 9$ ft.) [From $x = 0$ to $x = 2.7$ m.] Note, however, that we are ignoring the weight of the beam.

Next, consider a section 10 ft [3.0 m] from R_1. The reaction to the left of this section is 450 lb [2 kN] and the load to the left is 1800 lb [8 kN]. Then, because the shear equals the reactions minus the loads,

$$V_{(x=10)} = 450 - 1800 = -1350 \text{ lb}$$

$$[V_{(x=3.0)} = 2 - 8 = -6 \text{ kN}]$$

Because this beam has but one load, the value of the shear at any section between the load and the right reaction is −1350 lb [6 kN]. It should be noted here that for *simple* beams the value of the shear at the support is equal to the magnitude of the reaction.

Hence the maximum shear will have the value of the greater reaction.

5-8. Use of Shear Values

We compute the value of the shear in beams for two principal reasons. First, we must know the value of the maximum shear to be sure that there is ample material in the beam to prevent it from failing by shear. The second reason for computing the shear values is that *the greatest tendency for the beam to fail by bending is at the section of the beam at which the value of the shear is zero.* In the foregoing illustration we found both positive and negative values for the shear and we also found that values on the left of the 1800 lb [8 kN] load were positive and those on the right were negative. The critical section for bending is under the concentrated load, where the shear changes from positive to negative. This is explained further when shear diagrams are constructed.

5-9. Shear Diagrams

In the preceding article we saw that the value of the shear may be computed readily at any section along the beam's length. When designing beams it will be found that diagrams that show the variation in shear along the span are extremely helpful.

To construct a shear diagram first make a drawing of the beam to scale, showing the loads and their positions. Next, compute the reactions as explained in Art. 5-6. Below the beam draw a horizontal *base line* to represent zero shear. Then the values of the shear at various sections of the beam may be computed and plotted to a suitable scale vertically from the base line, positive values above and negative values below. The following examples illustrate the construction of shear diagrams for beams under various conditions of loading:

Example 1. A simple beam 15 ft [4.5 m] long has two concentrated loads located as shown in Fig. 5-11. Construct the shear diagram and note the value of the maximum shear and the section of the beam at which the shear passes through zero.

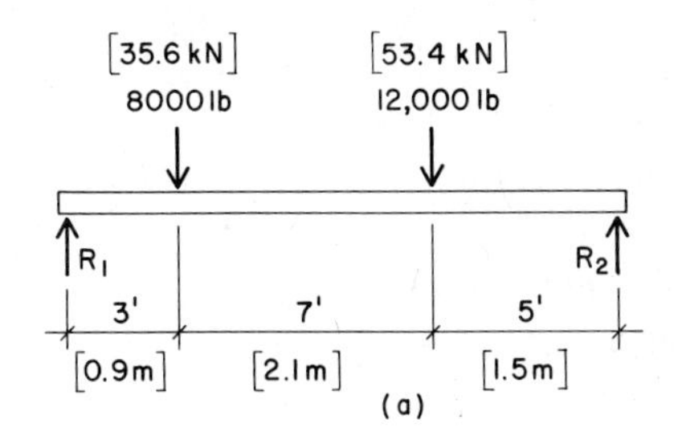

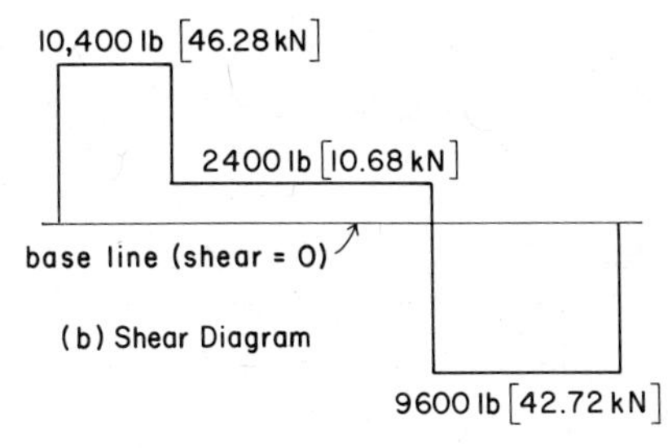

FIGURE 5-11

Solution: Taking R_2 as the center of moments,

$$15R_1 = (8000 \times 12) + (12{,}000 \times 5)$$

$$15R_1 = 156{,}000 \quad \text{and} \quad R_1 = 10{,}4000 \text{ lb}$$

$$[4.5R_1 = (35.6 \times 3.6) + (53.4 \times 1.5)]$$

$$[4.5R_1 = 208.26 \quad \text{and} \quad R_1 = 46.28 \text{ kN}]$$

With R_1 as the center of moments,

$$15R_2 = (8000 \times 3) + (12{,}000 \times 10)$$

$$15R_2 = 14{,}000 \quad \text{and} \quad R_2 = 9600 \text{ lb}$$

$$[4.5R_2 = (35.6 \times 0.9) + (53.4 \times 3.0)]$$

$$[4.5R_2 = 192.24 \quad \text{and} \quad R_2 = 42.72 \text{ kN}]$$

Now let us compute the value of the shear at various sections:

$$V_{(x=1)} = 10{,}400 - 0 = 10{,}400 \text{ lb}$$

$$V_{(x=4)} = 10{,}400 - 8000 = 2400 \text{ lb}$$

$$V_{(x=11)} = 10{,}400 - (8000 + 12{,}000) = -9600 \text{ lb}$$

$$\left[\begin{array}{r} V_{(x=0.3)} = 46.28 - 0 = 46.28 \text{ kN} \\ V_{(x=1.2)} = 46.28 - 35.6 = 10.68 \text{ kN} \\ V_{(x=3.3)} = 46.28 - (35.6 + 53.4) = -42.72 \text{ kN} \end{array}\right]$$

Because there are only two concentrated loads, it is unnecessary to compute the shear for other sections; for instance, the value of the shear at any section between the two loads is 2400 lb [10.68 kN].

We may now construct the shear diagram shown in Fig. 5-11*b* by plotting the three values of shear just computed, using the base line as the reference for zero shear. Positive values are plotted above the base line and negative values below the base line. Observing from the beam and its loads that the shear is a constant value between the points of location of the reactions and the loads, we can complete the shear diagram by drawing horizontal lines through the plotted points.

The value of the maximum shear is 10,400 lb [46.28 kN], the magnitude of the left reaction. On inspecting the shear diagram, we see that the shear passes through zero (the base line) at $x = 10$ ft [3.0 m] directly under the 12,000 lb [53.4 kN] load. Later on we shall see in the design of beams that it is important to know the location of this section. It is here that the greatest bending stresses occur.

So far we have considered a beam with only concentrated loads; now let us make a shear diagram for a uniformly distributed load.

Example 2. The beam shown in Fig. 5-12 has a span of 18 ft [5.4 m] and a uniformly distributed load of 500 lb per ft [7.3 kN/m]. Construct the shear diagram by noting the maximum shear and the section at which the shear passes through zero.

Solution: The total load is 500 × 18 = 9000 lb [7.3 × 5.4 = 39.42 kN]. Each of the reactions is equal to one-half the total load, or 4500 lb [19.71 kN], which is also the value of the maximum shear at the end of the beam. At a section 1 ft to the right of the left reaction the shear is equal to the reaction minus the increment of

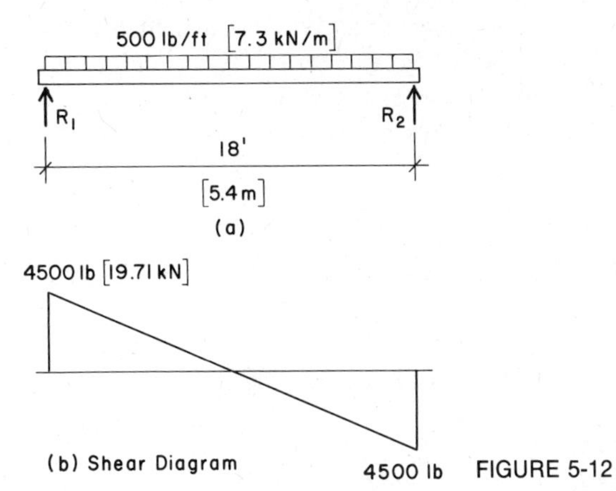

FIGURE 5-12

load on the 1-ft length of beam. Thus

$$V_{(x=1)} = 4500 - (500 \times 1) = 4000 \text{ lb}$$

$$[V_{(x=0.3)} = 19.71 - (7.3 \times 0.3) = 17.52 \text{ kN}]$$

The shears at some additional sections are as follows:

$$V_{(x=2)} = 4500 - (500 \times 2) = 3500 \text{ lb}$$

$$V_{(x=3)} = 4500 - (500 \times 3) = 3000 \text{ lb}$$

$$V_{(x=9)} = 4500 - (500 \times 9) = 0$$

$$V_{(x=12)} = 4500 - (500 \times 12) = -1500 \text{ lb}$$

$$V_{(x=18)} = 4500 - (500 \times 18) = -4500 \text{ lb}$$

$$\left[\begin{array}{c} V_{(x=0.6)} = 19.71 - (7.3 \times 0.6) = 15.33 \text{ kN} \\ V_{(x=0.9)} = 19.71 - (7.3 \times 0.9) = 13.14 \text{ kN} \\ V_{(x=2.7)} = 19.71 - (7.3 \times 2.7) = 0 \\ V_{(x=3.6)} = 19.71 - (7.3 \times 3.6) = -6.57 \text{ kN} \\ V_{(x=5.4)} = 19.71 - (7.3 \times 5.4) = -19.71 \text{ kN} \end{array}\right]$$

A plotting of these points, as shown in Fig. 5-12*b*, indicates that the shear diagram consists of a sloping straight line that passes through zero at the center of the span.

Example 3. Draw the shear diagram and determine the value of the maximum shear and the section at which the shear passes through zero for the beam shown in Fig. 5-13.

Solution: To compute the reactions first take R_2 as the center of moments. Thus

$$16R_1 = (6000 \times 10) + (200 \times 16 \times 8)$$

$$16R_1 = 85{,}600 \quad \text{and} \quad R_1 = 5350 \text{ lb}$$

$$\left[\begin{array}{l} 4.8R_1 = (26.7 \times 3.0) + (2.92 \times 4.8 \times 2.4) \\ 4.8R_1 = 113.7384 \quad \text{and} \quad R_1 = 23.70 \text{ kN} \end{array}\right]$$

With R_1 as the center of moments,

$$16R_2 = (6000 \times 6) + (200 \times 16 \times 8) =$$
$$16R_2 = 61{,}600 \quad \text{and} \quad R_2 = 3850 \text{ lb}$$

$$\left[\begin{array}{l} 4.8R_2 = (26.7 \times 1.8) + (2.92 \times 4.8 \times 2.4) \\ 4.8R_2 = 81.6984 \quad \text{and} \quad R_2 = 17.02 \text{ kN} \end{array}\right]$$

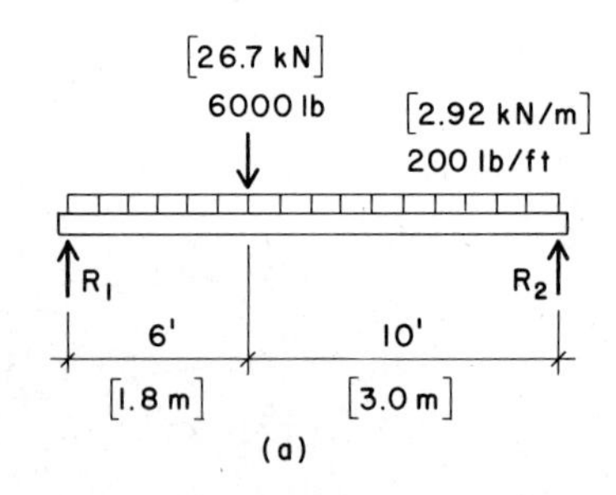

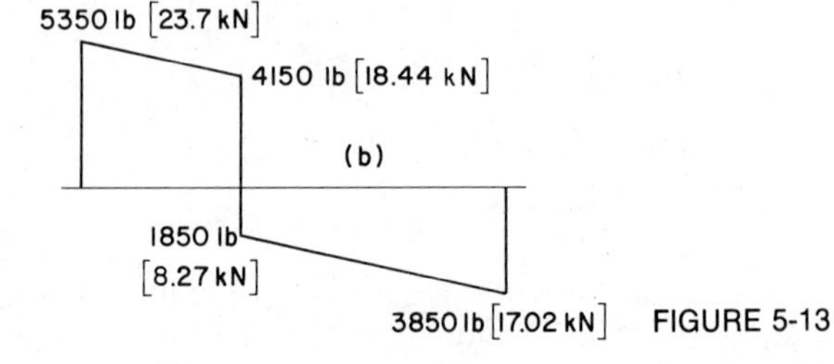

FIGURE 5-13

We know that the shear at R_1 is 5350 lb [23.70 kN]. Now let us compute the shear at a section infinitely close to, and to the left of, the concentrated load. We call this distance ($x = 6-$) from R_1. Then

$$V_{(x=6-)} = 5350 - (200 \times 6) = 4150 \text{ lb}$$

$$[V_{(x=1.8-)} = 23.70 - (2.92 \times 1.8) = 18.44 \text{ kN}]$$

We next consider the section just to the right of the concentrated load; call it ($x = 6+$) from R_1. Then

$$V_{(x=6+)} = 5350 - (200 \times 6) - 6000 = -1850 \text{ lb}$$

$$[V_{(x=1.8+)} = 23.70 - (2.92 \times 1.8) - 26.7 = -8.27 \text{ kN}]$$

Using these two computed values and the known values of shear at the ends of the beam, we can construct the diagram, as shown in Fig. 5-13*b*, noting that the maximum shear is at R_1 and that the shear passes through zero under the concentrated load.

Example 4. Construct the shear diagram for the beam shown in Fig. 5-14*a*.

Solution: With R_2 as the center of moments,

$$20R_1 = (800 \times 10 \times 15) + (4000 \times 6)$$

$$20R_1 = 144{,}000 \quad \text{and} \quad R_1 = 7200 \text{ lb}$$

$$\begin{bmatrix} 6.0R_1 = (11.67 \times 3.0 \times 4.5) + (17.8 \times 1.8) \\ 6.0R_1 = 189.585 \quad \text{and} \quad R_1 = 31.60 \text{ kN} \end{bmatrix}$$

With R_2 as the center of moments,

$$20R_2 = (800 \times 10 \times 5) + (4000 \times 14)$$

$$20R_2 = 96{,}000 \quad \text{and} \quad R_2 = 4800 \text{ lb}$$

$$\begin{bmatrix} 6.0R_2 = (11.67 \times 3.0 \times 1.5) + (17.8 \times 4.2) \\ 6.0R_2 = 127.275 \quad \text{and} \quad R_2 = 21.21 \text{ kN} \end{bmatrix}$$

We observe that the value of the shear at the left end is the same as R_1 and that at the right end is the same as R_2. Computing the

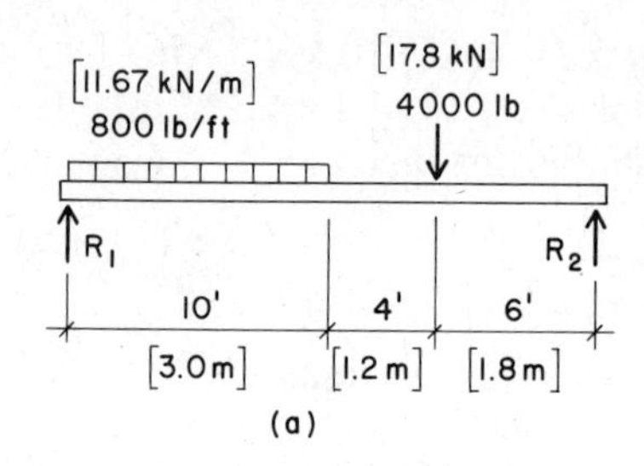

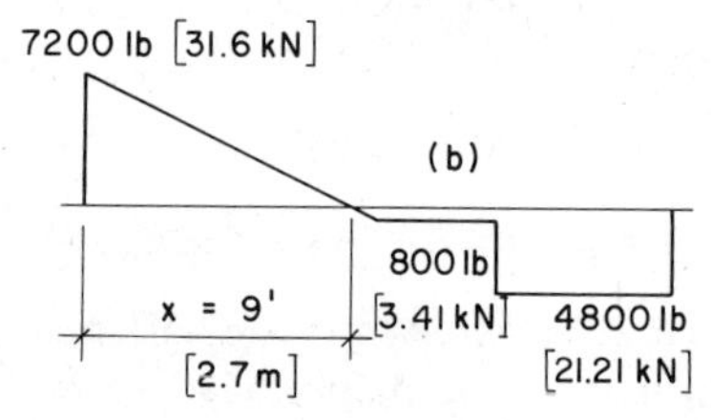

FIGURE 5-14

shears at other critical points, we obtain

$$V_{(x=10)} = 7200 - (800 \times 10) = -800 \text{ lb}$$

$$V_{(x=14-)} = 7200 - (800 \times 10) = -800 \text{ lb}$$

$$V_{(x=14+)} = 7200 - (800 \times 10) - 4000 = -4800 \text{ lb}$$

$$\left[\begin{array}{c} V_{(x=3.0)} = 31.60 - (11.67 \times 3.0) = -3.41 \text{ kN} \\ V_{(x=4.2-)} = 31.60 - (11.67 \times 3.0) = -3.41 \text{ kN} \\ V_{(x=4.2+)} = 31.60 - (11.67 \times 3.0) - 17.8 = -21.21 \text{ kN} \end{array}\right]$$

These various points are plotted; for the shear diagram see Fig. 5-14*b*.

An inspection of the shear diagram shows that the shear passes through zero at some point between R_1 and the end of the distributed load. We call this distance x and write an equation for the shear at this distance. Thus

$$V = 0 = 7200 - (800 \times x)$$

$$800x = 7200 \quad \text{and} \quad x = 9 \text{ ft}$$

$$\left[\begin{array}{c} 0 = 31.60 - (11.67 \times x) \\ x = 31.60/11.67 = 2.7 \text{ m} \end{array}\right]$$

Example 5. Construct the shear diagram for the beam shown in Fig. 5-15*a*.

Solution: With R_2 as the center of moments,

$$20R_1 + (6000 \times 4) = (4000 \times 26) + (8000 \times 14)$$

$$20R_1 = 192{,}000 \quad \text{and} \quad R_1 = 9600 \text{ lb}$$

$$\left[\begin{array}{c} 6.0R_1 + (26.7 \times 1.2) = (17.8 \times 7.8) + (35.6 \times 4.2) \\ 6.0R_1 = 256.32 \quad \text{and} \quad R_1 = 42.72 \text{ kN} \end{array}\right]$$

With R_1 as the center of moments,

$$20R_2 + (4000 \times 6) = (8000 \times 6) + (6000 \times 24)$$

$$20R_2 = 168{,}000 \quad \text{and} \quad R_2 = 8400 \text{ lb}$$

$$\left[\begin{array}{c} 6.0R_2 + (17.8 \times 1.8) = (35.6 \times 1.8) + (26.7 \times 7.2) \\ 6.0R_2 = 224.28 \quad \text{and} \quad R_2 = 37.38 \text{ kN} \end{array}\right]$$

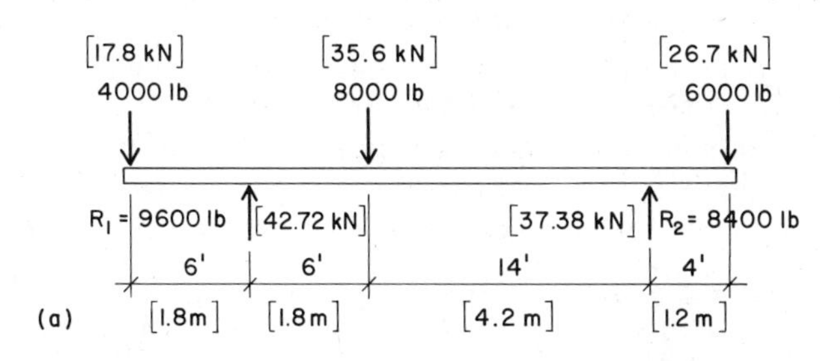

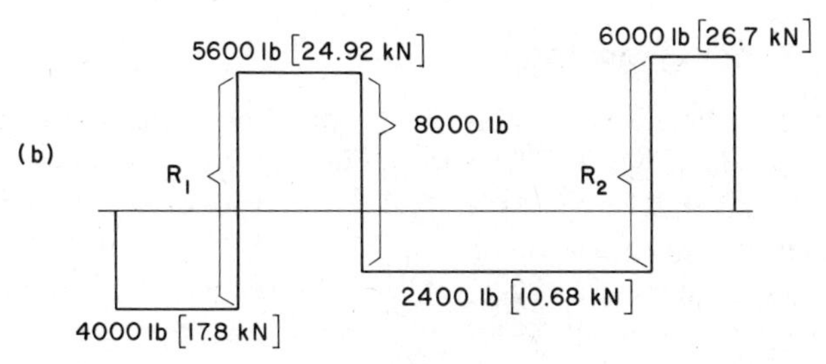

FIGURE 5-15

Then

$$V_{(x=1)} = -4000 \text{ lb}$$

$$V_{(x=7)} = 9600 - 4000 = 5600 \text{ lb}$$

$$V_{(x=13)} = 9600 - 4000 - 8000 = -2400 \text{ lb}$$

$$V_{(x=27)} = 9600 + 8400 - 4000 - 8000 = 6000 \text{ lb}$$

$$\left[\begin{array}{c} V_{(x=0.3)} = -17.8 \text{ kN} \\ V_{(x=2.1)} = 42.72 - 17.8 = 24.92 \text{ kN} \\ V_{(x=3.9)} = 42.72 - 17.8 - 35.6 = -10.68 \text{ kN} \\ V_{(x=8.1)} = 42.72 + 37.38 - 17.8 - 35.6 = 26.7 \text{ kN} \end{array}\right]$$

The shear at other sections might be computed, but we have sufficient information for our purpose. The values computed are now plotted with respect to the base line and the shear diagram shown in Fig. 5-15*b* results.

This diagram differs somewhat from those previously constructed. For simple beams the maximum shear is the value of the greater reaction; but this is an overhanging beam and we see that the maximum shear has a value of 6000 lb [26.7 kN]. We observe also that in this beam the shear passes through zero at three different points; at the two supports and under the 8000 lb [35.6 kN] load. The significance is explained later.

Problems 5.9.A, B, C, D,* E,* F. Construct the shear diagrams for the beams shown in Fig. 5-16*a, b, c, d, e,* and *f*. In each case note the magnitude of the maximum shear and the section at which the shear passes through zero.

5-10. Bending Moments

As noted earlier, a beam deforms by bending under the action of applied loads. Figure 5-1*a* shows the deformation curve for a simple beam that supports a concentrated load at midspan; the deformation curve for the beam in Fig. 5-17*a* would have the same general form. At any point between the left reaction and the 4000-lb [17.8-kN] load the tendency for the beam to bend is mea-

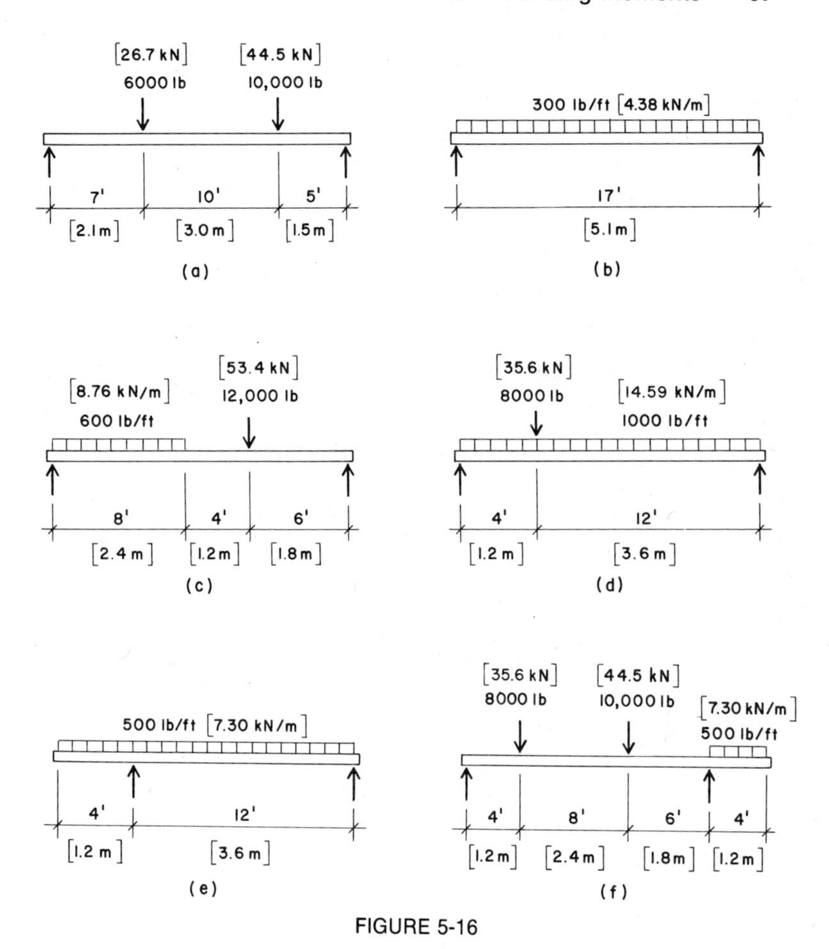

FIGURE 5-16

sured by the moment of the left reaction about the point in question; for example, consider a point 5 ft [1.5 m] to the right of the left support. The moment of the reaction about this point as the center of moments is 1500 × 5 or 7500 ft-lb [6.675 × 1.5 = 10 kN-m]. This moment is called the *bending moment* and its value may be plotted as a point on a moment diagram (Fig. 5-17*c*). The magnitude of the bending moment varies at different sections

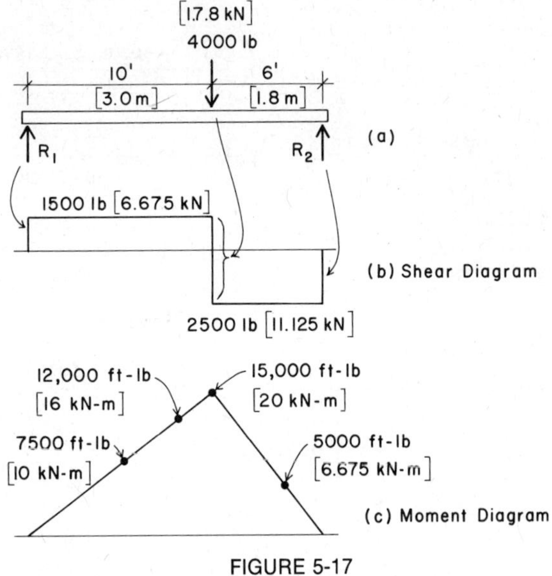

FIGURE 5-17

along the beam span. Thus

$$M_{(x=5)} = (1500 \times 5) = 7500 \text{ ft-lb}$$

$$M_{(x=8)} = (1500 \times 8) = 12{,}000 \text{ ft-lb}$$

$$M_{(x=10)} = (1500 \times 10) = 15{,}000 \text{ ft-lb}$$

$$\left[\begin{aligned} M_{(x=1.5)} &= (6.675 \times 1.5) = 10 \text{ kN-m} \\ M_{(x=2.4)} &= (6.675 \times 2.4) = 16 \text{ kN-m} \\ M_{(x=3.0)} &= (6.675 \times 3.0) = 20 \text{ kN-m} \end{aligned}\right]$$

Suppose we use the same procedure but take the reference from the other reaction. Then

$$M_{(x=2)} = (2500 \times 2) = 5000 \text{ ft-lb}$$

$$M_{(x=6)} = (2500 \times 6) = 15{,}000 \text{ ft-lb}$$

$$\left[\begin{aligned} M((_{x=0.6)} &= (11.125 \times 0.6) = 6.675 \text{ kN-m} \\ M((_{x=1.8)} &= (11.125 \times 1.8) = 20 \text{ kN-m} \end{aligned}\right]$$

The last value computed is the same as that obtained by proceeding from the other end. Thus we observe the following:

The bending moment at any section of a beam is equal to the algebraic sum of the moments of all forces on one side of the section.

For convenience call upward forces (the reactions) positive quantities, the downward forces (the loads) negative, and consider the forces to the left of the section. Then, in conformity with the foregoing statement, we can make the following rule:

The bending moment at any section of a beam is equal to the moments of the reactions minus the moments of the loads to the left of the section.

Actually this rule will also apply if the reactions and loads to the right of the section are used, as demonstrated in the previous computations. As we have already established the practice of proceeding from the left in the construction of the shear diagram, it will prove to be significant to do the same with moments.

5-11. Bending Moment Diagrams

Figure 5-17*c* is a bending moment diagram, four ordinates of which were computed in the preceding article. Bending moment diagrams are constructed quite like the shear diagrams. A base line ($M = 0$) is drawn and the values of the bending moments at various sections along the beam are plotted to scale. We compute the magnitudes of the moments in accordance with the rule given in Art. 5-10. The section at which the moment is to be computed is noted by using the subscripts $x = 1$, $x = 2$, etc. Then we write the moments of the reaction and subtract the moments of the loads to the left of the section. When we have established a sufficient number of values of bending moments we may construct the complete diagram.

In the design of steel beams we are particularly concerned with the maximum value of the bending moment. It will be noted that this value always occurs at a point at which the shear diagram passes through zero, which is why we have emphasized noting these points when shear diagrams are constructed.

5-12. Positive and Negative Bending Moments

Figure 5-18*a* illustrates the shape a simple beam tends to assume when it bends. The fibers in the upper surface of the beam are in compression; those in the lower surface are in tension. We say the beam is "concave upward" and define the bending moment under this condition as *positive*. Now refer to Fig. 5-18*b*, an overhanging beam. From the sketch we see that a portion of the beam in the vicinity of the right reaction has tension in the upper fibers and compression in the lower; here the beam is "concave downward" and the bending moment is negativc in sign. The section at which the bending moment changes from positive to negative is called the *inflection point*. As the following examples demonstrate, the inflection point corresponds to a value of zero moment on the moment diagram.

Problem 5.12.A.* Construct the bending moment diagram for the beam shown in Fig. 5-11*a* and note the value of the maximum bending moment.

Problem 5.12.B. Construct the bending moment diagram and note the value of the maximum bending moment for the beam in Fig. 5-7.

5-13. Concentrated Load at Center of Span

A condition that occurs frequently in practice is a concentrated load at the center of the span of a simple beam. In Fig. 5-19*a* let *P*

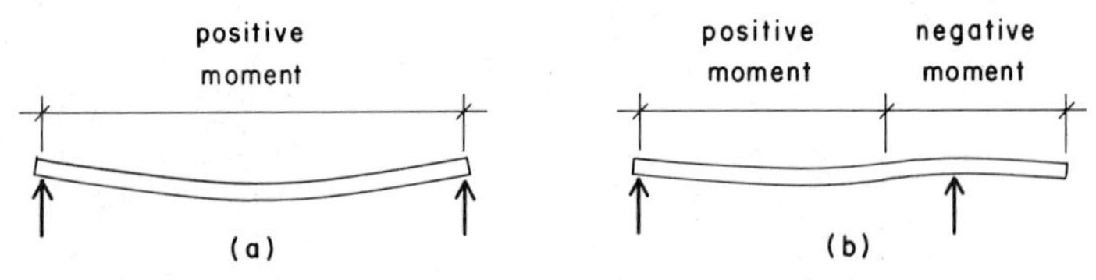

FIGURE 5-18. Positive and negative bending moments.

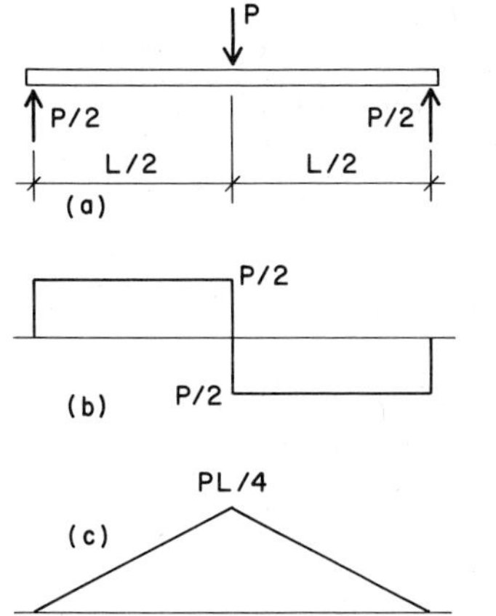

FIGURE 5-19

denote the concentrated load and L, the span length. Then, as the beam is symmetrical, each reaction is equal to $P/2$; see the shear diagram in Fig. 5-19*b*. This diagram shows that the shear passes through zero at the center of the span at $x = L/2$. At this section the value of the bending moment will be maximum. By following the rule given in Art. 4-10 its magnitude is readily found:

$$M_{(x=L/2)} = \frac{P}{2} \times \frac{L}{2} = \frac{PL}{4} \quad \text{(See Fig. 5-19}c\text{)}$$

This is a most useful expression; for instance, suppose we are to design a beam with a span of 24 ft [7.2 m] and a concentrated load of 18 kips [80 kN] at midspan. It will be necessary to compute the maximum bending moment, which is readily done for we have found that its value is $PL/4$. Thus

$$M = \frac{PL}{4} = \frac{18 \times 24}{4} = 108 \text{ k-ft}$$

$$\left[\frac{80 \times 7.2}{4} = 144 \text{ kN-m}\right]$$

If the bending moment is required in units of kip-in., as it frequently is, $M = 108 \times 12 = 1296$ k-in.

5-14. Simple Beam with Uniformly Distributed Load

Another condition that occurs perhaps more times than any other is represented by a simple beam with a uniformly distributed load (Fig. 5-20*a*). To develop a formula for its maximum bending moment let L be the span in feet and w the uniform load in lb per lin ft. Then the total load is wL, each reaction is $wL/2$, and the shear diagram takes the form shown in Fig. 5-20*b*. (See also Fig. 5-12.) The maximum bending moment is at the center of the span. Then by application of the rule given in Art. 5-10 the value of the bending moment at midspan is

$$M = \left(\frac{wL}{2} \times \frac{L}{2}\right) - \left(\frac{wL}{2} \times \frac{L}{4}\right) = \frac{wL^2}{8}$$

In this expression the load per lin foot is w lb and the total load is wL. If the total load is represented by W, we have

$$W = wL \quad \text{and} \quad M = \frac{wL^2}{8} = \frac{WL}{8}$$

which is another useful form of the equation. If moments are computed for other sections along the beam span, the values will plot to form a parabola with its apex at midspan (Fig. 5-20*c*).

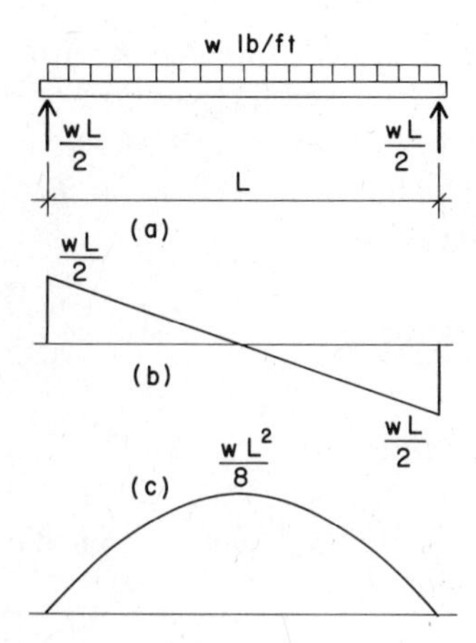

FIGURE 5-20

Example. A simple beam with a span of 22 ft [6.6 m] has a uniformly distributed load of 800 lb/ft [12 kN/m]. Compute the maximum bending moment.

Solution: Substituting in the formula that was previously derived, we obtain

$$M = \frac{wL^2}{8} = \frac{800 \times 22 \times 22}{8} = 48{,}400 \text{ ft-lb}$$

$$\left[M = \frac{12 \times 6.6 \times 6.6}{8} = 65.34 \text{ kN-m}\right]$$

This is the value of the bending moment at the center of the span. It is computed quickly by using the formula. If, however, we were required to construct the entire bending moment diagram, it would be necessary to compute the values at several other sections so that a smooth curve could be plotted. In this example R_1 is equal to half the total load on the beam, or 8800 lb [39.6 kN]; the bending moments at some other points are

$$M_{(x=4)} = (8800 \times 4) - (800 \times 4 \times 2) = 28{,}800 \text{ ft-lb}$$

$$M_{(x=8)} = (8800 \times 8) - (800 \times 8 \times 4) = 44{,}800 \text{ ft-lb}$$

$$M_{(x=11)} = (8800 \times 11) - (800 \times 11 \times 5.5) = 48{,}400 \text{ ft-lb}$$

$$M_{(x=16)} = (8800 \times 16) - (800 \times 16 \times 8) = 38{,}400 \text{ ft-lb}$$

$$\left[\begin{aligned} M_{(x=1.2)} &= (39.6 \times 1.2) - (12 \times 1.2 \times 0.6) = 38.88 \text{ kN-m} \\ M_{(x=2.4)} &= (39.6 \times 2.4) - (12 \times 2.4 \times 1.2) = 60.48 \text{ kN-m} \\ M_{(x=3.3)} &= (39.6 \times 3.3) - (12 \times 3.3 \times 1.65) = 65.34 \text{ kN-m} \\ M_{(x=4.8)} &= (39.6 \times 4.8) - (12 \times 4.8 \times 2.4) = 51.84 \text{ kN-m} \end{aligned}\right]$$

The reader should verify that a plot of these points will form a moment diagram like that in Fig. 5-20*c*.

Problem 5.14.A. A simple beam has a span of 16 ft [4.8 m] and a uniformly distributed load of 600 lb/ft [8.76 kN/m]. Compute the bending moment at the center and at the quarter points of the span and draw the shear and bending moment diagrams.

5-15. Maximum Bending Moments

To compute the magnitude of the maximum bending moment for a beam we must first know its position along the span. For an unsymmetrical loading it is generally necessary to construct the shear diagram. Then, as noted earlier, the section at which the shear passes through zero is the section at which maximum bending moment occurs. The following examples relate to unsymmetrically loaded beams and show the method of determining the maximum bending moments.

Example 1. The simple beam shown in Fig. 5-21*a* has three concentrated loads at the positions indicated. Compute the maximum bending moment.

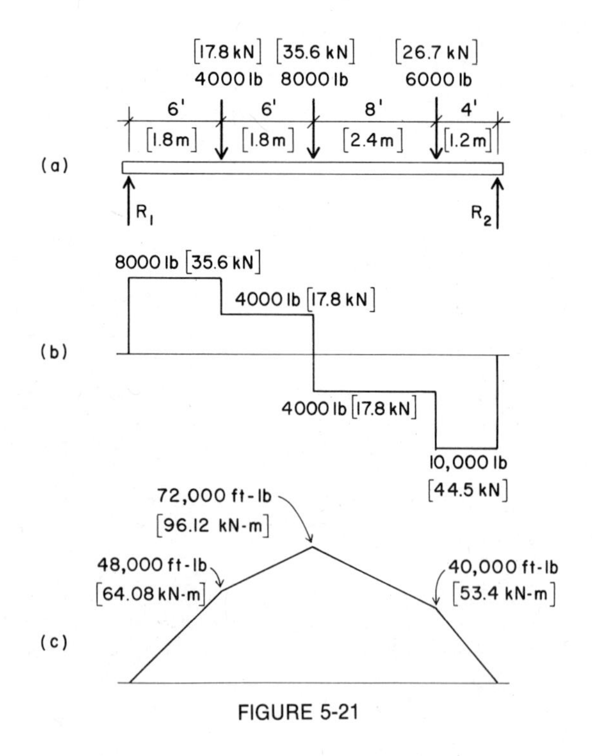

FIGURE 5-21

Solution: To find the reactions we may determine R_1 by writing a moment equation and R_2 by summing the vertical forces.

$$24R_1 = (4000 \times 18) + (8000 \times 12) + (6000 \times 4)$$

$$24R_1 = 192{,}000 \quad \text{and} \quad R_1 = 8000 \text{ lb}$$

$$4000 + 8000 + 6000 = 8000 + R_2$$

$$18{,}000 - 8000 = R_2 \quad \text{and} \quad R_2 = 10{,}000 \text{ lb}$$

$$\left[\begin{array}{c} 7.2R_1 = (17.8 \times 5.4) + (35.6 \times 3.6) + (26.7 \times 1.2) \\ 7.2R_1 = 256.32 \quad \text{and} \quad R_1 = 35.6 \text{ kN} \\ 17.8 + 35.6 + 26.7 = 35.6 + R_2 \\ 80.1 - 35.6 = R_2 \quad \text{and} \quad R_2 = 44.5 \text{ kN} \end{array}\right]$$

The value of the shear is computed at critical sections; the shear diagram is constructed as shown in Fig. 5-21*b*. In this diagram we find that the shear passes through zero under the 8000-lb [35.6-kN] load. The maximum bending moment is therefore

$$M_{(x=12)} = (8000 \times 12) - (4000 \times 6) = 72{,}000 \text{ ft-lb}$$

$$[M_{(x=3.6)} = (35.6 \times 3.6) - (17.8 \times 1.8) = 96.12 \text{ kN-m}]$$

To construct the bending moment diagram the expressions for moments under the other loads are

$$M_{(x=6)} = 8000 \times 6 = 48{,}000 \text{ ft-lb}$$

$$M_{(x=20)} = (8000 \times 20) - (4000 \times 14) - (8000 \times 8)$$
$$= 40{,}000 \text{ ft-lb}$$

$$\left[\begin{array}{l} M_{(x=1.8)} = 35.6 \times 1.8 = 64.08 \text{ kN-m} \\ M_{(x=6.0)} = (35.6 \times 6.0) - (17.8 \times 4.2) - (35.6 \times 2.4) \\ \quad = 53.40 \text{ kN-m} \end{array}\right]$$

The diagram in Fig. 5-21*c* may now be constructed.

Example 2. A simple beam with a span of 16 ft [4.8 m] sustains the loading shown in Fig. 5-22*a*. Compute the maximum bending moment.

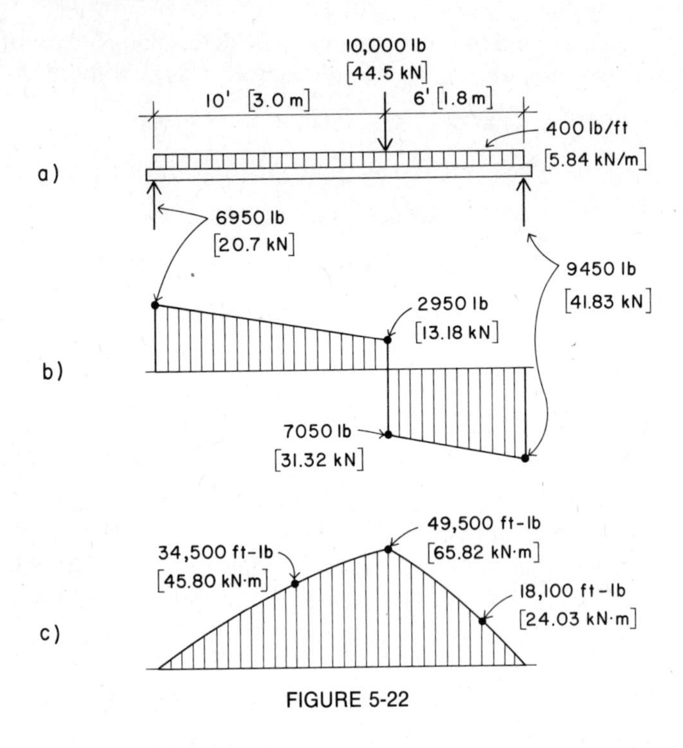

FIGURE 5-22

Solution: Determine R_1 and R_2 by using one moment equation and then summing the vertical forces:

$$16R_1 = (400 \times 16 \times 8) + (10{,}000 \times 6)$$

$$16R_1 = 111{,}200 \quad \text{and} \quad R_1 = 6950 \text{ lb}$$

$$(400 \times 16) + 10{,}000 = 6950 + R_2$$

$$16{,}400 - 6950 = R_2 \quad \text{and} \quad R_2 = 9450 \text{ lb}$$

$$\left[\begin{array}{c} 4.8R_1 = (5.84 \times 4.8 \times 2.4) + (44.5 \times 1.8) \\ 4.8R_1 = 147.3768 \quad \text{and} \quad R_1 = 30.70 \text{ kN} \\ (5.84 \times 4.8) + 44.5 = 30.70 + R_2 \\ 72.532 - 30.70 = R_2 \quad \text{and} \quad R_2 = 41.832 \text{ kN} \end{array}\right]$$

In constructing the shear diagram (Fig. 5-22*b*), we find that the maximum bending moment (point of zero shear) occurs under the concentrated load. Thus

$$M_{(x=10)} = (6950 \times 10) - (400 \times 10 \times 5) = 49{,}500 \text{ ft-lb}$$

$$[M_{(x=3.0)} = (30.7 \times 3.0) - (5.84 \times 3.0 \times 1.5) = 65.82 \text{ kN-m}]$$

To draw the bending moment diagram more accurately two additional values may be computed:

$$M_{(x=6)} = (6950 \times 6) - (400 \times 6 \times 3) = 34{,}500 \text{ ft-lb}$$

$$M_{(x=14)} = (6950 \times 14) - (400 \times 14 \times 7) - (10{,}000 \times 4) = 18{,}100 \text{ ft-lb}$$

$$\left[\begin{aligned} M_{(x=1.8)} &= (30.7 \times 1.8) - (5.84 \times 1.8 \times 0.9) - 45.80 \text{ kN-m} \\ M_{(x=4.2)} &= (30.7 \times 4.2) - (5.84 \times 4.2 \times 2.1) - (44.5 \times 1.2) \\ &= 24.03 \text{ kN-m} \end{aligned}\right]$$

Example 3. Compute the maximum bending moment for the beam shown in Fig. 5-14*a*.

Solution: Inspection of the loading diagram (Fig. 5-14*a*) gives no indication of the position of the maximum bending moment. However, after computing the reactions and drawing the shear diagram (Fig. 5-14*b*) it was found that the shear passed through zero at a section 9 ft [2.7 m] from the left support. The value of the bending moment at this section is

$$M_{(x=9)} = (7200 \times 9) - (800 \times 9 \times 4.5) = 32{,}400 \text{ ft-lb}$$

$$\left[\begin{aligned} M_{(x=2.7)} &= (31.6 \times 2.7) - (11.67 \times 2.7 \times 1.35) \\ &= 42.78 \text{ kN-m} \end{aligned}\right]$$

5-16. Overhanging Beams

For the simple beams previously discussed we have seen that the maximum shear has the same magnitude as the greater reaction. This is not true for overhanging beams; for example, we found that the beam shown in Fig. 5-15, Art. 5-9, had R_1 as the greater reaction, with a value of 9600 lb [42.72 kN]. The maximum shear, on the other hand, was 6000 lb [26.7 kN] and occurred just to the right of R_2. We noted also that the shear diagram for this beam passed through zero at three sections along its length; therefore it is necessary to determine at which of these sections the actual maximum bending moment occurs. This determination is accomplished by constructing the shear and moment diagrams, exercising particular care when plotting the positive and negative values. We take as an example the beam shown in Fig. 5-15 but restated here as Fig. 5-23.

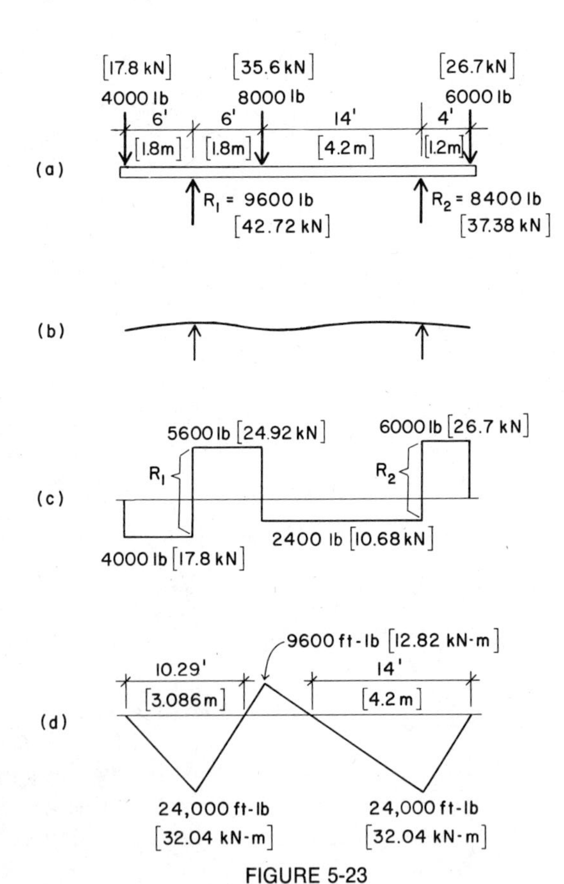

FIGURE 5-23

Example. Construct the shear and bending moment diagrams for the overhanging beam in Fig. 5-23*a* and note the values of the maximum shear and maximum bending moment.

Solution: The shear diagram is constructed as explained for Fig. 5-15*b* and shown here as Fig. 5-23*c*. The maximum shear is 6000 lb [26.7 kN].

When determining bending moments in overhanging beams it is helpful to sketch the approximate deformation curves. (See Fig. 5-23*b* and compare with Fig. 5-1*c*, in which the overhang occurs at one end only.) The curve for this beam and loading shows that we will encounter positive and negative bending moments, as defined in Art. 5-12. Now let us compute the bending moments at certain sections:

$$M_{(x=6)} = -(4000 \times 6) = -24{,}000 \text{ ft-lb}$$

$$M_{(x=12)} = (9600 \times 6) - (4000 \times 12) = +9600 \text{ ft-lb}$$

$$M_{(x=26)} = (9600 \times 20) - (4000 \times 26) - (8000 \times 14)$$
$$= -24{,}000 \text{ ft-lb}$$

$$\left[\begin{aligned} M_{(x=1.8)} &= -(17.8 \times 1.8) = -32.04 \text{ kN-m} \\ M_{(x=3.6)} &= (42.72 \times 1.8) - (17.8 \times 3.6) = +12.82 \text{ kN-m} \\ M_{(x=7.8)} &= (42.72 \times 6.0) - (17.8 \times 7.8) - (35.6 \times 4.2) \\ &= -32.04 \text{ kN-m} \end{aligned}\right]$$

Other values might be computed, but because there are no uniformly distributed loads it is unnecessary. The values just computed are plotted with respect to the base line: positive values above and negative values below. Figure 5-23*d* is the completed diagram. The points at which the bending moment diagram passes through zero are of special significance; they are discussed in the following article and the determination of their location is illustrated. The maximum value of the bending moment is 24,000 ft-lb [32.04 kN-m]; it is negative and occurs over both supports. (It is only coincidental that the moment has the same magnitude over both supports in this example.)

For beams with complicated loading patterns it is advantageous to compute bending moments from both ends to simplify the arithmetic. In a procedure that will always give the correct sign for the bending moment the moments of all upward forces are considered as positive and the moments of all downward forces, as negative. Using this convention, we compute the bending moment over the right reaction in Fig. 5-23:

$$M = -(6000 \times 4) = -24{,}000 \text{ ft-lb}$$

$$[M = -(26.7 \times 1.2) = 32.04 \text{ kN-m}]$$

which is the same value as that obtained, working from the left end, for M at $x = 26$ ft [7.8 m].

5-17. Inflection Point

The two points in Fig. 5-23*d*, where the bending moment diagram passes through zero, are called the *inflection points* or *points* of *contraflexure*. The inflection point is the section along the beam length at which the curvature of the beam changes and at which the value of the bending moment diagram is zero. On either side of this section the positions of the tension and compression stresses in the beam section reverse. This may be observed by inspecting the deformed shape (Fig. 5-23*b*).

Example 1. Determine the position of the inflection point to the right of the left reaction for the beam in Fig. 5-23.
Solution: Let x be the distance from the inflection point to the left end of the beam. The expression for bending moment at this point is

$$M = \{9600 \times (x - 6)\} - (4000 \times x)$$

$$[M = \{42.72 \times (x - 1.8)\} - (17.8 \times x)]$$

We know that the value of the moment at this section is zero. Then

$$0 = \{9600 \times (x - 6)\} - (4000 \times x)$$

$$9600x - 57{,}600 = 4000x$$

$$5600x = 57{,}600 \quad \text{and} \quad x = 10.29 \text{ ft}$$

$$\left[\begin{array}{c} 0 = \{42.72 \times (x - 1.8)\} - (17.8 \times x) \\ 42.72x - 76.896 = 17.8x \\ 24.92x = 76.896 \quad \text{and} \quad x = 3.086 \text{ m} \end{array}\right]$$

To find the position of the inflection point to the left of the right support it will simplify the mathematics if we consider the moments of the forces to the *right* of the section instead of the left. Let x be the distance from the inflection point to the right end of the beam. Then, as before,

$$0 = \{8400 \times (x - 4)\} - (6000 \times x)$$

$$8400x - 33{,}600 = 6000x$$

$$2400x = 33{,}600 \quad \text{and} \quad x = 14 \text{ ft}$$

$$\left[\begin{array}{c} 0 = \{37.38 \times (x - 1.2)\} - (26.7 \times x) \\ 37.38x - 44.856 = 26.7x \\ 10.68x = 44.856 \quad \text{and} \quad x = 4.2 \text{ m} \end{array}\right]$$

Example 2. The overhanging beam shown in Fig. 5-24*a* has a uniformly distributed load of 200 lb/ft [2.92 kN/m] over its entire length. Construct the shear and moment diagrams, note the values of maximum shear and maximum bending moment, and compute the position of the inflection point.

Solution: The curve the bent beam will take is approximated in Fig. 5-24*b*; we shall expect to find both positive and negative bending moments. By computing the reactions we obtain

$$18R_1 = 200 \times 24 \times 6$$

$$18R_1 = 28{,}800$$

$$R_1 = 1600 \text{ lb}$$

$$18R_2 = 200 \times 24 \times 12$$

$$18R_2 = 57{,}600$$

$$R_2 = 3200 \text{ lb}$$

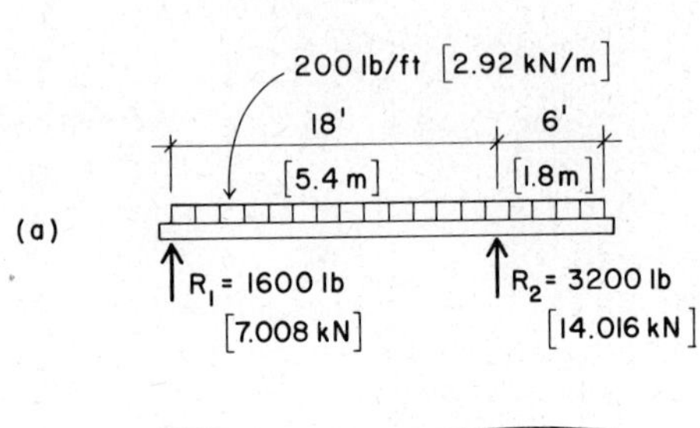

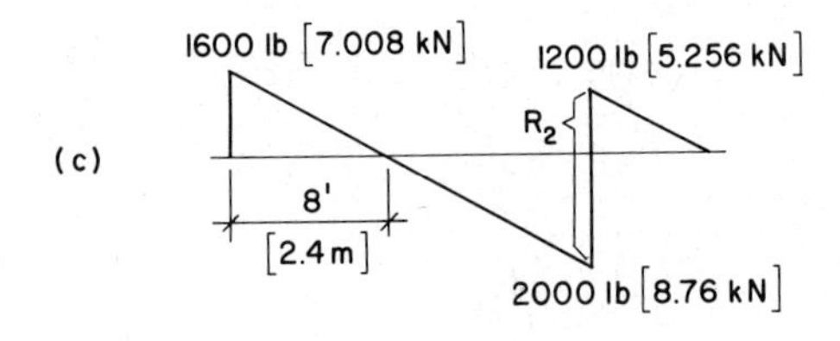

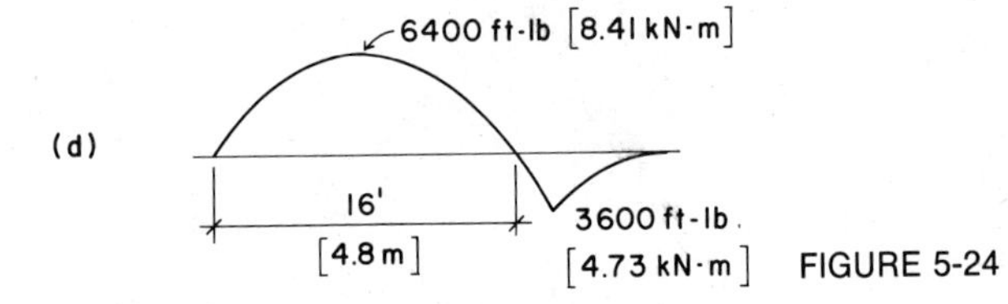

FIGURE 5-24

$$\left[\begin{aligned} 5.4R_1 &= 2.92 \times 7.2 \times 1.8 \\ 5.4R_1 &= 37.8432 \\ R_1 &= 7.008 \text{ kN} \\ 5.4R_2 &= 2.92 \times 7.2 \times 3.6 \\ 5.4R_2 &= 75.6864 \\ R_2 &= 14.016 \text{ kN} \end{aligned}\right]$$

Computing the values of the shears, we have

$$V \text{ at the left support} = 1600 \text{ lb}$$

$$V_{(x=18-)} = 1600 - (200 \times 18) = -2000 \text{ lb}$$

$$V_{(x=18+)} = 1600 - (200 \times 18) + 3200 = +1200 \text{ lb}$$

$$V_{(x=24)} = 1600 - (200 \times 24) + 3200 = 0$$

$$\left[\begin{array}{c} V \text{ at the left support} = 7.008 \text{ kN} \\ V_{(x=5.4-)} = 7.008 - (2.92 \times 5.4) = -8.760 \text{ kN} \\ V_{(x=5.4+)} = 7.008 - (2.92 \times 5.4) + 14.016 = +5.256 \\ V_{(x=7.2)} = 7.008 - (2.92 \times 7.2) + 14.016 = 0 \end{array}\right]$$

The shear diagram is plotted in Fig. 5-24*c* and the maximum shear value is 2000 lb [8.76 kN]. Because the shear passes through zero at two points, the bending moment diagram will have maximum values at two places; one, a positive moment, the other, a negative. The maximum negative moment is directly above the right support. To find the position of zero shear between the supports

$$0 = 1600 - (200 \times x)$$

$$200x = 1600 \quad \text{and} \quad x = 8 \text{ ft}$$

$$\left[\begin{array}{c} 0 = 7.008 - (2.92 \times x) \\ 2.92x = 7.008 \quad \text{and} \quad x = 2.4 \text{ m} \end{array}\right]$$

The bending moment will have values of zero at each end of the beam. Values for the other two critical sections are computed as follows:

$$M_{(x=8)} = (1600 \times 8) - (200 \times 8 \times 4) = 6400 \text{ ft-lb}$$

$$M_{(x=18)} = (1600 \times 18) - (200 \times 18 \times 9) = -3600 \text{ ft-lb}$$

$$\left[\begin{array}{c} M_{(x=2.4)} = (7.008 \times 2.4) - (2.92 \times 2.4 \times 1.2) = 8.4096 \text{ kN-m} \\ M_{(x=5.4)} = (7.008 \times 5.4) - (2.92 \times 5.4 \times 2.7) = -4.7304 \text{ kN-m} \end{array}\right]$$

The positive value is the maximum for the beam.

To find the position of the inflection point let x be its distance from the left support. The value of the bending moment at this section is zero. Then

$$(1600 \times x) - \left(200 \times x \times \frac{x}{2}\right) = 0$$

$$\frac{200x^2}{2} - 1600x = 0$$

$$100x^2 - 1600x = 0$$

$$x^2 - 16x = 0$$

$$\left[\begin{aligned} (7.008 \times x) - \left(2.92 \times x \times \frac{x}{2}\right) &= 0 \\ \frac{2.92x^2}{2} - 7.008x &= 0 \\ 1.46x^2 - 7.008x &= 0 \\ x^2 - 4.8x &= 0 \end{aligned}\right]$$

To complete the square

$$x^2 - 16x + 64 = 64$$

$$(x - 8)^2 = 64$$

$$\left[\begin{aligned} x^2 - 4.8x + 5.76 &= 5.76 \\ (x - 2.4)^2 &= 5.76 \end{aligned}\right]$$

By extracting the square root of both sides we obtain

$$x - 8 = 8 \quad \text{and} \quad x = 16 \text{ ft}$$

$$[x - 2.4 = 2.4 \quad \text{and} \quad x = 4.8 \text{ m}]$$

It may also be observed in this case that the moment diagram will be a simple symmetrical parabola and the distance to the inflection point will be twice that to the point of zero shear. This will not be true when any concentrated loads fall within this distance.

5-18. Cantilever Beams

The computations for shear and bending moments for the portion of cantilever beams that extends beyond the face of the support are quite simple. The vertical reaction at the support is equal to the sum of the loads on the beam. If the beam diagram is drawn with the support at the right, the rules given in Arts. 5-7 and 5-10 for computing shear and moments are used. The maximum values for both shear and bending moment are at the face of the support.

Example. The cantilever beam shown in Fig. 5-25*a* has a concentrated load at the free end and a uniformly distributed load applied over the first 6 ft [1.8 m] of length from the support. Construct the shear and bending moment diagrams and note the maximum values of each.

Solution: The value of the shear is computed at various sections:

$$V_{(x=1)} = -400 \text{ lb}$$

$$V_{(x=4)} = -400 \text{ lb}$$

$$V_{(x=6)} = -400 - (100 \times 2) = -600 \text{ lb}$$

$$V_{(x=10)} = -400 - (100 \times 6) = -1000 \text{ lb}$$

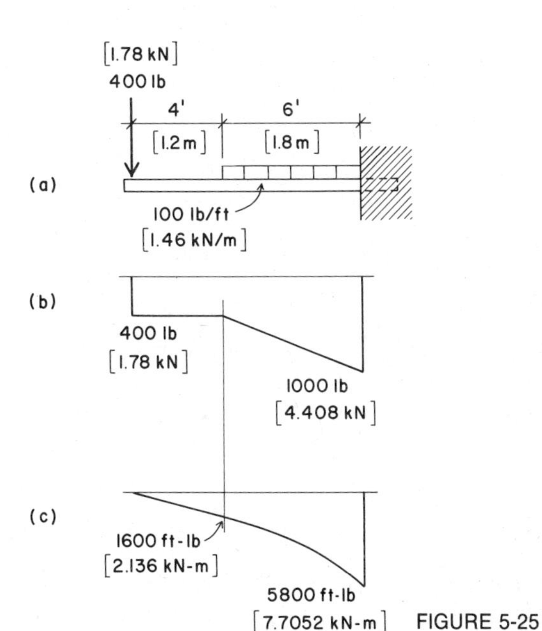

FIGURE 5-25

$$\begin{bmatrix} V_{(x=0.3)} = -1.78 \text{ kN} \\ V_{(x=1.2)} = -1.78 \text{ kN} \\ V_{(x=1.8)} = -1.78 - (1.46 \times 0.6) = -2.656 \text{ kN} \\ V_{(x=3.0)} = -1.78 - (1.46 \times 1.8) = -4.408 \text{ kN} \end{bmatrix}$$

The maximum value of the shear is at the support; the complete shear diagram is shown in Fig. 5-25*b*.

By computing the values of the bending moment at various sections we have

$$M_{(x=1)} = -(400 \times 1) = -400 \text{ ft-lb}$$

$$M_{(x=2)} = -(400 \times 2) = -800 \text{ ft-lb}$$

$$M_{(x=4)} = -(400 \times 4) = -1600 \text{ ft-lb}$$

$$M_{(x=8)} = -(400 \times 8) - (100 \times 4 \times 2) = -4000 \text{ ft-lb}$$

$$M_{(x=10)} = -(400 \times 10) - (100 \times 6 \times 3) = -5800 \text{ ft-lb}$$

$$\begin{bmatrix} M_{(x=0.3)} = -(1.78 \times 0.3) = -0.534 \text{ kN-m} \\ M_{(x=0.6)} = -(1.78 \times 0.6) = -1.068 \text{ kN-m} \\ M_{(x=1.2)} = -(1.78 \times 1.2) = -2.136 \text{ kN-m} \\ M_{(x=2.4)} = -(1.78 \times 2.4) - (1.46 \times 1.2 \times 0.6) = -5.323 \text{ kN-m} \\ M_{(x=3.0)} = -(1.78 \times 3.0) - (1.46 \times 1.8 \times 0.9) = -7.705 \text{ kN-m} \end{bmatrix}$$

The maximum bending moment is at the face of the support; its value is −5800 ft-lb [−7.705 kN-m]. The curve of the bending moment is a straight line from the free end up to the beginning of the uniformly distributed load and a portion of a parabola from that point to the support. In accordance with the convention established in Art. 5-12 the full length of the cantilever will be under a negative bending moment.

5-19. Typical Loads for Simple and Cantilever Beams

A simple beam with the load at the center of the span and a simple beam with a uniformly distributed load over its entire length are

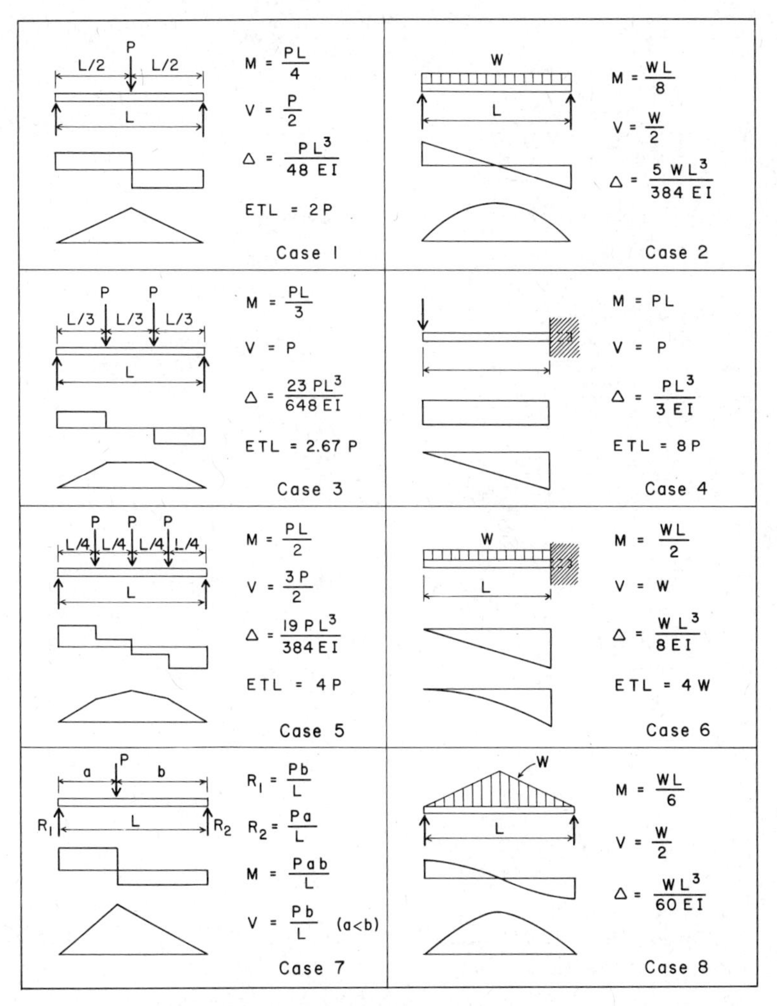

FIGURE 5-26. Values for typical beam loadings.

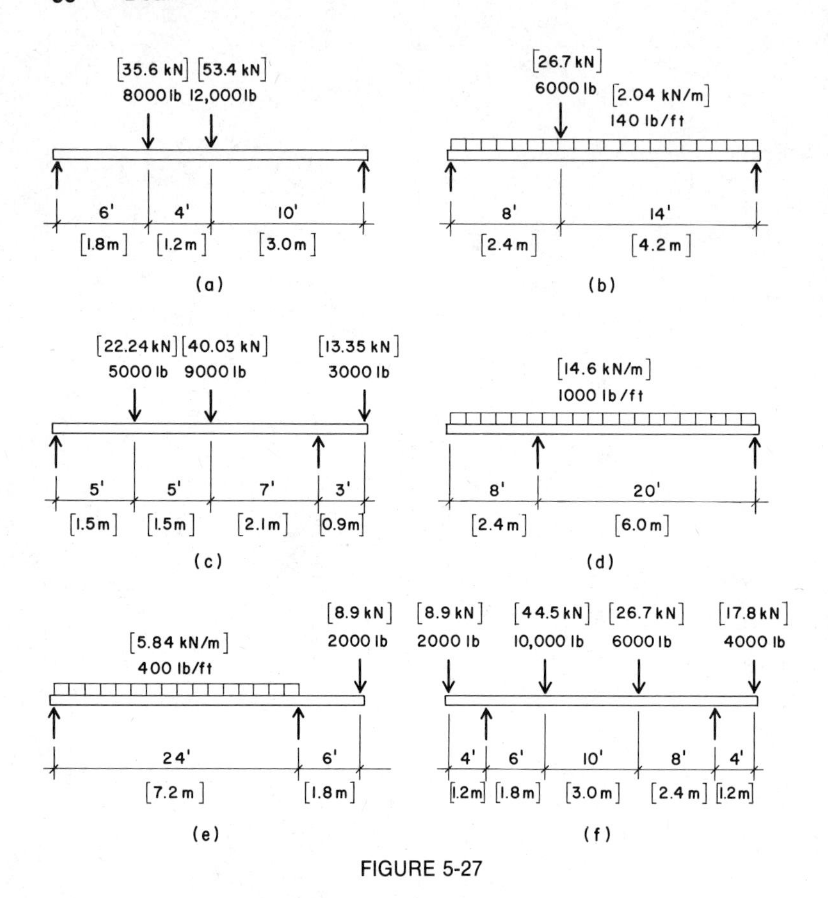

FIGURE 5-27

conditions that are frequently met in practice. These and other common types of loading are shown in Fig. 5-26. This figure reveals at a glance the values of the maximum shear, maximum bending moment, and maximum deflection (to be discussed later) for the conditions presented. The following notation is used in the figure:

P = the concentrated load in pounds
W = the total uniformly distributed load in pounds (in Case 8, W represents a triangular loading)

L = the span in feet
V = the maximum shear in pounds
M = the maximum bending moment in foot-pounds
Δ = the maximum deflection in inches
ETL = the equivalent tabular load (see Art. 9-10)

Problems 5.19.A, B. Construct the shear and bending moment diagrams for the beams shown in Fig. 5-27*a* and *b*. Note the magnitudes of the maximum shear and the maximum bending moment on the diagrams.

Problems 5.19.C,* D, E, F. For the beams shown in Fig. 5-27*c, d, e,* and *f* draw the shear and bending moment diagrams. For each beam give the value of the maximum shear and maximum bending moment and compute the position of the inflection points.

6

Theory of Bending and Properties of Sections

6-1. Resisting Moment

In the preceding chapter the concept of bending moment was developed as a means of measuring the tendency of the external loads on a beam to deform it by bending. It is now necessary to consider the action within the beam that resists bending and is called the *resisting moment*.

A rectangular beam subjected to bending is shown in Fig. 6-1*a*. If we consider any section along the beam, such as *X-X*, the external forces produce a bending moment at this section. By letting x be the distance of the section from the left reaction, as indicated in the enlarged detail of Fig. 6-1*b*, the value of the bending moment is $R_1 \times x$. This bending moment is resisted by internal stresses set up in the fibers of the beam.

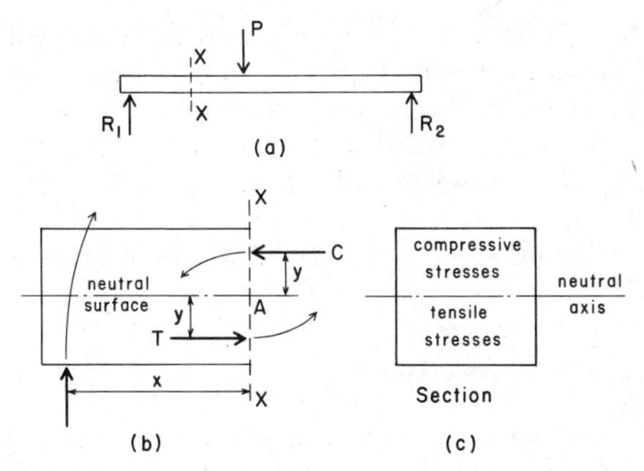

FIGURE 6-1. Development of bending stress in a beam.

As pointed out in Art. 3-4, the stresses above the neutral surface are compressive and those below are tensile. Let C be the resultant of all the compressive stresses, T, the resultant of all the tensile stresses, and y, their distances from the neutral axis, point A. Then $(C \times y) + (T \times y)$ equals the sum of the moments of all the internal stresses at section X-X. Referring to Fig. 6-1*b*, we note that the moments of the stresses in the fibers tend to cause a counterclockwise rotation about point A, whereas the bending moment generated by R_1 tends to produce a clockwise rotation about the same point. The sum of the moments of all the internal stresses about the neutral axis is called the *resisting moment* because it holds the bending moment in equilibrium.

At any section along the length of a beam the bending moment and the resisting moment are equal in magnitude; and using the foregoing terms, we may write

$$R_1 \times x = (C \times y) + (T \times y)$$

or

$$\text{bending moment} = \text{resisting moment}$$

The design of a beam consists primarily of (1) computing the maximum bending moment that will be developed by the design loading and (2) selecting a structural steel shape that will provide a potential resisting moment equal to or greater than the maximum bending moment. The study of Chapter 5 has enabled us to compute the maximum bending moment for most of the steel beams that will be encountered in practice; the second part of the procedure is accomplished by use of the *flexure formula* (frequently called the *beam formula*) developed in the following article.

6-2. The Flexure Formula

The flexure formula is an expression for the resisting moment that involves the size and shape of the beam cross section and the material of which the beam is made. It is used in the design of all homogeneous beams (i.e., beams made of one material only, such as steel, aluminum, or wood).

Figure 6-2 represents a portion of the side elevation and the cross section of a homogeneous beam subject to bending stresses. The cross section shown is unsymmetrical about the neutral axis, but this discussion applies to a cross section of any shape. In Fig. 6-2*a* let c be the distance of the most remote fiber from the neutral axis and let f be the unit stress on the fiber at c distance. If f, the greatest fiber stress, does not exceed the elastic limit of the material, the stresses in the other fibers will be directly proportional to their distances from the neutral axis. If c is in inches, the unit

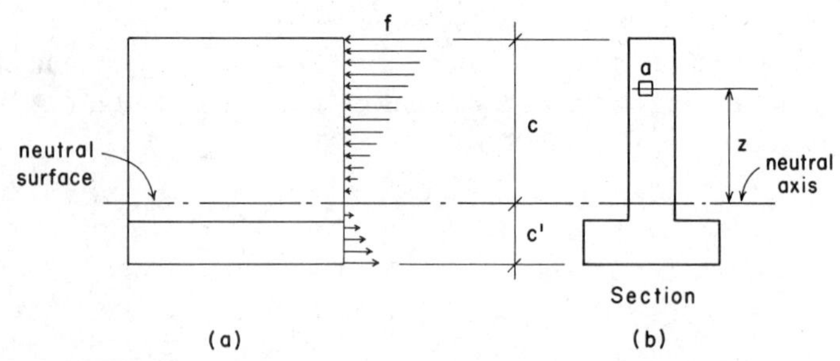

FIGURE 6-2. Distribution of bending stress on a beam section.

stress on a fiber at 1-in. distance is f/c. To explain this, suppose $c = 8$ in. and the stress f on the most remote fiber is 1600 psi. Then the stress on a fiber at 1 in. from the neutral axis is f/c, or 1600/8, or 200 psi; at 2 in. it is 2×200, or 400 psi; at 3 in. it is 3×200, or 600 psi, etc.

Now imagine an infinitesimally small area at z distance from the neutral axis and call this area a. Because the unit stress on a fiber at unity distance is f/c, the unit stress on the fiber at z distance must be $f/c \times z$, and because there are a square inches in this elementary area the stress on the fiber a will be $f/c \times z \times a$. The *moment* of the stress on fiber a about the neutral axis is the stress times its lever arm, or

$$\left(\frac{f}{c} \times z \times a\right) \times z \quad \text{or} \quad \frac{f}{c} \times a \times z^2$$

We know that an infinite number of these elementary areas are contained in the cross section, and if we use the symbol Σ to indicate the sum of an infinite number $\Sigma(f/c) \times a \times z^2$ will represent the sum of the moments of all the stresses in the cross section (both above and below the neutral axis) with respect to the neutral axis. This sum, we know from Art. 6-1, is the *resisting moment*. Because we know that the bending and resisting moments at any section of a beam under load are equal in magnitude, we may write

$$M = M_R \quad \text{or} \quad M = \frac{f}{c}\,\Sigma az^2$$

The term Σaz^2 is called the moment of inertia of the cross section with respect to the neutral axis and is discussed in more detail in Art 6-5. It is usually designated by the letter I. Using this symbol, we may write the above expression as

$$M = \frac{f}{c} \times I \quad \text{or} \quad M = \frac{fI}{c}$$

It is called the *beam* or *flexure formula* and may be simplified further by substituting S for I/c, called the *section modulus*, a term that is described more fully in Art. 6-6. By making this substitution the formula becomes

$$M = fS$$

Use of the flexure formula in the design and investigation of beams is explained in Chapter 7.

6-3. Properties of Sections

Each of the terms, moment of inertia and section modulus, used in deriving the flexure formula represents a *property* of a particular beam cross section. Other properties are the area of the section, the radius of gyration, and the position of the centroid of the cross-sectional area. Although these properties and other useful design constants are tabulated for the commonly used structural shapes, a knowledge of their significance will be of great value to the designer. Several of the properties listed in the tables are discussed in this chapter and others are considered in Chapter 9 under the design of beams. Table 6-1 contains formulas that may be used to determine the properties of simple geometric shapes.

6-4. Centroids

The term *centroid* is related to plane areas in the same manner that *center of gravity* is related to solids. *The centroid of a plane surface is a point that corresponds to the center of gravity of a very thin homogeneous plate of the same area and shape*. It can be shown that the neutral axis of a beam cross section passes through its centroid; consequently it is necessary to know its exact position. For symmetrical sections such as wide flange shapes and I-beams it can be seen by inspection that the centroid lies at a point midway between the flanges, at the intersection of the X-X and Y-Y axes shown at the head of Appendix Tables A-1 and A-2. The distance c from the extreme fiber to the neutral axis is therefore half the depth of such sections. This is also true for standard channels used as beams when bending takes place about the X-X axis, although the position of the centroid with respect to its distance from the back of the web cannot be determined by

TABLE 6-1. Properties of Geometric Shapes

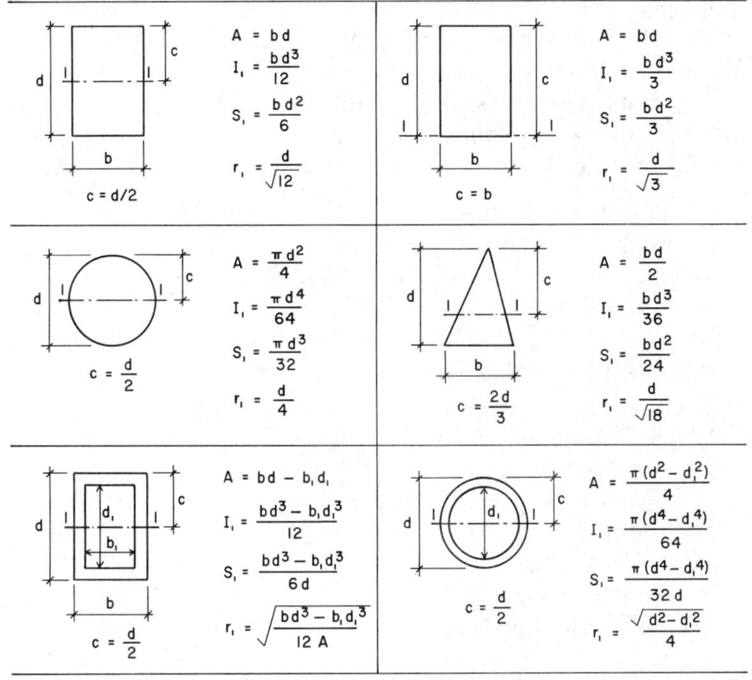

A = Area I = Moment of inertia S = Section modulus = $\frac{I}{c}$ r = Radius of gyration = $\sqrt{\frac{I}{A}}$

inspection and must be computed. This situation is shown in the sketch at the head of Appendix Table A-4 and computed values of $\bar{x}$ (read x bar) are given in the table; for example, the centroid of a C 10 × 20 is located 5 in. from the backs of the flanges and 0.606 in. from the back of the web (the distance $\bar{x}$).

For unsymmetrical sections such as angles the centroid may be found by referring to Appendix Table A-6. As an illustration, consider an L 6 × 4 × $\frac{1}{2}$. The table shows that the centroid is 1.99 in. from the back of the short leg and 0.987 in. from the back of the long leg. In the sketch at the head of the table these distances are indicated as y and x, respectively. As Appendix Table A-9 shows, the x and y distances are equal for equal leg angles.

The procedure for computing the position of centroids uses the principle of *statical moments*. A statical moment of a plane area with respect to an axis is the area multiplied by the perpendicular distance from the centroid to the axis. If a plane area is divided into a number of parts, the sum of the statical moments of the parts is equal to the statical moment of the entire area. The following example illustrates how readily this principle is applied.

Example. Compute the distance of the centroid from the back of the short leg of an L $6 \times 4 \times \frac{1}{2}$.

Solution: A diagram that shows the dimensions of the angle is made first (Fig. 6-3); it is then divided into the two areas indicated by the diagonals and marked *A* and *B*, but any convenient division will serve our purpose. The value sought is denoted c'. Area *A* is 6×0.5, or 3 sq in. [150×13, or 1950 mm^2] and Area *B* is 3.5×0.5, or 1.75 sq in. [87×13, or 1131 mm^2]. The area of the entire cross section is $3 + 1.75$, or 4.75 sq in. [$1950 + 1131$, or 3081 mm^2]. Now, *with respect to an axis that passes through the back of the short leg* we may write the sum of the statical moments of the two parts and equate it to the statical moment of the entire area. Noting that the distances of the centroids of rectangular areas *A* and *B* to the axis selected are 3 and 0.25 in. [75 and 6.5 mm], respectively, we obtain

$$(3 \times 3) + (1.75 \times 0.25) = 4.75 \times c'$$

$$9.438 = 4.75c' \quad \text{and} \quad c' = 1.99 \text{ in.}$$

$$\begin{bmatrix} (1950 \times 75) + (1131 \times 6.5) = 3081 \times c' \\ 153{,}600 = 3081c' \quad \text{and} \quad c' = 49.85 \text{ mm} \end{bmatrix}$$

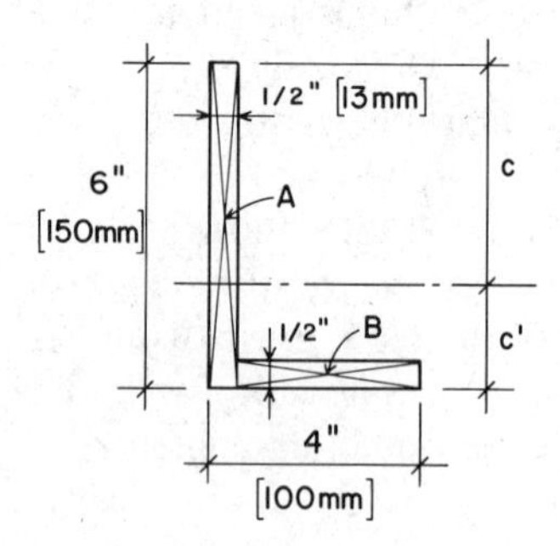

FIGURE 6-3

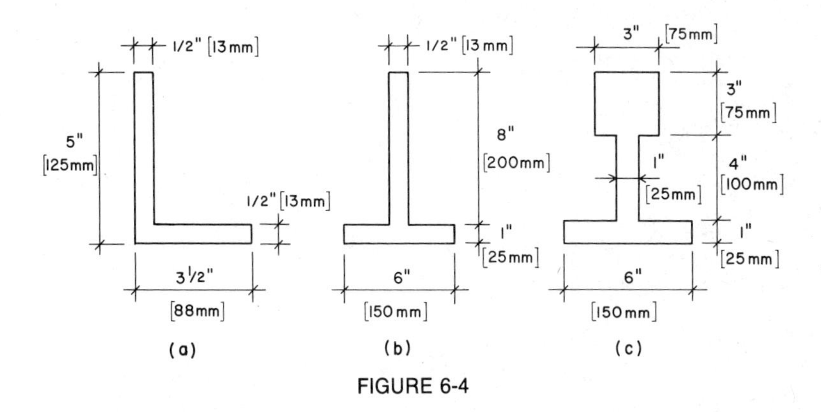

FIGURE 6-4

This is the same value given as y for this angle in Appendix Table A-6.

It should be observed that if this angle were used as a beam with the long leg in the vertical position the neutral axis would be located 1.99 in. [49.85 mm] above the bottom of the cross section. The distance of the *most remote fiber* from the neutral axis, however, would be 6 − 1.99, or 4.01 in. [150 − 49.85, or 100.15 mm]. This would be the distance c in the flexure formula. (See also the discussion of the *shear center* in Art. 6-9.)

Problems 6.4.A,* B, C. Compute the values of c with respect to the horizontal axes of the beam cross sections shown in Fig. 6-4*a, b,* and *c*.

6-5. Moment of Inertia

When developing the flexure formula (Art. 6-2) we called the term Σaz^2 the moment of inertia of the beam cross section and denoted it by the symbol I. Moment of inertia may be defined as *the sum of the products obtained by multiplying all the elementary areas of a cross section by the squares of their distances from a given axis.* It may be found for any axis, but the axis commonly used is the neutral axis of a beam cross section, which, as we know, passes through the centroid of the section. Because the elements of area are expressed in square inches and the distances in inches, the units for moment of inertia will be inches raised to the fourth power, written *inches*^{4}.

In order to understand more clearly the significance of moment

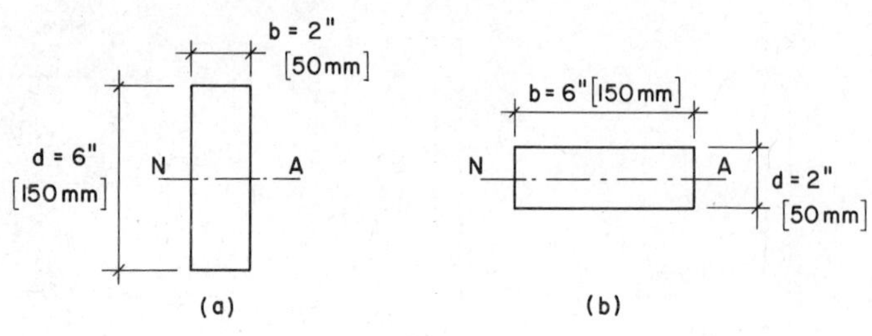

FIGURE 6-5

of inertia, we may observe from the expression Σaz^2 that I varies with the *shape* of the cross section as well as with the area. It is, of course, a matter of common observation that a wood plank, say 2 in. by 6 in. [50 × 150 mm] in cross section, is much stiffer when used on a given span with its 6-in. [150-mm] side vertical than when the same surface is placed flat. In the former situation (Fig. 6-5*a*) the term Σaz^2 is larger than in the latter (Fig. 6-5*b*) because the average value of z is larger. It follows, therefore, that it is desirable to place as much material as far as practicable from the neutral axis when seeking to maximize the value of I.

The moment of inertia of any section may be computed by use of the calculus or other summation method. However, I values for structural steel shapes are listed in tables of the properties of structural sections. As an example, refer to Appendix Table A-1 which contains data that relate to wide flange shapes. Under W 18 × 40 we find that the I of this section with respect to the X-X axis is 612 in.^{4} If we also look up the moment of inertia of a W 16 × 40 (a shape with the same weight per linear foot and cross-sectional area), we find that its $I_{X\text{-}X}$ is 518 in.^{4} In a comparison of these two values the effect of the 2-in. difference in depth on the moments of inertia is readily apparent. As an aid to fixing in mind the concept of moment of inertia, it is recommended that the reader make similar comparisons between other sections listed in the tables.

6-6. Section Modulus

As noted in Art. 6-2, the term I/c in the flexure formula is called the *section modulus*. It is defined as the moment of inertia divided by the distance of the most remote fiber from the neutral axis and is denoted by the symbol S. Because I and c always have the same values for any given cross section, values of S may be computed and tabulated for structural shapes. With I expressed in inches to the fourth power and c, a linear dimension in inches, S is in units of inches to the third power, written in.^{3} Appendix Table A-1 lists the values of the section modulus with respect to the two major axes of wide flange shapes; for example, a W 12 × 40 has an $S_{X\text{-}X}$ of 51.9 in.^{3} and an $S_{Y\text{-}Y}$ of 11.0 in.^{3} We are concerned mostly with $S_{X\text{-}X}$ because structural shapes are normally used as beams with the X-X axis as the axis about which bending takes place.

Like moment of inertia, the section modulus is related to the size and shape of the beam cross section, and shapes with a high percentage of material placed in the flanges will have larger values of S than other configurations of the same cross-sectional area. This is made apparent by a comparison of the wide flange shape W 12 × 50 (Appendix Table A-1) and the Standard I-beam S 12 × 50 (Appendix Table A-3). Both sections have approximately the same depth, weight per linear foot, and cross-sectional area. The former however, has a section modulus of 64.7 in.^{3} and the latter, a section modulus of 50.8 in.^{3} Further scanning of the tables reveals that the flange of the W shape is 8.08 in. wide and 0.640 in. thick, making the area of one flange 5.17 in.^{2} The flange of the S shape is 5.477 in. wide and 0.659 in. thick, giving a flange area of 3.61 in.^{2} Therefore, considering both flanges, the W shape has 2 × 5.17 = 10.34 in.^{2} in the flanges in the total cross-sectional area of 14.7 in.,^{2} whereas the S shape has only 2 × 3.61 = 7.22 in.^{2} in the flanges in the same total area. As noted in Art. 2-2, this means that the W shapes are more efficient in bending resistance than the Standard I-beams.

The section modulus is one of the most common properties of cross sections; it is used often in the design of beams and is a measure of the ability to resist bending stresses. The example and two problems that follow should serve to fix in mind the relationship between it and the moment of inertia.

Example. Verify the tabulated value of the section modulus of a W 18 × 46 with respect to the X-X axis.

Solution: Appendix Table A-1 gives the moment of inertia of this section as 712 in.^{4} [296 × 10^6 mm^4]. Because it is a symmetrical section, $c = d/2$, or 18.06/2 = 9.03 in. [459/2 = 229.5 mm]. Therefore

$$S = \frac{I}{c} = \frac{712}{9.03} = 78.85 \text{ in}^3$$

$$\left[S = \frac{296 \times 10^6}{229.5} = 1290 \times 10^3 \text{ mm}^3\right]$$

which checks with the value given in the table.

Problem 6.6.A. Verify the value of $S_{X\text{-}X}$ listed in Appendix Table A-4 for the American Standard Channel C 15 × 50.

Problem 6.6.B. Verify the value of $S_{X\text{-}X}$ given in Appendix Table A-6 for the unequal leg angle $L \times 5 \times 3 \times \frac{1}{2}$.

6-7. Radius of Gyration

This property of a cross section is related to the design of compression members rather than beams and is discussed in more detail under the design of columns in Chapter 11. Because it is listed in the tables of properties, however, it is considered here briefly.

Just as the section modulus is a measure of the resistance of a beam section to bending, the radius of gyration (which is also related to the size and shape of the cross section) is an index of the stiffness of a structural section when used as a column or other compression member. The radius of gyration is found by the formula

$$r = \sqrt{\frac{I}{A}}$$

and is expressed in inches because the moment of inertia is in inches^4 and the cross-sectional area is in square inches. For W, S, and C shapes (Appendix Tables A-1 through A-5) r is tabulated for both X-X and Y-Y axes; for angles an additional value is given with respect to an oblique axis through the centroid marked Z-Z.

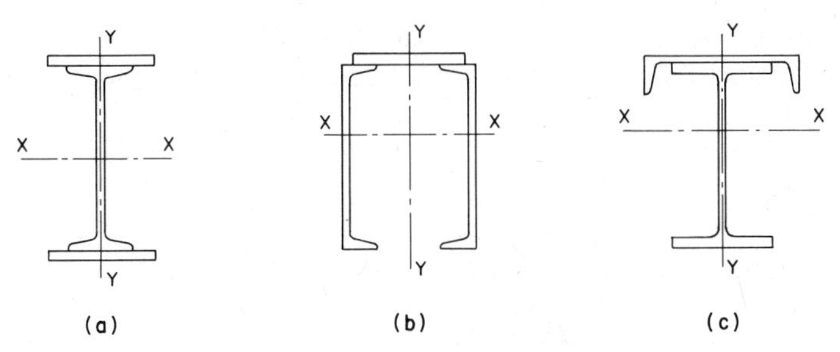

FIGURE 6-6. Built-up sections.

When used in the design of columns the *least* radius of gyration is generally used.

6-8. Properties of Built-Up Sections

When rolled steel shapes are combined to form built-up structural sections such as those illustrated in Fig. 6-6 their properties for use in design must be computed. The first step in such computations is to determine the moment of inertia of the built-up section about its neutral axis. This usually requires transferring the moments of inertia of the cross sections of the individual parts from one axis to another because the moment of inertia of a built-up section about its neutral axis is equal to the sum of the moments of inertia of the individual parts about the same axis.

The simple built-up section shown in Fig. 6-7 consists of an S 10×25.4 with two 12 in. by $\frac{1}{2}$ in. [300 by 13 mm] plates welded to

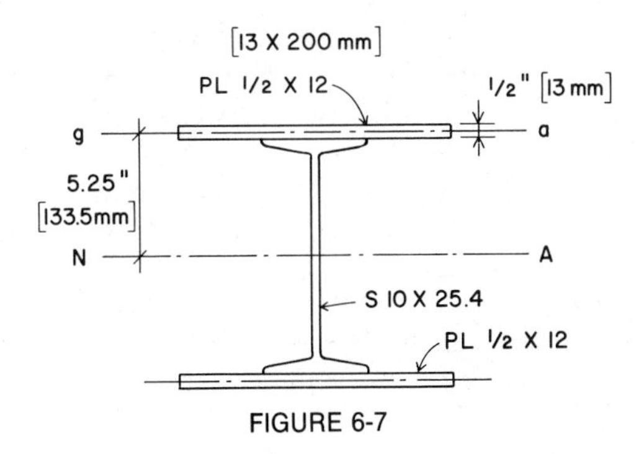

FIGURE 6-7

its flanges. Because the X-X axis of the S shape coincides with the neutral axis NA of the built-up section, its I with respect to NA may be found in Appendix Table A-3 directly and is 124 in.^{4} [51.6×10^6 mm^4]. The moment of inertia of one of the plates about an axis through its centroid and parallel to its longer side (called its *gravity axis* and marked ga in Fig. 6-7) may be found from the formula for rectangular cross sections (Table 6-1). This is

$$I = \frac{bd^3}{12} = \frac{12 \times 0.5^3}{12} = 0.125 \text{ in.}^4$$

$$\left[I = \frac{300 \times 13^3}{12} = 54925 \text{ mm}^4\right]$$

In order to find the I of the plate about the NA of the built-up section, it is necessary to use the *transfer formula*, which may be stated as follows:

The moment of inertia of a cross section about any axis parallel to an axis through its own centroid is equal to the moment of inertia of the cross section about its own gravity axis plus its area times the square of the distance between the two axes.

Expressed mathematically,

$$I = I_o + Az^2$$

In this formula I = the moment of inertia of the cross section about the required axis,
I_o = the moment of inertia of the cross section about its own gravity axis parallel to the required axis,
A = the area of the cross section,
z = the distance between the two parallel axes.

To apply this equation to the *two* plates shown in Fig. 6-7, we note that the area of each plate is $12 \times 0.5 = 6$ in^2 [$300 \times 13 = 3900$ mm^2] and that the distance between ga and NA is 5.25 in.

[133.5 mm]. Then

$$
\begin{aligned}
I &= 2(I_o + Az^2) \\
&= 2\{0.125 + (6 \times 5.25 \times 5.25)\} \\
&= 2(0.125 + 165.375) = 2 \times 165.5 = 331 \text{ in.}^4
\end{aligned}
$$

$$
\left[
\begin{aligned}
I &= 2\{54925 = (3900 \times 133.5 \times 133.5)\} \\
&= 2(54925 + 69{,}506{,}775) = 2 \times 69.5 \times 10^6 \\
&= 139 \times 10^6 \text{ mm.}^4
\end{aligned}
\right]
$$

The moment of inertia of the entire built-up section about the neutral axis *NA* is then 124 + 331 = 445 in.^{4} [$(51.6 + 139) \times 10^6 = 190.6 \times 10^6$ mm^4].

Because this built-up section is symmetrical with respect to *NA*, its section modulus will be equal to *I*, divided by half the overall depth:

$$S = \frac{I}{c} = \frac{455}{5.5} = 82.7 \text{ in.}^3$$

$$S = \frac{190.6 \times 10^6}{140} = 1361 \times 10^3 \text{ mm}^3$$

6-9. Unsymmetrical Built-Up Sections

In the preceding article the location of the neutral axis of the built-up section shown in Fig. 6-7 was determined by inspection. This was possible because the cross section is symmetrical about *NA*; consequently its centroid lies at the midpoint of the overall depth. When working with unsymmetrical sections, however, it is necessary to locate the centroid by using the principle of statical moments discussed in Art. 6-4. The application of this principle to unsymmetrical built-up sections is explained in the following example:

Example. The built-up section shown in Fig. 6-8 consists of a W 14 × 30 to which a C 10 × 15.3 is welded. This section is sometimes used when the beam has a long length with no side-

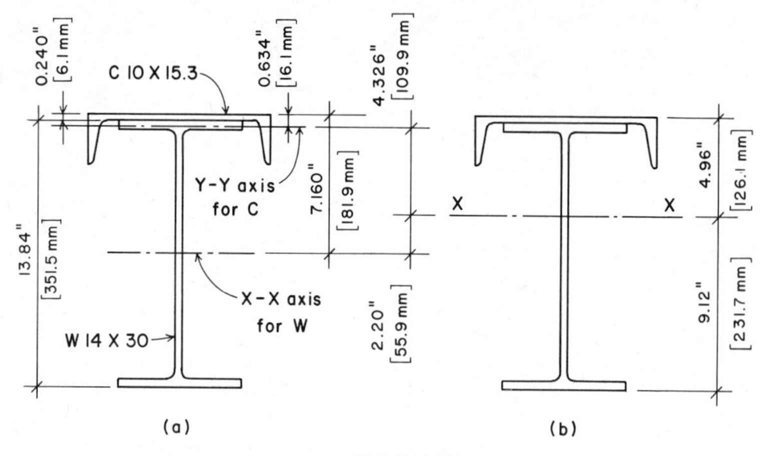

FIGURE 6-8

ways bracing or is subjected to a significant sideways loading in addition to a vertical one. Determine the section modulus for the built-up section with respect to the horizontal neutral axis *X-X*.

Solution: The first step is to locate the *X-X* axis. This is accomplished by finding the distance of the centroid from either the top or bottom of the section; we use the top, or back of the channel, as our reference.

By referring to Appendix Table A-1 we find the following for the W 14 × 30:

$$\begin{aligned} I_x &= 291 \text{ in.}^4 & &[121.1 \times 10^6 \text{ mm}^4] \\ A &= 8.85 \text{ in.}^2 & &[5710 \text{ mm}^2] \\ d &= 13.84 \text{ in.} & &[351.5 \text{ mm}] \end{aligned}$$

and from Appendix Table A-3, for the C 10 × 15.3,

$$\left[\begin{aligned} I_y &= 2.28 \text{ in.}^4 & &0.949 \times 10^6 \text{ mm}^4 \\ A &= 4.49 \text{ in.}^2 & &2897 \text{ mm}^2 \\ t_w &= 0.240 \text{ in.} & &6.10 \text{ mm} \\ \bar{x} &= 0.634 \text{ in.} & &16.1 \text{ mm} \end{aligned}\right]$$

The combined area of the built-up section is 8.85 + 4.49 = 13.34 in.^{2} [5710 + 2897 = 8607 mm^2].

By writing an equation for the statistical moments about the back of the channel as an axis and letting c be the distance of the centroid of the entire section from this location we obtain the following equation:

$$13.34 \times c = (8.85 \times 7.160) + (4.49 \times 0.634)$$

$$13.34c = 66.213 \quad \text{and} \quad c = 4.96 \text{ in.}$$

$$\left[\begin{aligned} 8607 \times c &= (5710 \times 181.9) + (2897 \times 16.1) \\ 8607c &= 1{,}085{,}291 \quad \text{and} \quad c = 126.1 \text{ mm} \end{aligned}\right]$$

This dimension is recorded in Fig. 6-8b.

Having found the value of c, the next step is to transfer the moments of inertia of the W and C shapes to the X-X axis of the combined section. Using the data previously given, the moment of inertia of the channel is as follows:

$$\begin{aligned} I_X &= I_o + Az^2 \\ &= (2.28) + (4.49 \times 4.326 \times 4.326) \\ &= 2.28 + 84.027 = 86.307 \text{ in.}^4 \end{aligned}$$

$$\left[\begin{aligned} I_X &= (0.949 \times 10^6) + (2897 \times 110 \times 110) \\ &= (0.949 \times 10^6) + (35.05 \times 10^6) = 36.0 \times 10^6 \text{ mm}^4 \end{aligned}\right]$$

For the wide flange shape

$$\begin{aligned} I_X &= I_o + Az^2 \\ &= (291) + (8.85 \times 2.20 \times 2.20) \\ &= 291 + 42.834 = 333.834 \text{ in.}^4 \end{aligned}$$

$$\left[\begin{aligned} I_X &= (121.1 \times 10^6) + (5708 \times 55.8 \times 55.8) \\ &= (121.1 \times 10^6) + (17.8 \times 10^6) = 138.9 \times 10^6 \text{ mm}^4 \end{aligned}\right]$$

The combined moment of inertia is

$$I_X = 86.424 + 333.834 = 420.258 \text{ in.}^4$$

$$I_X = (138.9 + 36)(10^6) = 174.9 \times 10^6 \text{ mm}^4$$

Although there is only a single value for the moment of inertia, there are two values for the section modulus with respect to the *X-X* axis of the combined section. One refers to the top of the section (c = 4.96 in. [126.1 mm]) and the other, to the bottom of the section (c' = 9.12 in. [231.5 mm]). If the maximum stress is desired, the lower value is used. However, if we wish to know the maximum values for tension and compression stresses due to bending, both values are used. We determine both values as follows:

For stress at the uppermost fiber

$$S_1 = \frac{I}{c} = \frac{420.258}{4.96} = 84.73 \text{ in.}^3$$

$$\left[S_1 = \frac{174.9 \times 10^6}{126.1} = 1387 \times 10^3 \text{ mm}^3\right]$$

For stress at the bottommost fiber

$$S_2 = \frac{I}{c} = \frac{420.258}{9.12} = 46.08 \text{ in.}^3$$

$$\left[S_2 = \frac{174.9 \times 10^6}{231.5} = 755 \times 10^3 \text{ mm}^3\right]$$

6-10. Shear Center

Loadings that produce bending can also have other effects on beams. In Fig. 6-9*a* a concentrated load is shown on the end of a cantilevered beam that has a simple rectangular cross section. With the loading placed as shown, the effect produced is one of simple bending in which the neutral axis is the centroidal axis *X-X* and the beam assumes the deformed shape shown in Fig. 6-9*c*. If, however, the load is moved off center (Fig. 6-9*b*), the beam is also subjected to a torsional twist and assumes the deformed shape shown in Fig. 6-9*d*.

To avoid the torsional effect illustrated in Fig. 6-9*d*, it is necessary that the plane of the loading (or the plane in which the bending moment occurs) coincide with the location of the *shear center* of the beam cross section. For beam shapes with biaxial

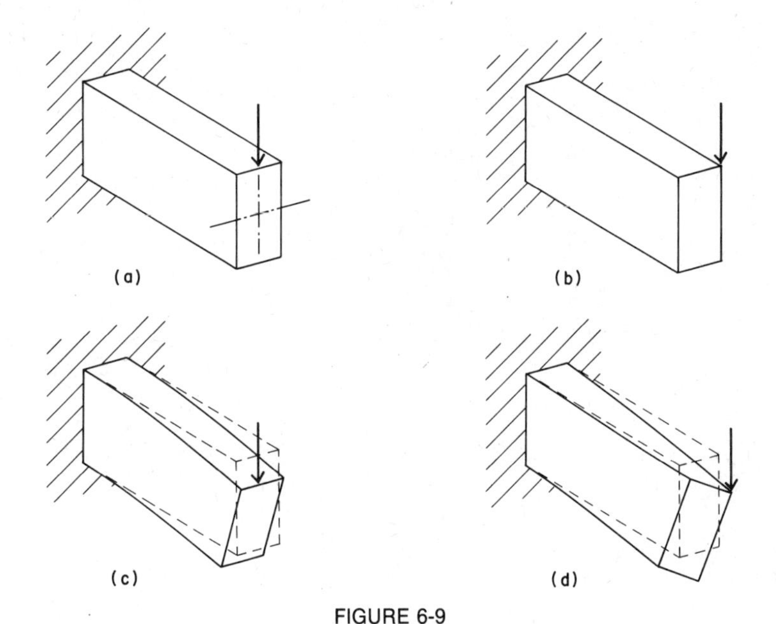

FIGURE 6-9

symmetry (symmetrical about both centroidal axes) the shear center will coincide with the centroid of the cross section. Thus for the beam in Fig. 6-9 the twisting effect is avoided if the plane of the bending moment coincides with the *Y-Y* centroidal axis (Fig. 6-9*a*). The same relationship exists for the doubly symmetrical H shape shown in Fig. 6-10*a*.

For sections that have no axis of symmetry parallel to the plane of the bending moment, such as the C shape in Fig. 6-10*c*, the shear center is at a location separate from the centroid of the section. In this case loading through the centroid causes twisting. For the C shape the shear center is located on the major axis of symmetry (*X-X*) but slightly in back of the C web, as shown in the illustration. The location of the shear center for rolled steel C shapes is given as the value e_o in Appendix Tables A-4 and A-5.

Among the common rolled shapes C, L, and T are those that have a separate location for the shear center. When these sections are loaded through their centroids twisting may occur (Fig.

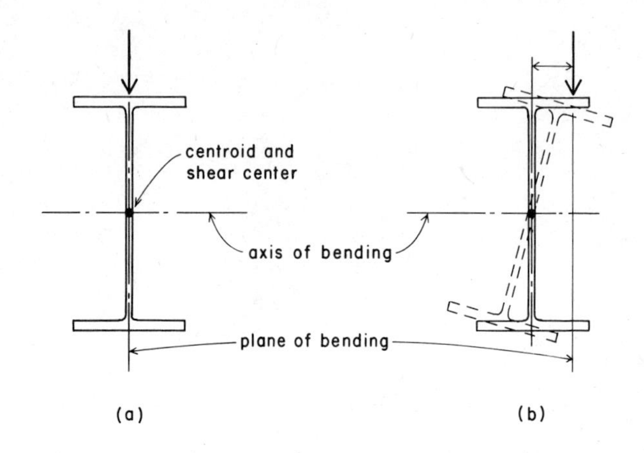

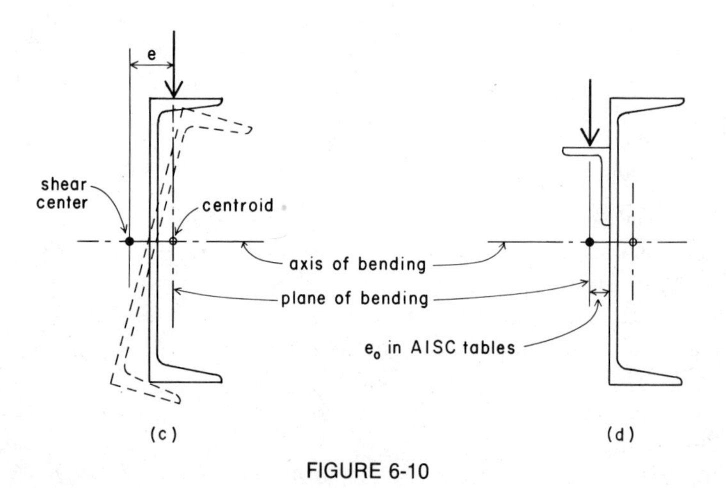

FIGURE 6-10

6-11*a*). If C and T shapes are loaded in the plane of their axes of symmetry, however, there is no twisting (see Fig. 6-11*c*). Also, when a symmetrical combined section is produced with matched shapes, as shown in Fig. 6-11*d*, centroidal loading does not produce twisting.

Analysis for torsional effects is beyond the scope of this book. Designers are cautioned to be careful when using unsymmetrical

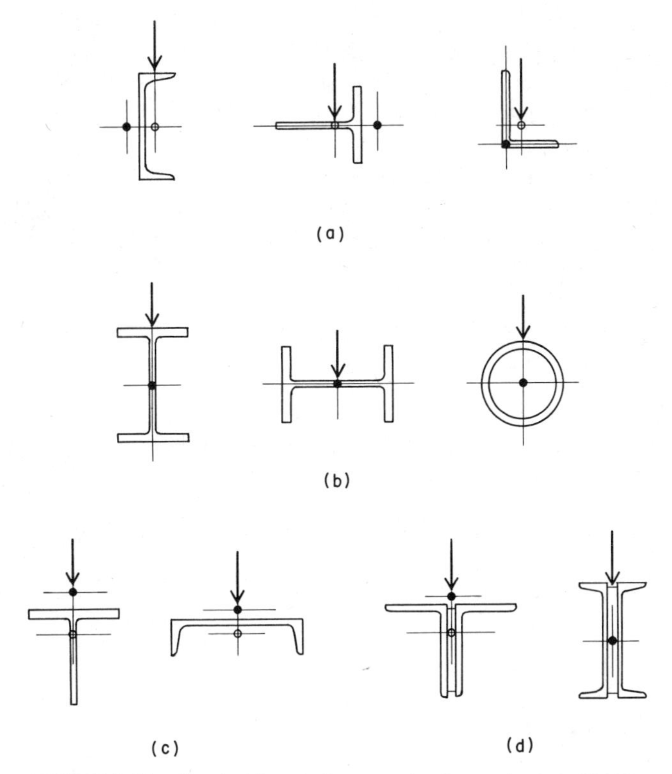

FIGURE 6-11. Centroid and shear center for various sections.

sections for beams and, if possible, to load them in a manner that will avoid the torsional effects. When this is not possible, an alternative solution may be to use other elements of the construction to brace the beam in a manner that will prevent torsional rotation.

Review Problems

Problem 6.9.A.* The built-up section shown in Fig. 6-6*c* consists of a W 18 × 50 with a C 12 × 20.7 welded to the top flange. Determine the moment of inertia and the section modulus with respect to the *X-X* axis.

Problem 6.9.B. Compute the value of I and S with respect to the Y-Y axis of the combined section described in the preceding problem.

Problem 6.9.C.* For the angle section shown in Fig. 6-4*a* compute the moment of inertia, the section modulus, and radius of gyration with respect to an axis through the centroid and parallel to the short leg.

Problem 6.9.D. An S 12 × 31.8 has an 8 by $\frac{1}{2}$ in. [200 by 13 mm] plate welded to its upper flange. Compute the moment of inertia and section modulus for this combined section with respect to an axis through its centroid and parallel to the plate.

7

Use of the Beam Formula

7-1. Forms of the Equation

The expression $M = fI/c$ or fS, developed in Art. 6-2, may be stated in three different forms, depending on the information sought. These are given below, using the AISC general nomenclature, which makes a distinction between allowable bending stress (F_b) and computed bending stress (f_b):

$$(1)\ M = \frac{F_b I}{c} \quad (2)\ f_b = \frac{Mc}{I} \quad (3)\ \frac{I}{c} = \frac{M}{F_b}$$

Letting $I/c = S$,

$$(1)\ M = F_b S \quad (2)\ f_b = \frac{M}{S} \quad (3)\ S = \frac{M}{F_b}$$

Form (1) gives the maximum potential resisting moment when the section modulus of the beam and the maximum allowable bending stress are known. Form (2) gives the computed bending stress when the maximum bending moment due to the loading and the section modulus of the beam are known. These are the two forms used in investigation.

Form (3) is the one used in design. It gives the *required* section modulus when the maximum bending moment and the allowable bending stress are known. When the required section modulus has been determined a beam with an S equal to or greater than the computed value is selected from tables that list the properties of the various structural shapes.

When the beam formula is used care must be exercised with respect to the units in which the terms are expressed. Bending stress values F_b and f_b may be written in pounds per square inch (psi) or kips per square inch (ksi); S is stated in inches^3 and I in inches^4; M, therefore, must be in inch-pounds or kip-inches; M, as customarily computed from the loads and reactions, is expressed in foot-pounds or kip-feet and must be converted to inch-pounds or kip-inches by multiplying its value by 12 before it is used in the formula. In SI units stresses are usually in megapascals (MPa), S in a value times 10^3 and measured in mm^3, I in a value times 10^6 and measured in mm^4, loads and reactions, in kN, and bending moments, in kN-m. The challenge is in keeping track of the decimal point or the powers of 10.

7-2. Investigation of Beams

The process of determining whether a beam of given size and span can safely support a proposed loading is called investigation. Use of the beam formula in making determinations is illustrated in the following examples:

Example 1. A W 10 × 26 is proposed to carry a total uniformly distributed load of 30 kips [133 kN], including an allowance for the beam's weight, on a span of 13 ft [4 m]. (See Fig. 7-1.) If the allowable bending stress is 24 ksi [165 MPa], determine whether the beam is safe (a) by comparing the maximum resisting moment of the section with the maximum bending moment developed by the loading and (b) by comparing the allowable bending stress with that actually produced by the loading.

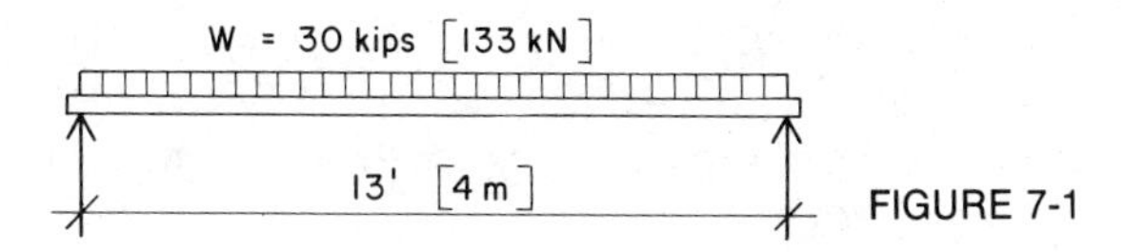

FIGURE 7-1

Solution (a): By referring to Appendix Table A-1, we find that the section modulus for the W 10 × 26 is 27.9 in.^{3} [457 × 10^3 mm^3]. Then

$$M = F_b S = 24 \times 27.9 = 670 \text{ kip-in.}$$

or

$$M = \frac{670}{12} = 55.8 \text{ kip-ft}$$

$$\left[M = F_b S = \frac{165 \times 457 \times 10^3}{10^6} = 75.4 \text{ kN-m} \right]$$

From Art. 5-14 and Fig. 5-26 (Case 2) we know that the maximum bending moment for the proposed loading occurs at midspan and may be found with the formula M = WL/8. Then

$$M = \frac{WL}{8} = \frac{30 \times 13}{8} = 48.8 \text{ kip-ft}$$

$$\left[M = \frac{133 \times 4}{8} = 66.5 \text{ kN-m} \right]$$

The beam is safe as long as the bending moment developed by the loading is less than the permissible resisting moment.
Solution (b): The maximum bending stress will occur at the top and bottom surfaces (of this symmetrical section) at the location of the largest bending moment. Then

$$f_b = \frac{M}{S} = \frac{48.8 \times 12}{27.9} = 20.99 \text{ ksi}$$

$$\left[f_b = \frac{66.5 \times 10^6}{457 \times 10^3} = 146 \text{ MPa} \right]$$

This equation verifies that the beam is safe because the actual extreme fiber stress (20.99 ksi [146 MPa]) is less than the allowable (24 ksi [165 MPa]). Note that the bending moment was multiplied by 12 to convert it to kip-in.

Example 2. An S × 12 × 31.8 has a span of 14 ft [4.3 m]. If the allowable bending stress is 22,000 psi [152 MPa], find the maximum concentrated load it will support at midspan (Fig. 7-2).
Solution: By referring to Appendix Table A-3 we find that the section modulus for the beam is 36.4 in.^{3} [597 × 10^3 mm^3]. The maximum resisting moment of the beam is

$$M = F_b S = 22{,}000 \times 36.4 = 800{,}800 \text{ in.-lb}$$

or

$$M = \frac{800{,}800}{12} = 66{,}733 \text{ ft-lb}$$

$$\left[M = \frac{152 \times 597 \times 10^3}{10^6} = 90.7 \text{ kN-m}\right]$$

From Art. 5-13 and Fig. 5-26 (Case 1) we find that the maximum bending moment for this loading occurs at midspan and is given by the formula $M = PL/4$. Before this equation is solved for P, however, the bending moment due to the beam weight must be deducted from the maximum resisting moment. The moment due to the beam weight of 31.8 lb/ft [0.464 kN/m] also occurs at midspan and is found from the expression $M = wL^2/8$ (Art. 5-14). Then

$$M = \frac{wL^2}{8} = \frac{31.8 \times (14)^2}{8} = 779 \text{ ft-lb}$$

$$\left[M = \frac{0.464 \times (4.3)^2}{8} = 1.07 \text{ kN-m}\right]$$

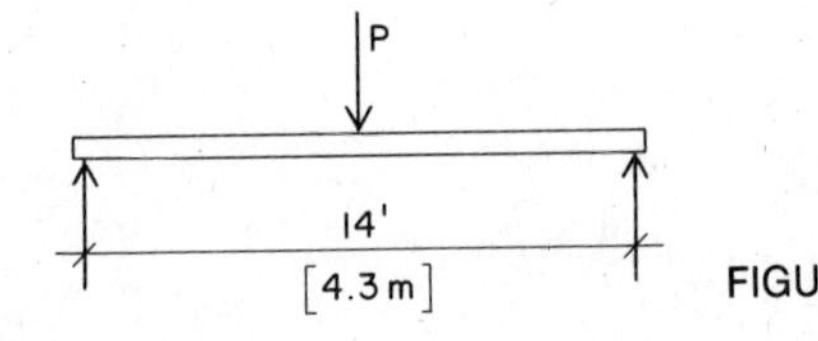

FIGURE 7-2

and the resisting moment available to support the proposed loading is

$$M = 66{,}733 - 779 = 65{,}954 \text{ ft-lb}$$

$$[M = 90.7 - 1.1 = 89.6 \text{ kN-m}]$$

Therefore the maximum safe load at the center of the span is

$$P = \frac{4M}{L} = \frac{4 \times 65{,}954}{14} = 18{,}844 \text{ lb}$$

$$\left[P = \frac{4 \times 89.6}{4.3} = 83.3 \text{ kN}\right]$$

Example 3. Determine the maximum resisting moment for the unsymmetrical built-up section shown in Fig. 6-8 if the allowable bending stress in 24 ksi [165 MPa].

Solution: In the example in Art. 6-9 the least section modulus of this built-up shape was 46.08 in.^{3} [755×10^3 mm^3]; therefore its maximum resisting moment is

$$M = F_b \times S = 24 \times 46.08 = 1106 \text{ kip-in}$$

$$\frac{1106}{12} = 92.17 \text{ kip-ft}$$

$$\left[M = \frac{165 \times 755 \times 10^3}{10^6} = 124.6 \text{ kN-m}\right]$$

Problem 7.2.A.* An S 10 × 25.4 has a span of 10 ft [3 m] with a uniformly distributed load of 36 kips [160 kN] in addition to its own weight. The allowable bending stress is 24 ksi [165 MPa]. Is the beam safe with respect to bending stresses?

Problem 7.2.B. A W 16 × 45 has a loading consisting of 10 kips [45 kN] at each of the quarter points of a 24 ft [7.2 m] span (Fig. 5-26, Case 5) and a uniformly distributed load of 5.2 kips [23 kN] including the beam weight. If the allowable bending stress is 24 ksi [165 MPa], is the beam safe with respect to bending stresses?

Problem 7.2.C.* Two $5 \times 3\frac{1}{2} \times \frac{1}{2}$ in. angles fastened together back-to-back are to be used as a beam on a span of 5 ft [1.5 m]. The allowable bending stress is 22 ksi [152 MPa]. Find the total permissible uniformly distributed load (a) when the long legs are placed vertically back-to-back and (b) when the short legs are so placed.

7-3. Design of Beams for Bending

As noted in Art. 7-1, form (3) of the beam formula is the one used in design. It gives the required section modulus after the maximum bending moment has been computed from the loading and the allowable bending stress, determined from the specifications. The complete design of a beam includes consideration of additional items such as deflection, shear, web crippling, and lateral support of the compression flange against buckling. These items are dealt with in succeeding chapters, but the cardinal principle underlying beam design is that the beam cross section shall have a potential resisting moment equal to or greater than that developed by the loading. The beam formula provides the mechanism for establishing this basic condition.

Example. Design a simply supported beam (Art. 5-2) to carry a superimposed load of 2 kips per ft [29.2 kN/m] over a span of 24 ft [7.3 m]. (The term *superimposed load* is used to denote any load other than the weight of a structural member itself.) The allowable bending stress is 24 ksi [165 MPa].

Solution: The bending moment due to the superimposed load is

$$M = \frac{wL^2}{8} = \frac{2 \times (24)^2}{8} = 144 \text{ kip-ft}$$

$$\left[M = \frac{29.2 \times (7.3)^2}{8} = 195 \text{ kN-m}\right]$$

The required section modulus for this moment is

$$S = \frac{M}{F_b} = \frac{144 \times 12}{24} = 72.0 \text{ in.}^3$$

$$\left[S = \frac{195 \times 10^6}{165} = 1182 \times 10^3 \text{ mm}^3\right]$$

Scanning Appendix Table A-1, we find a W 16 × 45 with a section modulus of 72.7 in.^{3} [1192 × 10^3 mm^3]. This value, however, is so close to that required that almost no margin is provided for the effect of the beam weight. Further scanning of the table reveals a W 16 × 50 with an S of 81.0 in.^{3} [1328 × 10^3 mm^3] and a W

18 × 46 with an S of 78.8 in.^{3} [1291 × 10^3 mm^3]. In the absence of any known restriction on the beam depth we try the lighter section. The bending moment at the center of the span with this beam is

$$M = \frac{wL^2}{8} = \frac{46 \times (24)^2}{8} = 3312 \text{ ft-lb} \quad \text{or} \quad 3.3 \text{ kip-ft}$$

$$\left[M = \frac{0.67 \times (7.3)^2}{8} = 4.46 \text{ kN-m}\right]$$

Thus the total bending moment at midspan is

$$M = 144 + 3.3 = 147.3 \text{ kip-ft}$$

$$[M = 195 + 4.5 = 199.5 \; kN\text{-}m]$$

The section modulus required for this moment is

$$S = \frac{M}{F_b} = \frac{147.3 \times 12}{24} = 73.7 \text{ in.}^3$$

$$\left[S = \frac{199.5 \times 10^6}{165} = 1209 \times 10^3 \text{ mm}^3\right]$$

Because this required value is less than that of the W 18 × 46, this section is acceptable.

7-4. Use of Section Modulus Tables

Selection of rolled shapes on the basis of required section modulus may be achieved by the use of the tables in the AISC Manual (Ref. 1) in which beam shapes are listed in descending order of their section modulus values. Material from these tables is presented in Appendix B of this book. Note that certain shapes have their designations listed in boldface type. These are sections that have an especially efficient bending moment resistance, indicated by the fact that there are other sections of greater weight but the same or smaller section modulus. Thus for a savings of material cost these *least-weight* sections offer an advantage. Consideration of other beam design factors, however, may sometimes make this a less important concern.

Data are also supplied in the tables in Appendix B for the consideration of compact sections and lateral support for beams of A36 steel with F_y = 36 ksi [248 MPa]. Shapes that are noncompact for A36 steel are noted by a mark in the listings of designations. For consideration of lateral support the values are given for the two limiting lengths L_c and L_u. If a calculation has been made by assuming the maximum allowable stress of 24 ksi [165 MPa], the required section modulus obtained will be proper only for sections not indicated as noncompact and beams in which the lateral unsupported length is equal to or less than L_c.

A second method of using the tables in Appendix B for beams of A36 steel omits the calculation of a required section modulus and refers directly to the listed values for the maximum bending resistance of the sections, given as M_R in the tables. Although the condition of the noncompact section may be noted, in this case the M_R values have taken the reduced values for bending stress into account.

Example. Rework the problem in the example in Art. 7-3 by using the tables in Appendix B.

Solution: As before, we determine that the bending moment due to the superimposed loading is 144 kip-ft [195 kN-m]. Noting that some additional M_R capacity will be required because of the beam's own weight, we scan the tables for shapes with an M_R of slightly more than 144 kip-ft [195 kN-m]. Thus we find

Shape	M_R (kip-ft)	M_R (kN-m)
W 21 × 44	163	221
W 16 × 50	162	220
W 18 × 46	158	214
W 12 × 58	156	212
W 14 × 53	156	212

Although the W 21 × 44 is the least-weight section, other design considerations, such as restricted depth, may make any of the other shapes the appropriate choice.

It should be noted that not all the available W shapes listed in Appendix Table A-1 are included in the tables in Appendix B. Specifically excluded are the shapes that are approximately square (depth equal to flange width) and are ordinarily used for columns rather than beams.

7-5. Structural Design Methods

Two different methods are used in the design of steel beams for bending stresses. The first, called *allowable stress design,* applies to this chapter and to the major part of the book. It is based on the idea of using F_b, an allowable extreme fiber stress, as a certain fraction of the yield stress (Art. 4-5). These allowable stresses fall below the elastic limit of the material, and we speak of them as conforming to the elastic behavior of the material. This approach to the design of steel members has been standard practice for many years.

The second method, known as *plastic design,* is a more recent development and is based on the idea of computing an ultimate load and using a portion of the reserve strength, after initial yield stress has been reached, as part of the factor of safety (Art. 3-8). This method was introduced into the AISC Specification in 1963 and forms Part 2 of the 1978 Specification. A brief explanation of plastic design theory is given in Chapter 15.

The following problems involve design for bending stresses only. A36 steel with an allowable bending stress of 24 ksi [165 MPa] is to be used. Full lateral support (as mentioned in Art. 7-3) is to be assumed.

Problem 7.5.A. A simple beam has a span of 12 ft [3.66 m] with a uniformly distributed load, including its own weight, of 28 kips [124.5 kN]. What size wide flange section should be used?

Problem 7.5.B.* Two concentrated loads of 20 kips [89 kN] each are placed at the third points of a simple beam with a span of 24 ft [7.32 m] (Fig. 5-26, Case 3). What is the size of the beam?

Problem 7.5.C.* A simple beam has a span of 14 ft [4.27 m] with a concentrated load of 15 kips [66.7 kN] applied at midspan. In addition, there is a superimposed load of 1 kip/ft [14.6 kN/m] extending over the entire span. Design the beam.

8

Deflection of Beams

8-1. Deflection

The deformation that accompanies the bending of a beam is called *deflection*. If the beam is in a horizontal position, it is the vertical distance moved from a horizontal line. The deflection of a beam may not be apparent visually but it is, nevertheless, always present. Figure 8-1 illustrates the deflection of a simple beam; we are principally concerned with its maximum value, which in this instance occurs at the center of the span.

A beam may be strong enough to withstand the bending stresses without failure, but the curvature may be so great that cracks will appear in suspended plaster ceilings, water will collect in low spots on roofs, and the general lack of stiffness will result in an excessively springy floor. When designing floor construction for buildings in which machinery will be used, particular care should be given to the deflection of beams because excessive deflection may be the cause of inordinate vibrations or misalignments. Another fault that results from excessive deflection occurs

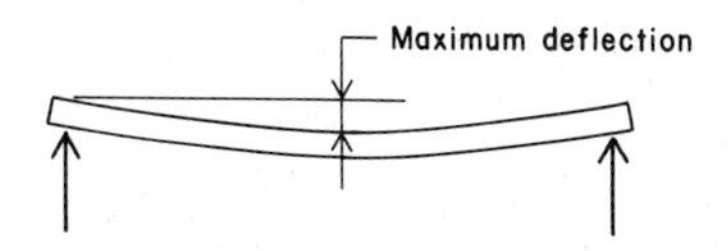

FIGURE 8-1

where floor beams frame into girders. At these points there is a tendency for cracks to develop in the flooring directly over the girder. This is illustrated in Fig. 8-2.

If, in the design of a beam, the deflection is computed to be excessive, the remedy is to select a deeper beam—a beam with a greater moment of inertia. For a given span and loading the deflection of a beam varies directly as the fiber stress (bending stress) and inversely as the depth. For this reason it is preferable to select a beam that has the greatest practicable depth so that the deflection will be a minimum. When ample headroom is available a deeper beam is preferable to a shallow beam with the same section modulus.

In general, the procedure is to design a beam of ample dimensions to resist bending stresses (design for strength) and then to investigate the beam for deflection.

8-2. Allowable Deflection

There is common agreement that deflection of beams should be limited, but authorities differ with respect to the maximum degree of deflection to be permitted. Some codes require that deflection be limited to 1/360 of the span but say nothing of the kind of load that produces the deflection. The AISC Specification requires

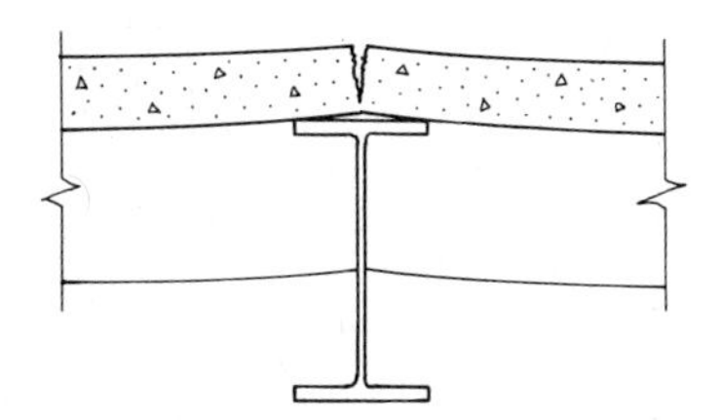

FIGURE 8-2

that beams and girders that support plastered ceilings have a maximum deflection of 1/360 of the span due to the live load. (See discussion of loads in Chapter 10.) For actual design work the local building code should be consulted for any deflection limits.

8-3. Deflection for Uniformly Distributed Loads

By referring to Fig. 5-26 we find formulas that can be used to compute the maximum deflection for the several types of loading that occur most frequently. The notation in the formulas is explained in Art 5-19. The Greek letter delta (Δ) represents the maximum deflection.

A number of terms make up these formulas and their solution may be a tedious task. There are, however, simplified methods of computing deflections. Consider the following example:

A designer is engaged in determining the size of steel floor beams; an allowable bending stress of 24 ksi has been chosen. The immediate problem is a simple beam with a span of 16 ft [4.88 m] and a uniformly distributed load of 34 kips [151 kN]. The computations show that a W 12 × 26 is large enough to resist the bending stresses and the next step is to compute the maximum deflection. To do this the designer uses the formula

$$\Delta = \frac{0.02483L^2}{d}$$

$$\left[\frac{0.1719L^2}{d}\right]$$

where L = the span length of the beam in feet [meters],
d = the depth of the beam in inches [meters],
Δ = the maximum deflection of the beam in inches [millimeters].

Referring to Appendix Table A-1, we find that d for the section is 12.22 in. Then

$$\Delta = \frac{0.02483 \times (16)^2}{12.22} = 0.520 \text{ in.}$$

$$\left[\Delta = \frac{0.1719 \times (4.88)^2}{0.3104} = 13.2 \text{ mm}\right]$$

Assume that the allowable deflection is 1/360 of the span, or

$$\Delta = \frac{16 \times 12}{360} = 0.533 \text{ in.}$$

$$\left[\frac{4880}{360} = 13.56 \text{ mm}\right]$$

The actual deflection does not exceed the allowable.

Now let us see how the foregoing formula, identified later as Fr. 8.3.3, is determined.

The maximum bending moment for this type of loading is $M = wl^2/8$. In Art. 7-1 we find

$$f = \frac{Mc}{I}$$

Consequently by substituting the value of M we obtain

$$f = \frac{wl^2c}{8I}$$

The maximum deflection for a simple beam with a uniformly distributed load is given in Case 2, Fig. 5-26:

$$\Delta = \frac{5}{384} \times \frac{Wl^3}{EI}$$

or, since $W = wl$,

$$\Delta = \frac{5}{384} \times \frac{wl^4}{EI} = \left(\frac{wl^2c}{8I}\right)\left(\frac{5l^2}{48Ec}\right)$$

$$= (f)\left(\frac{5l^2}{48Ec}\right) = \frac{5fl^2}{48Ec} \qquad \text{(Formula 8.3.1)}$$

This is a basic formula; it can be used for any material by substituting the appropriate values for f and E.

For sections symmetrical with respect to a horizontal axis through the centroid, such as rectangular and I-sections, $c = d/2$. Then by substituting this value of c in Fr. 8.3.1 we have

$$\Delta = \frac{5}{48}\left(\frac{f}{E}\right)\left(\frac{2l^2}{d}\right) \qquad \text{(Formula 8.3.2)}$$

Now, let f = 24 ksi, E = 29,000 ksi, and convert the span to feet (l = 12L). Then

$$\Delta = \frac{5}{48}\left(\frac{24}{29{,}000}\right)\left\{\frac{2\,(12L)^2}{d}\right\}$$

$$\Delta = \frac{0.02483L^2}{d} \qquad \text{(Formula 8.3.3)}$$

In SI units, with f = 165 MPa and E = 200 GPa,

$$\Delta = \frac{5}{48}\left(\frac{165}{200{,}000}\right)\left(\frac{2L^2}{d}\right)$$

$$\Delta = \frac{0.0001719L^2}{d} \quad \text{(in meters)}$$

$$\Delta = \frac{0.1719L^2}{d} \quad \text{(in millimeters)} \quad \text{(Formula 8.3.4)}$$

Remember that these equations apply only to simple beans with uniformly distributed loading and only when the stress is 24 ksi [165 MPa]. Because both stress and deflection are directly proportional to the magnitude of the load, the deflection that accompanies some other value of stress may be directly proportioned as shown in the examples.

Example 1. A simple beam has a span of 20 ft [6.10 m] and a uniformly distributed load of 39 kips [173.5 kN]. The section used for this load is a W 14 × 34 and the extreme fiber stress is 24 ksi [165 MPa]. Compute the deflection.
Solution: Formula 8.3.3 is appropriate without modification. Referring to Appendix Table A-1, we find d = 13.98 in. [0.355 m]. Then

$$\Delta = \frac{0.02483L^2}{d} = \frac{0.02483(20)^2}{13.98} = 0.710 \text{ in.}$$

$$\left[\Delta = \frac{0.1719L^2}{d} = \frac{0.1719(6.1)^2}{0.355} = 18.0 \text{ mm}\right]$$

If we had not known that f = 24 ksi, we might have used the formula for deflection given in Fig. 5-26, Case 2. Thus

$$\Delta = \frac{5}{384} \times \frac{Wl^3}{EI}$$

Appendix Table A-1 shows that I for the W 14 × 34 is 340 in.^{4} [141.5 × 10^6 mm^4]. Then

$$\Delta = \frac{5}{384} \times \frac{39 \times (20 \times 12)^3}{29{,}000 \times 340} = 0.712 \text{ in.}$$

$$\left[\Delta = \frac{5}{384} \times \frac{173.5 \times (6100)^3}{200{,}000 \times 141.5 \times 10^6} = 18.1 \text{ mm}\right]$$

Example 2. A W 12 × 26 is used as a simple beam on a span of 19 ft [5.79 m]. It supports a uniformly distributed load and the computed maximum bending stress is 20 ksi [138 MPa]. Find the deflection.

Solution: Referring to Appendix Table A-1, we find d = 12.22 in. [0.310 m]. To use Formula 8.3.3 we must adjust for the actual stress of 20 ksi [138 MPa], which is done by a simple proportion:

$$\Delta = \frac{20}{24} \times \frac{0.02483\,(19)^2}{12.22} = 0.611 \text{ in.}$$

$$\left[\Delta = \frac{138}{165} \times \frac{0.1719\,(5.79)^2}{0.310} = 15.5 \text{ mm}\right]$$

8-4. Deflections Found by Coefficients

A method of computing deflections preferred by some designers consists in using the coefficients found in Table 8-1. This table is applicable for simple beams with uniformly distributed loads and for symmetrical sections such as I-beams and wide flange sections.

To use this table find the coefficient corresponding to the beam span and f, the bending stress. This coefficient divided by the depth of the beam in inches gives the deflection in inches:

$$\text{Deflection} = \frac{\text{coefficient in table}}{\text{depth of beam}}$$

TABLE 8-1. Deflection Coefficients for Uniformly Distributed Loads[a]

Span (ft)	Coeff.	Span (ft)	Coeff.	Span (ft)	Coeff.	Span (m)	Coeff.	Span (m)	Coeff.
10	2.483	24	14.301	38	35.851	3	1.547	10	17.188
12	3.575	26	16.783	40	39.724	4	2.750	11	20.797
14	4.866	28	19.465	42	43.796	5	4.297	12	24.750
16	6.356	30	22.345	44	48.066	6	6.188	13	29.047
18	8.044	32	25.423	46	52.535	7	8.422	14	33.688
20	9.931	34	28.701	48	57.203	8	11.000	15	38.672
22	12.017	36	32.177	50	62.069	9	13.922	16	44.000

[a] Simple span, $f_b = 24$ ksi [165 MPa]. Deflection equals table coefficient divided by beam depth in inches or meters. Coefficient for concentrated load at center of span: 0.8 of table values.

Table 8-1 has been prepared for *uniformly distributed* loads. For other types of loading the coefficients are multiplied by the various factors given in the footnote to the table.

Example 1. A W 14 × 34 supports a distributed load of 39 kips [173 kN] on a span of 20 ft [6 m]. The bending stress is 24 ksi [165 MPa]. Determine the deflection by the use of Table 8-1 (See Example 1, Art. 8-3.)

Solution: From Appendix Table A-1 we find that the depth of the beam is 13.98 in. [355 mm] and in Table 8-1 we see that the coefficient is 9.931 [6.188]. Therefore

$$\Delta = \frac{9.931}{13.98} = 0.710 \text{ in.}$$

$$\left[\frac{6.188}{0.355} = 17.4 \text{ mm}\right]$$

It may happen that the extreme fiber stress is not 24 ksi. in which case the value for the deflection may be determined by proportion, as shown in the following example:

Example 2. A simple beam with a section of W 16 × 36 has a span of 16 ft [5 m] and supports a uniformly distributed load of 40 kips [178 kN]. Compute the deflection.

Solution: From Appendix Table A-1 we find that $d = 15.86$ in. [403 mm] and $S = 56.5$ in^3 [926×10^3 mm^3] for this section. Then

$$M = \frac{WL}{8} = \frac{40 \times 16}{8} = 80 \text{ kip-ft}$$

$$\left[\frac{178 \times 5}{8} \; 111 \text{ kN-m}\right]$$

$$F_b = \frac{M}{S} = \frac{80 \times 12}{56.5} = 16.99 \text{ ksi}$$

$$\left[\frac{111 \times 10^3}{926} = 120 \text{ MPa}\right]$$

and, using the coefficient for 24 ksi [165 MPa] from Table 8-1,

$$\Delta = \frac{16.99}{24} \times \frac{6.356}{15.86} = 0.284 \text{ in.}$$

$$\left[\frac{120}{165} \times \frac{4.297}{0.403} = 7.75 \text{ mm}\right]$$

Problems 8.4.A,* B, C. Find the maximum deflection of the following simple span beams with the indicated total uniformly distributed loadings. Find the deflection (a) by use of the formula from Fig. 5-26 and (b) by use of the coefficient from Table 8.1.

		Span		Load	
	Beam	(ft)	(m)	(kips)	(kN)
A	W 10 × 33	16	5	32	140
B	W 12 × 45	20	6	40	175
C	W 18 × 40	32	10	30	130

9

Beam Design Procedures

9-1. General

The complete design of a beam includes consideration of bending strength, shear resistance, deflection, lateral support, web crippling, and support details. Design for bending is treated in Chapter 7 and methods for computing deflections are presented in Chapter 8. In this chapter the remaining items are considered and design procedures are established.

The current AISC Specification includes special factors that influence design for bending. The availability of higher strength steels and attendant higher allowable stresses has led to the classification of structural shapes for beams as *compact* or *noncompact* sections. Consequently the shape, the laterally unsupported length of the span, and the grade of steel must be known in order to establish allowable stresses for bending. Unless otherwise stated, the examples and problems presented in this book are based on the assumption that A36 steel is used. The design procedures and specification formulas, however, are applicable to any grade of steel by selecting the appropriate values for the yield stress F_y or the ultimate strength F_u.

9-2. Compact and Noncompact Sections

To qualify for use of the maximum allowable bending stress of 0.66 F_y a beam consisting of a rolled section must satisfy several qualifications, principal among which are the following:

The beam section must be symmetrical about its minor (*Y-Y*) axis and the plane of the loading must coincide with the plane of this axis. (See discussion of shear center in Art. 6-10.)

The web and flanges of the section must have width-to-thickness ratios that qualify the section as *compact*.

The compression flange of the beam must be adequately braced against lateral buckling.

The criteria for determining whether a section is compact includes as a variable the F_y of the steel. It is therefore not possible to identify sections for this condition strictly on the basis of their geometric properties. The yield stress limit for qualification as a compact section is given as the value F_y' in Appendix Tables A-1, A-2, and A-3 for W, M, and S shapes. When members are subjected to combined bending and compression the qualifications for consideration as compact get even tougher and the value in the tables given as F_y''' indicates the limit for this condition.

When sections do not qualify as compact the allowable bending stress must be reduced by using the formulas in Section 1.5.1.4 of the AISC Specification. In some cases these reductions have been incorporated into the design aids in the AISC Manual (Ref. 1) for A36 steel, as described in the example problems.

9-3. Lateral Support of Beams

A beam may fail by sideways buckling of the top (compression) flange when lateral deflection is not prevented. The tendency to buckle increases as the compressive bending stress in the flange increases and as the unbraced length of the span increases. Consequently the full value of the allowable extreme fiber stress $F_b = 0.66\ F_y$ can be used only when the compression flange is adequately braced. As for the compact section, the value of this required length includes the variable of the F_y of the beam steel.

When the compression flanges of compact beams are supported laterally at intervals not greater than L_c the full allowable stress of $0.66\ F_y$ may be used. For lateral unsupported lengths greater than L_c, but not greater than L_u, the allowable bending stress is reduced to $0.60\ F_y$. (In certain instances the AISC Specification permits a proportionate reduction in the allowable stress between the two limits; however, in the examples in this book we assume that the drop occurs totally as L_c is exceeded.)

When the laterally unsupported length exeeds L_u the specifications provide a formula for the determination of the allowable stress, based on the specific value of the unsupported length. The design of beams based on these requirements is not a simple matter and the AISC Manual (Ref. 1) contains supplementary charts to aid the designer. These beam charts include "Allowable Moments in Beams" as a function of the laterally unsupported length and provide a workable approach to this otherwise rather cumbersome problem. Reproductions of the charts for beams of A36 steel appear in Appendix C of this book.

After determining the maximum bending moment in kip-ft and noting the longest unbraced length of the compression flange these two coordinates are located on the sides of the chart and are projected to their intersection. Any beam whose curve lies above and to the right of this intersection point satisfies the bending stress requirement. The nearest curve that is a solid (versus dashed) line represents the most economical, or least-weight, section in terms of beam weight, a relationship similar to that discussed in Art. 7-4 with regard to selection from the section modulus tables in Appendix B. It should be noted that selection from the charts incorporates the considerations of bending stress, compact sections, and lateral support; however, deflection, shear, and other factors may also have to be considered.

Example. A simple beam carries a total uniform load of 19.6 kips [87.2 kN], including its own weight, over a span of 20 ft [6.10 m]. It has no lateral support except at the ends of the span. Assuming A36 steel, select from the charts of Appendix C a beam that will meet bending strength requirements and not deflect more than 0.75 in. [19 mm] under the full load.

Solution: For this loading the maximum bending moment is

$$M = \frac{WL}{8} = \frac{19.6 \times 20}{8} = 49 \text{ kip-ft}$$

$$\left[M = \frac{87.2 \times 6.10}{8} = 66.5 \text{ kN-m}\right]$$

Entering the appropriate chart with the values for the moment and the unsupported length, we locate the critical intersection point. The nearest solid line curve above and to the right of this point is that for a W 8 × 31.

For consideration of the deflection limit we may now find the actual deflection for the W 8 × 31 for the given span and load. If this is excessive, we then return to the chart to read the other shapes whose curves are above and to the right of the critical intersection point and examine the corresponding deflections until we can make an acceptable choice. An alternative to this pick-and-try method is a separate calculation for the required moment of inertia of the beam with a transformed version of the formula for maximum deflection for the beam (Fig. 5-26, Case 2). Thus

$$I = \frac{5}{384} \times \frac{Wl^3}{E\Delta} = \frac{5 \times 19.6 \times (20 \times 12)^3}{384 \times 29{,}000 \times 0.75} = 162 \text{ in}^4$$

$$\left[I = \frac{5 \times 87.2 \times (6.1)^3}{384 \times 200{,}000 \times 0.019} = 0.06782 \text{ m}^4 \quad \text{or} \quad 67.82 \times 10^6 \text{ mm}^4\right]$$

For the W 8 × 31, from Appendix Table A-1, I = 110 in.^{4} Because this is less than that required, we return to the chart and select the W 10 × 33 for which I = 170 in.^{4} The W 10 × 33 is an acceptable selection for the criteria established.

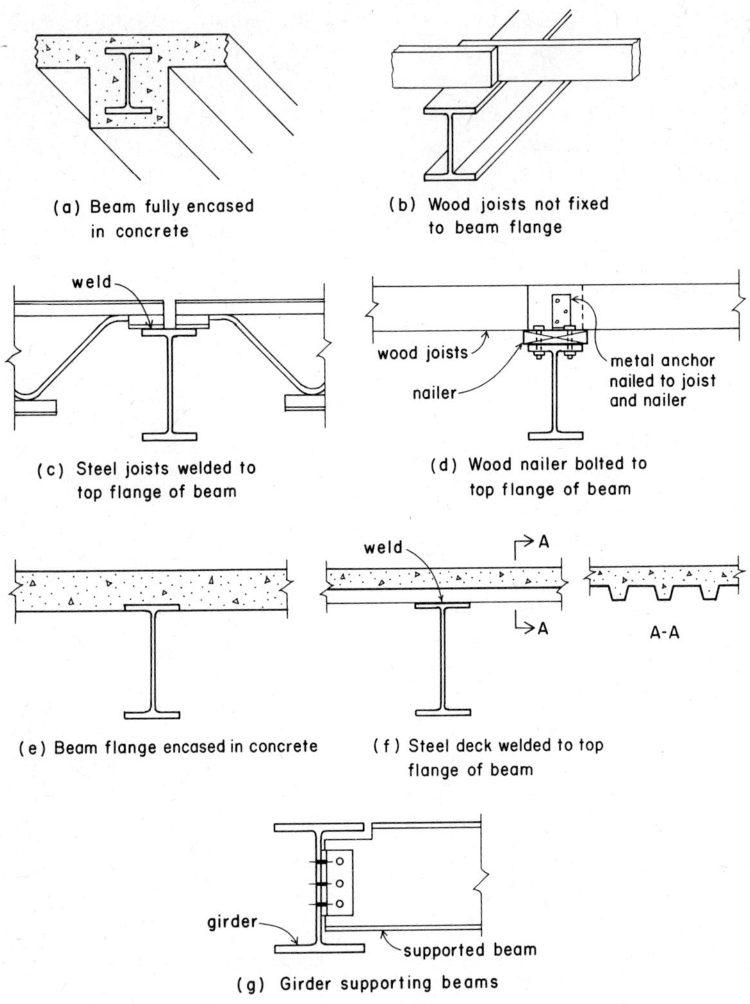

FIGURE 9-1. Lateral bracing of beams.

It is not always a simple matter to decide that a beam is laterally supported. In cases such as that shown in Fig. 9-1*a* lateral support is supplied to beams by the floor construction; it is evident from the figure that lateral deflection of the top flange is prevented by the concrete slab. On the other hand, the type of floor construction shown in Fig. 9-1*b*, where wood joists simply rest on the top flange of a steel beam, offers no resistance to sideways buckling. Floor systems of the types indicated in Figs. 9-1*c*, *d*, *e*, and *f* usually furnish adequate lateral support of the top flange. However, metal or precast floor systems held in place by clips generally have insufficiently rigid connections to the flange to provide adequate lateral bracing. If a beam acts as a girder and supports other beams with connections like those shown in Fig. 9-1*g*, lateral bracing is provided at the connections and the laterally unsupported length of the top flange to be checked against L_c or L_u becomes the distance between the supported beams.

Problem 9.3.A.* A W 21 × 83 is used as a simple beam to carry a total uniformly distributed load, including the beam weight, of 53 kips [236 kN] on a span of 36 ft [11 m]. Lateral support is provided only at the ends and at the midspan. Is the section an adequate choice for these conditions?

Problem 9.3.B.* A simple beam is required for a span of 32 ft [9.75 m]. The load, including the beam weight, is uniformly distributed and consists of a total of 100 kips [445 kN]. If lateral support is provided only at the ends and midspan and the deflection under full load is limited to 0.6 in. [15.2 mm], find an adequate rolled shape for the beam.

9-4. Torsional Effects

In various situations steel beams may be subjected to torsional twisting effects in addition to the primary conditions of shear and bending. These effects may occur when the beam is loaded in a plane that does not coincide with the shear center of the section; this problem of special concern for channel and angle shapes is discussed in Art. 6-10. Even for the doubly symmetrical W, M, and S shapes loadings may produce torsion when applied off-center, as shown in Fig. 9-2.

A special torsional effect is that of the rotational effect known as torsional buckling. Beams that are weak on their minor axes

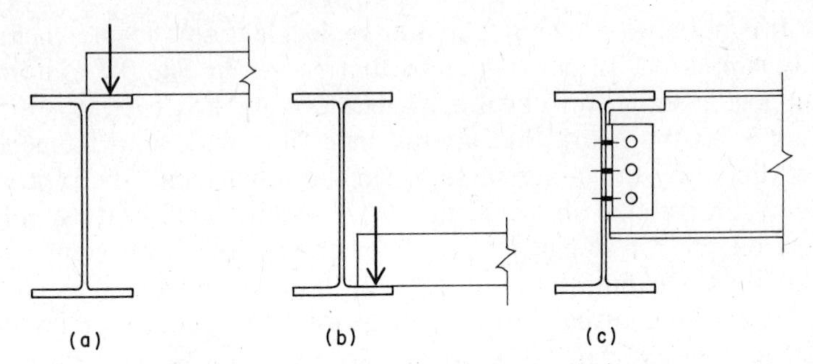

FIGURE 9-2. Torsion produced by off-center loading.

are subject to this effect, which occurs at the points of support or at the location of concentrated loads (Fig. 9-3*a*).

When potential torsional effects threaten they may be dealt with in one of two ways. In the first we simply compute the torsional moments and design the beam to resist them or we determine the torsional buckling and reduce the allowable stress accordingly. The analysis required, however, is complex and beyond the scope of this book. In the second method, which is usually the preferred, adequate bracing is provided for the beam

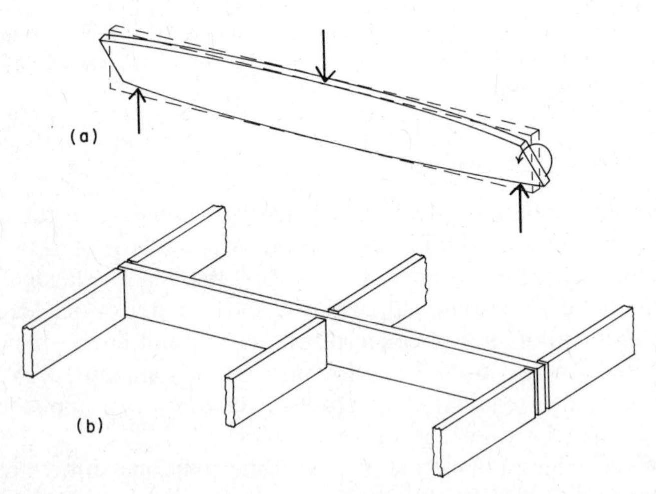

FIGURE 9-3. Torsional (rotational) buckling (a) of an unbraced beam and (b) prevented by framing.

to prevent the potential torsional rotation, in which case the torsional effect is essentially avoided.

In many cases the ordinary details of construction will result in adequate bracing against torsion. In the detail shown in Fig. 9-2*c*, for example, although the supported beam applies a slightly off-center load, the connection between the beam and girder may be adequately stiff to prevent torsional twisting of the girder. Some judgment must be exercised, of course, with regard to the relative sizes of girder, beam, and the connection itself.

In Fig. 9-3*b* the torsional buckling of the beam shown in Fig. 9-3*a* is prevented by the framing that is attached at right angles to the beam.

9-5. Shear Effects

Shear stress in a steel beam is seldom a factor in determining its size. It is customary to determine first the size of the beam to resist bending stresses. Having done this, the beam is then investigated for shear, which means that we compute the actual maximum unit shear stress to see that it does not exceed the allowable stress. The AISC Specification gives F_v, the allowable shear stress in beam webs, as $0.40F_y$ on the gross section of the web, which is computed as the product of the web thickness and the overall beam depth. For A36 steel $F_v = 0.40 \times 36$ or 14.4 ksi. This value is rounded off at $F_v = 14.5$ ksi [100 MPa], as shown in Table 4-3.

As noted in Art. 3-5, shearing stresses in beams are not distributed uniformly over the cross section but are zero at the extreme fibers, with the maximum value occurring at the neutral surface. Consequently the material in the flanges of wide flange sections, I-beams, and channels has little influence on shearing resistance, and the working formula for determining shearing stress is taken as

$$f_v = \frac{V}{A_w}$$

where f_v = the unit shearing stress,
V = the maximum vertical shear,
A_w = the gross area of the web (actual depth of section times the web thickness, or $d \times t_w$).

The shearing stresses in beams are seldom excessive. If, however, the beam has a relatively short span with a large load placed near one of the supports, the bending moment is relatively small and the shearing stress becomes comparatively high. This situation is demonstrated in the following example:

Example. A simple beam of A36 steel is 6 ft [1.83 m] long and has a concentrated load of 36 kips [160 kN] applied 1 ft [0.3 m] from the left end. It is found that a W 8 × 24 is large enough to support this load with respect to bending stresses (required S = 15 in.^{3} [246 × 10^3 mm^3]). Investigate the beam for shear, neglecting the weight of the beam.
Solution: The two reactions are computed by the methods explained in Art. 5-6; we find that the left reaction is 30 kips [133 kN] and the right reaction is 6 kips [27 kN]. The maximum vertical shear is thus 30 kips [133 kN].

From Appendix Table A-1 we find that d = 7.93 in. [201.4 mm] and t_w = 0.245 in. [6.22 mm] for the W 8 × 24. Then

$$A_w = d \times t_w = 7.93 \times 0.245 = 1.94 \text{ in.}^2$$

$$[A_w = 201.4 \times 6.22 = 1253 \text{ mm}^2]$$

and

$$f_v = \frac{V}{A_w} = \frac{30}{1.94} = 15.4 \text{ ksi}$$

$$\left[f_v = \frac{133 \times 10^3}{1253} = 106 \text{ MPa}\right]$$

Because this exceeds the allowable value of 14.5 ksi [100 MPa], the W 8 × 24 is not acceptable.

Recalling from Art. 2-3 that S shapes have somewhat thicker webs than W shapes, we find in Appendix Table A-3 that an S 8 × 23 has a depth of 8 in. [203.2 mm] and a web thickness of 0.441 in. [11.2 mm]. Then

$$A_w = d \times t_w = 8 \times 0.441 = 3.53 \text{ in.}^2$$

$$[A_w = 203.2 \times 11.2 = 2276 \text{ mm}^2]$$

and

$$f_v = \frac{V}{A_w} = \frac{30}{3.53} = 8.50 \text{ ksi}$$

$$\left[f_v = \frac{1.33 \times 10^3}{2276} = 58.4 \text{ MPa}\right]$$

which is less than the allowable stress. It should be observed that both W and S shapes are adequate for bending stress. Thus both sections are acceptable for bending resistance, but only the S 8 × 23 will provide adequate shearing resistance. In real design situations often many additional matters must be considered, such as concern for support or framing details.

Problem 9.5.A. Compute the maximum permissible web shears for the following beams of A36 steel: (a*) S 12 × 40.8; (b) W 12 × 40; (c) W 10 × 22; (d) C 10 × 20.

9-6. Crippling of Beam Webs

An excessive end reaction on a beam or an excessive concentrated load at some point along the interior of the span may cause crippling or localized yielding of the beam web. The AISC Specification requires that end reactions or concentrated loads for beams without stiffeners or other web reinforcement shall not exceed the following (Fig. 9-4):

$$\text{Maximum end reaction} = 0.75F_y t(N + k)$$

$$\text{Maximum interior load} = 0.75F_y t(N + 2k)$$

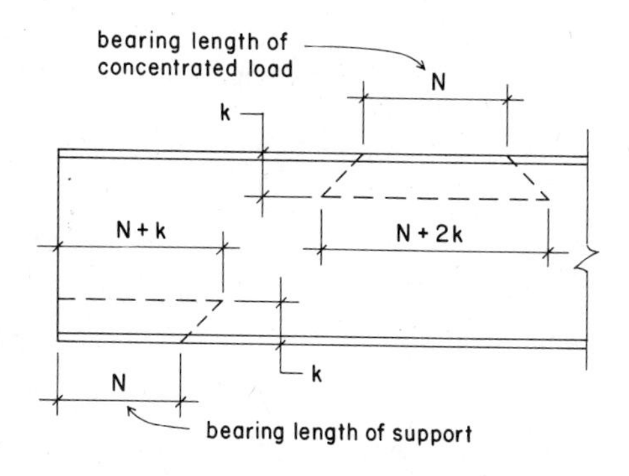

FIGURE 9-4. Determination of effective length for computation of web crippling.

where t = thickness of beam *web,* in inches,
N = length of bearing or length of concentrated load (not less than k for end reactions), in inches,
k = distance from outer face of flange to web toe of fillet, in inches,
$0.75F_y$ = 27 ksi for A36 steel [186 MPa].

When these value are exceeded the webs of the beams should be reinforced with stiffeners, the length of bearing increased, or a beam with a thicker web selected.

Example 1. A W 21 × 57 beam of A36 steel has an end reaction that is developed in bearing over a length of N = 10 in. [254 mm]. Check the beam for web crippling if the reaction is 44 kips [196 kN].
Solution: In Appendix Table A-1 we find that k = 1.375 in. [35 mm] and the web thickness is 0.405 in. [10 mm]. To check for web crippling we find the maximum end reaction permitted and compare it with the actual value for the reaction. Thus

$$R = F_p \times t \times (N + k) = 27 \times 0.405 \times (10 + 1.375)$$

$$= 124 \text{ kips (the allowable reaction)}$$

$$\left[R = \frac{186 \times 10 \times (254 + 35)}{10^3} = 538 \text{ kN}\right]$$

Because this is greater than the actual reaction, the beam is not critical with regard to web crippling.

Example 2. A W 12 × 26 of A36 steel supports a column load of 70 kips [311 kN] at the center of the span. The bearing length of the column on the beam is 10 in. [254 mm]. Investigate the beam for web crippling under this concentrated load.
Solution: In Appendix Table A-1 we find that k = 0.875 in. [22 mm] and the web thickness is 0.230 in. [5.84 mm]. The allowable load that can be supported on the given bearing length is

$$P = F_p \times t \times (N + 2k) = 27 \times 0.230 \times \{10 + (2 \times 0.875)\}$$

$$= 73 \text{ kips}$$

$$\left[P = \frac{186 \times 5.84}{10^3} \times \{254 + (2 \times 22)\} = 324 \text{ kN}\right]$$

which exceeds the required load. Because the column load is less than this value, the beam web is safe from web crippling.

Problem 9.6.A. Compute the maximum allowable reaction with respect to web crippling for a W 14 × 30 of A36 steel with an 8 in. [200 mm] bearing-plate length.

Problem 9.6.B. A column load of 81 kips [360 kN] with a bearing-plate length of 11 in. [279 mm] is placed on top of the beam in the preceding problem. Are web stiffeners required to prevent web crippling?

9-7. Beam Bearing Plates

Beams that are supported on walls or piers of masonry or concrete usually rest on steel bearing plates. The purpose of the plate is to provide an ample bearing area. The plate also helps to seat the beam at its proper elevation. Bearing plates provide a level surface for a support and, when properly placed, afford a uniform distribution of the beam reaction over the area of contact with the supporting material.

By reference to Fig. 9-5 the area of the bearing plate is $B \times N$. It is found by dividing the beam reaction by F_p, the allowable bearing value of the supporting material. Then

$$A = \frac{R}{F_p}$$

where $A = B \times N$, the area of the plate in sq in.,
R = reaction of beam in pounds or kips,
F_p = allowable bearing pressure on the supporting material in psi or ksi (see Table 9-1).

The thickness of the wall generally determines N, the dimension of the plate parallel to the length of the beam. If the load from the beam is unusually large, the dimension B may become excessive. For such a condition one or more shallow-depth I-beams, placed parallel to the wall length, may be used instead of a plate. The dimensions B and N are usually in even inches and a great variety of thicknesses is available.

The thickness of the plate is determined by considering the projection n (Fig. 9-5*b*) as an inverted cantilever; the uniform bearing pressure on the bottom of the plate tends to curl it upward about the beam flange. The required thickness may be computed

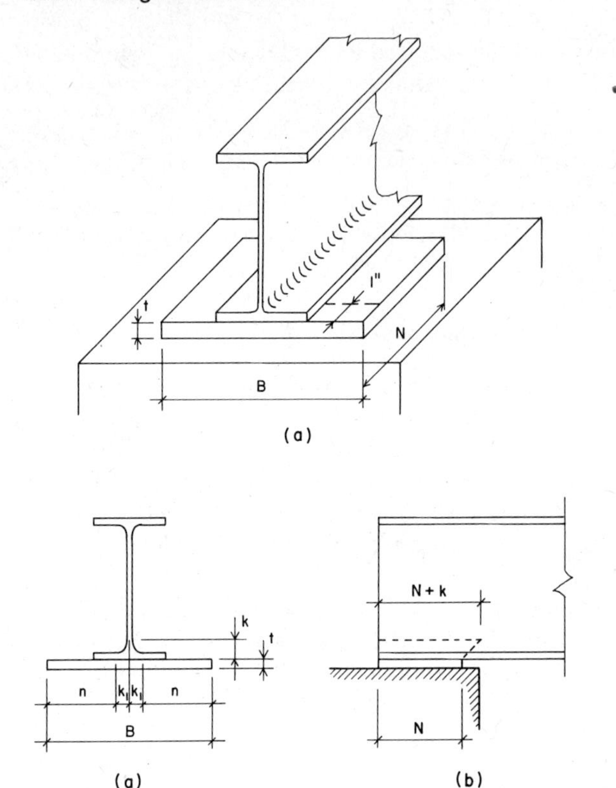

FIGURE 9-5. Reference dimensions for beam end bearing plates.

readily by the following formula, which does not involve direct computation of bending moment and section modulus:

$$t = \sqrt{\frac{3f_p n^2}{F_b}}$$

where t = thickness of plate in inches,

f_p = *actual* bearing pressure of the plate on the masonry, psi or ksi,

F_b = allowable bending stress in the plate (the AISC Specification gives the value of F_b as $0.75F_y$; for A36 steel F_y = 36 ksi; therefore $F_b = 0.75 \times 36 = 27$ ksi),

TABLE 9-1. Allowable Bearing Pressure on Masonry and Concrete

Type of material and conditions	Allowable unit stress in bearing, F_p (psi)	(kPa)
Solid brick, unreinforced, type S mortar		
$f'_m = 1500$ psi	170	1200
$f'_m = 4500$ psi	338	2300
Hollow unit masonry, unreinforced, type S mortar, $f'_m = 1500$ psi (on net area of masonry)	225	1500
Concrete[a]		
(1) Bearing on full area of support		
$f'_c = 2000$ psi	500	3500
$f'_c = 3000$ psi	750	5000
(2) Bearing on $\frac{1}{3}$ or less of support area		
$f'_c = 2000$ psi	750	5000
$f'_c = 3000$ psi	1125	7500

[a] Stresses for areas between these limits may be determined by direct proportion.

$n = \frac{B}{2} - k_1$, in inches (see Fig. 9-5b),

k_1 = the distance from the center of the web to the toe of the fillet; values of k_1 for various beam sizes may be found in the appendix tables.

The foregoing formula is derived by considering a strip of plate 1 in. wide (Fig. 9-5a) and t in. thick, with a projecting length of n inches, as a cantilever. Because the upward pressure on the steel strip is f_p, the bending moment at distance n from the edge of the plate is

$$M = f_p n \times \frac{n}{2} = \frac{f_p n^2}{2}$$

For this strip with rectangular cross section

$$\frac{I}{c} = \frac{bd^2}{6} \qquad \text{(Art. 6-5)}$$

and because $b = 1$ in. and $d = t$ in.

$$\frac{I}{c} = \frac{1 \times t^2}{6} = \frac{t^2}{6}$$

Then, from the beam formula,

$$\frac{M}{F_b} = \frac{I}{c} \qquad \text{(Art. 7-1)}$$

Substituting the values of M and I/c determined above,

$$\frac{f_p n^2}{2} \times \frac{1}{F_b} = \frac{t^2}{6}$$

and

$$t^2 = \frac{6 f_p n^2}{2F_b} \quad \text{or} \quad t = \sqrt{\frac{3 f_p n^2}{F_b}}$$

When the dimensions of the bearing plate are determined the beam should be investigated for web crippling on the length $(N + k)$ shown in Fig. 9-5*c*. This is explained in Art. 9-6.

Example 1. A W 21 × 57 of A36 steel transfers an end reaction of 44 kips [196 kN] to a wall built of solid brick by means of a bearing plate of A36 steel. Assume type S mortar and a brick with $f'_m = 1500$ psi. The N dimension of the plate (see Fig. 9-5) is 10 in. [254 mm]. Design the bearing plate.

Solution: In Appendix Table A-1 we find that k_1 for the beam is 0.875 in. [22 mm]. From Table 9-1 the allowable bearing pressure F_p for this wall is 170 psi [1200 kPa]. The required area of the plate is then

$$A = \frac{R}{F_p} = \frac{44{,}000}{170} = 259 \text{ in.}^2$$

$$\left[A = \frac{196 \times 10^6}{1200} = 163{,}333 \text{ mm}^2\right]$$

Then, because N = 10 in. [254 mm],

$$B = \frac{259}{10} = 25.9 \text{ in.}^2$$

$$\left[= \frac{163{,}333}{254} = 643 \text{ mm}\right]$$

which is rounded off to 26 in. [650 mm].

With the true dimensions of the plate we now compute the true bearing pressure:

$$f_p = \frac{R}{A} = \frac{44{,}000}{10 \times 26} = 169 \text{ psi}$$

$$\left[f_p = \frac{196 \times 10^6}{254 \times 650} = 1187 \text{ kPa}\right]$$

To find the thickness we first determine the value of n:

$$n = \frac{B}{2} - k_1 = \frac{26}{2} - 0.875 = 12.125 \text{ in.}$$

$$\left[n = \frac{650}{2} - 22 = 303 \text{ mm}\right]$$

Then

$$t = \sqrt{\frac{3f_p n^2}{F_b}} = \sqrt{\frac{3 \times 169 \times (12.125)^2}{27{,}000}} = \sqrt{2.760} = 1.66 \text{ in.}$$

$$\left[t = \sqrt{\frac{3 \times 1187 \times (303)^2}{186{,}000}} = \sqrt{1758} = 42 \text{ mm}\right]$$

The complete design for this problem would include a check of the web crippling in the beam. This has already been done as Example 1 in Art. 9-6.

In the event that a comparatively light reaction or a high allowable bearing pressure reduces the bearing area required so that a beam may be supported without a bearing plate, the beam flange should be checked for bending induced by the bearing pressure to make certain that it does not exceed F_b. This may be accomplished by use of the formula

$$f_b = \frac{3f_p n^2}{t^2}$$

where f_b = the actual bending stress in the beam flange,
f_p = the actual bearing pressure of the beam flange on the supporting structure,
n = (flange width/2) − k_1,
t = the thickness of the flange.

Example 2. A W 12 × 53 of A36 steel transfers a load of 16 kips [71 kN] to a brick wall laid up with type S mortar and brick with f'_m of 4500 psi. The beam has an 8-in. [203-mm] bearing length (dimension N) on the wall. If a bearing plate is not required, compute the maximum bending stress in the beam flange. Does the bending stress exceed 27,000 psi [186 MPa], the allowable in the flange when acting as a bearing plate?

Solution: In Appendix Table A-1 we find that the flange width is 9.995 in. [254 mm], the flange thickness, 0.575 in. [14.6 mm], and the dimension k_1, 0.8125 in. [20.6 mm]. From Table 9-1 we find that the allowable bearing pressure on the brick wall is 338 psi [2300 kPa].

The bearing area of the flange on the wall is

$$A = 8 \times 10 = 80 \text{ in.}^2$$

$$[203 \times 254 = 51{,}562 \text{ mm}^2]$$

and because the reaction is 16 kips [71 kN] the actual bearing pressure on the wall is

$$f_p = \frac{R}{A} = \frac{16{,}000}{80} = 200 \text{ psi}$$

$$\left[f_p = \frac{71}{0.05156} = 1377 \text{ kPa}\right]$$

Because this value is less than the allowable pressure, no bearing plate is required unless the bending stress in the beam flange exceeds the allowable. Investigating the beam flange for bending stress, we obtain

$$n = \frac{10}{2} - 0.8125 = 4.1875 \text{ in.}$$

$$\left[n = \frac{254}{2} - 20.6 = 106.4 \text{ mm}\right]$$

and

$$f_b = \frac{3f_p n^2}{t^2} = \frac{3 \times 200 \times (4.1875)^2}{(0.575)^2} = 31{,}822 \text{ psi}$$

$$\left[f_b = \frac{3 \times 1377 \times (106.4)^2}{(14.6)^2} = 219{,}400 \text{ kPa}\right]$$

which is greater than that allowable.

Although the bearing stress is not critical, in this case a bearing plate must be used unless the bearing length can be increased.

Problem 9.7.A.* A W 14 × 30 with a reaction of 20 kips [89 kN] rests on a brick wall with brick of $f'_m = 1500$ psi and type S mortar. The beam has a bearing length of 8 in. [203 mm] parallel to the length of the beam. If the bearing plate is A36 steel, determine its dimensions.

Problem 9.7.B. A wall of brick with f'_m of 1500 psi and type S mortar supports a W 18 × 50 of A36 steel. The beam reaction is 25 kips [111 kN] and the bearing length N is 9 in. [229 mm] Design the beam bearing plate.

Problem 9.7.C. A W 12 × 65 of A36 steel with a reaction of 4 kips [17.8 kN] is supported by a wall of brick with f'_m of 4500 psi and type S mortar. The bearing length of the beam on the wall (dimension N) is 3.5 in. [89 mm]. Determine whether a bearing plate is required. If not, does the bending stress in the beam flange exceed the allowable value?

9-8. Beam Connections

When steel beams are supported at their ends by steel columns or by other steel beams a common form of connection is that shown in Fig. 9-6. To achieve this connection a pair of steel angles is used with one angle placed on each side of the web of the sup-

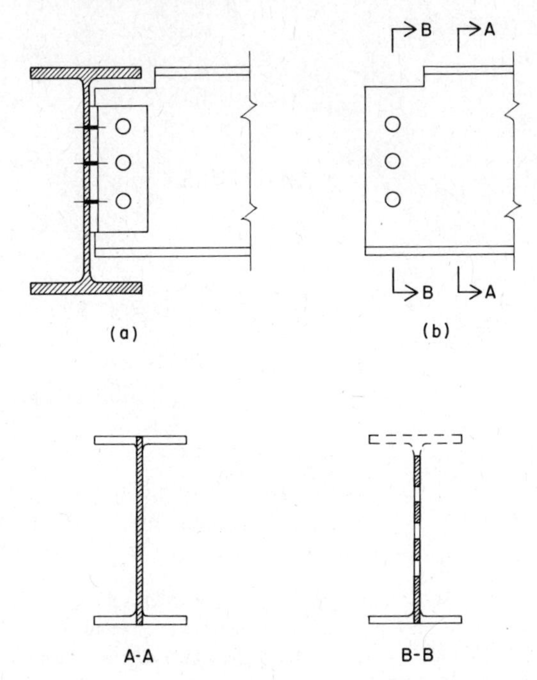

FIGURE 9-6. Shear at reduced sections of beams.

ported beam and the other angle legs outstanding at right angles to the beam web. The outstanding legs are then attached to the web of the supporting beam or to the side of the steel column.

The attachment of the angles for this connection may be achieved by welding or bolting. Most often the angles are attached to the supported beam in the fabricating shop by welding or bolting. The beam is then placed in position in the field and attached to the supporting structure by bolting the outstanding angle legs.

For floor and roof framing it is usual to have all the beam tops at a common level, but this presents the problem of how to attach the supported beam to the web of the supporting beam because the flange of the supporting beam is in the way. One solution is to cut away part of the top of the supported beam as shown in Fig. 9-6*a*.

If the angles are bolted to the supported beam and the top is cut back (Fig. 9-6*b*), there may be a substantial reduction in the web cross section. This is illustrated by the two sections *A-A* and *B-B*. Section *A-A* (Fig. 9-6) shows the unreduced web as it is considered for resistance to shear (beam depth times web thickness). Section *B-B* (Fig. 9-6) is the net usable area for shear at the connection. This net area must be capable of resisting the end reaction or the connection cannot be used. In some cases, for example, when a beam is supported at the face of a column (Fig. 9-7*a*), it is not necessary to cut back the beam flange. A really critical situation occurs, however, when a beam is supported by another beam of the same depth (Fig. 9-7*b*), in which case both flanges must be cut back. The problems of design of bolted connections, including the issue of reduced cross sections, are dealt with in Chapter 12.

9-9. Safe Load Tables

The simple span beam loaded entirely with uniformly distributed load occurs so frequently in steel structural systems that it is useful to have a rapid design method for quick selection of shapes for a given load and span condition. In preceding articles we have demonstrated the use of the section modulus tables and the charts for laterally unsupported beams which are in the AISC Manual

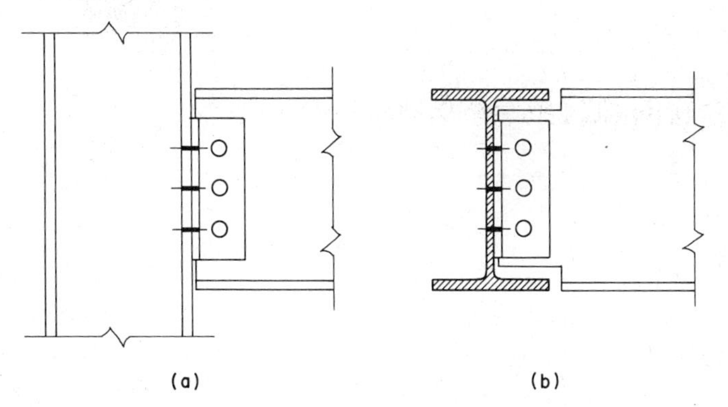

FIGURE 9-7. Use of framing connections (a) at a column and (b) at a supporting beam.

(Ref. 1). Use of the tables in Appendix D allows an even simpler procedure when design conditions permit their use.

For the simple beam with uniformly distributed load we note the maximum bending moment to be $WL/8$. If we equate this to the resisting moment of the beam, expressed as $S \times F_b$, we obtain the following expression for the limiting load on the beam:

$$W = \frac{8SF_b}{L}$$

where W = the total uniformly distributed load,
S = the section modulus of the beam (S_x),
F_b = the allowable bending stress,
L = the beam span.

If we assume a doubly symmetrical shape (W, M, or S), a compact section, and a laterally unsupported length not greater than L_c, we may use the maximum allowable bending stress of $0.66\,F_y$. Then, for a given shape and span, the allowable value of W can be computed for a given grade of steel.

The tables in Appendix D have been developed by this process, assuming A36 steel. Use of the tables requires only the determination of the span and total load if bending stress is the single concern. If the distance between points of lateral bracing exceeds the value of L_c for a given shape, the charts in Appendix C should be used instead of the tables in Appendix D. For a check the values of L_c are given in the tables for each shape. The loads in the tables will not result in excessive shear stress on the beam webs if the full section is available for resistance. Shear stress becomes increasingly critical as the span becomes shorter and should be investigated as described in Chapter 12 if the web section is reduced to form end connections.

Deflection due to the loads in the table may be determined by use of the deflection factors given at the top of the table for each span. Deflections for loads less than those tabulated may be found by proportion, as demonstrated in previous examples. As the span increases, deflection becomes increasingly critical, and the point at which the load will cause a deflection greater than 1/360 of the span is noted in the table by a heavy vertical line.

The following examples illustrate the use of the tables in Appendix D for some common design situations.

Example 1. A simple span beam of A36 steel is required to carry a total uniformly distributed load of 40 kips [178 kN] on a span of 30 ft [9.14 m]. Find (a) the lightest shape permitted and (b) the shallowest shape permitted.
Solution: From Table D.1 we find the following:

Shape	Allowable load (kips)
W 21 × 44	43.5
W 18 × 46	42.0
W 16 × 50	43.2
W 14 × 53	41.5

Thus the lightest section is the W 21 × 44 and the shallowest is the W 14 × 53.

Example 2. A simple span beam of A36 steel is required to carry a total uniformly distributed load of 25 kips [111 kN] on a span of 24 ft [7.32 m] while sustaining a maximum deflection of 1/360 of the span. Find the lightest shape permitted.
Solution: In Table D-1 we find that the lightest shape that will carry this load is the W 16 × 26. For this beam the deflection will be

$$\Delta = \frac{25}{25.6} \times \frac{14.3}{16} = 0.873 \text{ in.}$$

which exceeds the allowable of (24 × 12)/360 = 0.80 in.

The next heaviest beam in the table is a W 16 × 31, for which the deflection will be

$$\Delta = \frac{25}{31.5}\frac{14.3}{16} = 0.709 \text{ in.}$$

which is less than the limit; therefore the W 16 × 31 is the lightest choice.

Problems 9.9 A,* B, C, D, E,* F. For each of the following conditions find (a) the lightest permitted shape and (b) the shallowest permitted shape of A36 steel.

	Span	Total uniformly distributed load (kips)	Deflection limited to 1/360 of the span
A	16	10	No
B	20	30	No
C	36	40	No
D	18	16	Yes
E	32	20	Yes
F	42	50	Yes

9-10. Equivalent Tabular Loads

The safe loads shown in the tables in Appendix D are uniformly distributed loads on simple beams. By the use of coefficients we can convert other types of loading to equivalent uniform loads and thereby greatly extend the usefulness of the tables.

The maximum bending moments for typical loadings are shown in Fig. 5-26. For a simple beam with a uniformly distributed load $M = WL/8$. For a simple beam with equal concentrated loads at the third points $M = PL/3$. These values are shown in Fig. 5-26, Cases 2 and 3, respectively. By equating these values

$$\frac{WL}{8} = \frac{PL}{3} \quad \text{and} \quad W = 2.67 \times P$$

which shows that if the value of one of the concentrated loads (in Case 3) were multiplied by the coefficient 2.67 we would have an equivalent distributed load that would produce the same bending moment as the concentrated loads. The coefficients for finding equivalent uniform loads for other beams and loadings are given in Fig. 5-26. Because of their use with safe load tables, equivalent uniform loads are usually called *equivalent tabular loads,* abbreviated *ETL*.

It is important to remember that an *ETL* does not include the weight of the beam, for which an estimated amount should be added. Beams found by this method should be investigated for shear and deflection; it is assumed that they are adequately supported laterally. Also, when recording the beam reactions they must be determined from the *actual* loading conditions without regard to the *ETL*.

9-11. General Design Procedure

The following general procedure describes the usual steps required for the complete design of beams. As one becomes proficient in structural work, the procedure may be abbreviated, for the designer will develop insights with respect to situations in which shear, deflection, or lateral support tests may be omitted.

Step 1: Determine the beam span, support conditions, distance between lateral supports, deflection limits, and any other design restrictions.

Step 2: Compute the loads.

Step 3: Compute the values for the reactions, the maximum shear, and the maximum bending moment. For ordinary loadings use the data from Fig. 5-26.

Step 4: Find the beam shape required for the maximum bending moment. This may be done by any of the following procedures:

(a) Determine the allowable bending stress and compute the required section modulus: $S_x = M/F_b$. Scan the tables in Appendix A or use the section modulus tables in Appendix B.

(b) Using the value for the maximum bending moment, select directly from the section modulus tables in Appendix B by finding a shape with a resisting moment (M_R) equal to or higher than that required.

(c) Using the computed maximum bending moment and the length between lateral supports, select a shape from the charts in Appendix C.

(d) Using only the total uniformly distributed load (or the ETL as explained in Art. 9-10) and the span, select a shape from the tables in Appendix D if the beam has a simple span.

Step 5: Check for shear on the beam web.

Step 6: Check the deflection. Compute from the appropriate formula in Fig. 5-26, use the coefficient from Table 8-1, or use the factor from the tables in Appendix D. If deflection is excessive, use the deflection formula to derive the required moment of inertia (I_x) and reselect a shape from the alternatives that satisfy Step 4.

Step 7: Check for web crippling if the end support is the bearing type or a concentrated load is placed on top of the beam.

Step 8: Design the end connections, bearing plates, and any other details required for the beam.

Commentary on Procedure. Step 1 is the definition of the beam design problem. All pertinent data and restrictions should be noted. Additional considerations may include the type of steel to be used, dimensional limits on beam size, type of shape preferred, type of connection to be used, and any special requirements for the attachment of decking or support of other parts of the construction.

Determination of loads may include consideration of various load combinations. Most beams carry both dead and live loads and for some design purposes a separate load tabulation may be necessary. If wind or seismic loads are a problem, the possible combinations that must be considered may be numerous. (See discussions in Chapters 10 and 14 regarding design load combinations.)

Although the design of the beam may not actually require Step 3 if, for example, the tables in Appendix D can be used, it is generally useful to know these values. Because the reactions for the beam will constitute loads on some other structural element—a column, a wall, or another beam—it will be necessary to know their values for the design of those elements.

Allowance must be made for the beam weight in Steps 2, 3, and 4. This may be done by adding an estimated uniformly distributed load in Step 2, which can then be verified for accuracy once the beam size is chosen. An alternative is to ignore the weight and simply choose a section with some extra capacity in Step 4; more S_x, M_R, or W.

As already discussed, if all other considerations are satisfied, the desired beam is usually the lightest one that will do the job. Use of the tables in Appendix B or the charts in Appendix C simplifies this determination. If a section is found in the tables in Appendix D or by simply scanning the tables in Appendix A, some care must be exercised to find the lightest shape.

The following examples illustrate the design of various types of beam. Most examples are not carried through all the steps in the general procedure but rather are used to illustrate some particular variation.

Example 1. A simple beam has a span of 10 ft [3.05 m] and carries a uniformly distributed load of 22 kips [98 kN], including its own weight. Deflection is limited to 1/360 of the span for the total load. The beam is supported laterally for its entire length. Design the beam.

Solution: The beam and loading are shown in Fig. 9-8. No allowance is made for the beam weight because it is included in the given load. Since the beam is symmetrical, the reactions and the maximum shear are equal to one-half the total load. The maximum bending moment is given by the formula for Case 2, Fig. 5-26:

$$M = \frac{WL}{8} = \frac{22 \times 10}{8} = 27.5 \text{ kip-ft}$$

$$\left[M = \frac{98 \times 3.05}{8} = 37.36 \text{ kN-m}\right]$$

With full lateral support the allowable bending stress is F_b = 24 ksi [165 MPa] if a compact section is used. Then

$$S = \frac{M}{F_b} = \frac{27.5 \times 12}{24} = 13.75 \text{ in.}^3$$

$$\left[S = \frac{37.36 \times 10^6}{165} = 226 \times 10^3 \text{ mm}^3\right]$$

By referring to Appendix B we find that a W 12 × 14 with S_x of 14.9 in.^{3} is the lightest acceptable shape.

In this situation we could have omitted the determination of

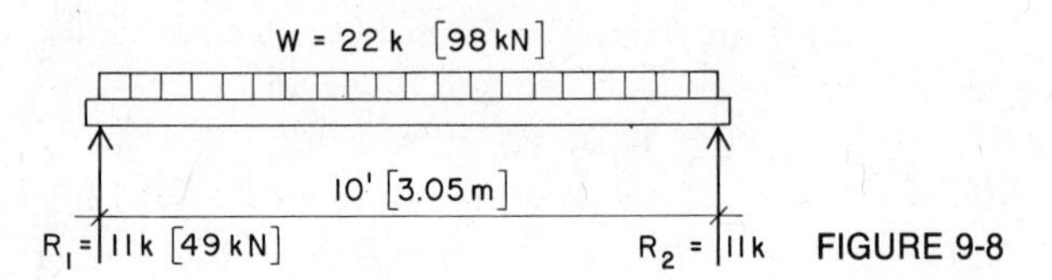

FIGURE 9-8

the required section modulus and used only the load and span to find a section in the tables in Appendix D that shows an allowable load of 23.8 kips for the W 12 × 14.

Checking the shear stress requires reference to Appendix Table A-1, which shows that d = 11.91 in. [302 mm] and t_w = 0.200 in. [5.08 mm] for the W 12 × 14. Then for the maximum shear stress

$$f_v = \frac{V}{A_w} = \frac{11}{11.91 \times 0.200} = 4.62 \text{ ksi}$$

$$\left[f_v = \frac{49 \times 10^3}{302 \times 5.08} = 31.9 \text{ MPa}\right]$$

This value is less than the allowable of 14.5 ksi [100 MPa]; therefore the section is acceptable for shear.

For consideration of deflection we have several alternatives. The formula given for Case 2 in Fig. 5-26 could be used. The slightly simpler formulas derived in Chapter 8 could be used. In this situation, however, the simplest procedure is to use the factor from the table in Appendix D:

Factor for the 10 ft span: 2.483

Then

$$\Delta = \frac{\text{(deflection factor)}}{\text{(beam depth)}} \times \frac{\text{(actual load on beam)}}{\text{(load from table)}}$$

$$\Delta = \frac{2.483}{11.91} \times \frac{22}{23.8} = 0.193 \text{ in.}$$

The specified limit for deflection is

$$\Delta = \frac{L}{360} = \frac{10 \times 12}{360} = 0.33 \text{ in.}$$

The deflection is not a critical factor.

Example 2. A simply supported girder has a span of 18 ft [5.49 m] with a concentrated load of 50 kips [222 kN] at the center of the span. The girder is braced laterally at midspan by beams that frame into it. Design the girder for bending stresses.

Solution: A diagram of the beam and loading is made as shown in Fig. 9-9. The lateral unsupported length is 9 ft [2.745 m]. We consider the matter of the beam weight in a later discussion.

The reactions and maximum shear are equal to one-half the superimposed load plus one-half the beam weight. For the superimposed load only

$$R_1 = R_2 = V = \frac{50}{2} = 25 \text{ kips}$$

$$M = \frac{PL}{4} = \frac{50 \times 18}{4} = 225 \text{ kip-ft}$$

$$S = \frac{M}{F_b} = \frac{225 \times 12}{24} = 112.5 \text{ in.}^3$$

$$\left[\begin{array}{c} \dfrac{222}{2} = 111 \text{ kN} \\ \dfrac{222 \times 5.49}{4} = 305 \text{ kN-m} \\ \dfrac{305 \times 10^6}{165} = 1848 \times 10^3 \text{ mm}^3 \end{array} \right]$$

In the tables in Appendix B the lightest section with the required S_x or M_R value is a W 24 × 55; however, its L_c is only 7.0 ft. Thus it is necessary to use the charts in Appendix C from

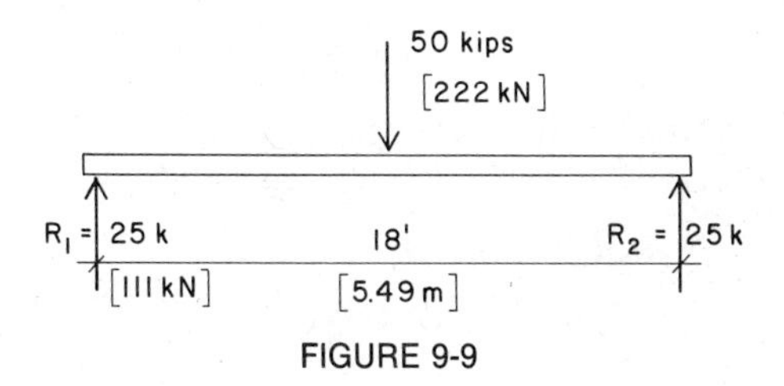

FIGURE 9-9

which we determine the following possible selections:

W 21 × 62 (the lightest choice)

W 12 × 87 (the shallowest choice

W 14 × 82

W 16 × 67

W 18 × 71

With other design considerations any of these sections may be the acceptable choice in a given situation. If we take the lightest beam, the effect of the beam weight is

$$R = V = 25{,}000 + (9 \times 62) = 25{,}558 \text{ lb}$$

$$M = 225 + \frac{0.062 \times (18)^2}{8} = 227.5 \text{ kip-ft}$$

These are insignificant changes and do not result in any change in the design choices.

Example 3. A simple beam has a span of 16 ft [4.88 m] with a uniformly distributed load of 1000 lb/ft [14.59 kN/m], including its own weight, over its entire length. In addition, a concentrated load of 8 kips [35.6 kN] is applied 4 ft [1.22 m] from the right reaction. The beam is laterally supported throughout its length and the maximum deflection is limited to 0.5 in. [13 mm]. Design the beam.

Solution: The beam diagram is drawn as shown in Fig. 9-10. The computation of the reactions for this irregular loading are explained in Art. 5-6:

$$16R_1 = (1000 \times 16 \times 8) + (8000 \times 4) \quad \text{and} \quad R_1 = 10{,}000 \text{ lb}$$

$$16R_2 = (1000 \times 16 \times 8) + (8000 \times 12) \quad \text{and} \quad R_2 = 14{,}000 \text{ lb}$$

$$\left[\begin{array}{l} 4.88R_1 = (14.59 \times 4.88 \times 2.44) \\ \qquad + (35.6 \times 1.22) \quad \text{and} \quad R_1 = 44.50 \text{ kN} \\ 4.88R_2 = (14.59 \times 4.88 \times 2.44) \\ \qquad + (35.6 \times 3.66) \quad \text{and} \quad R_2 = 62.30 \text{ kN} \end{array}\right]$$

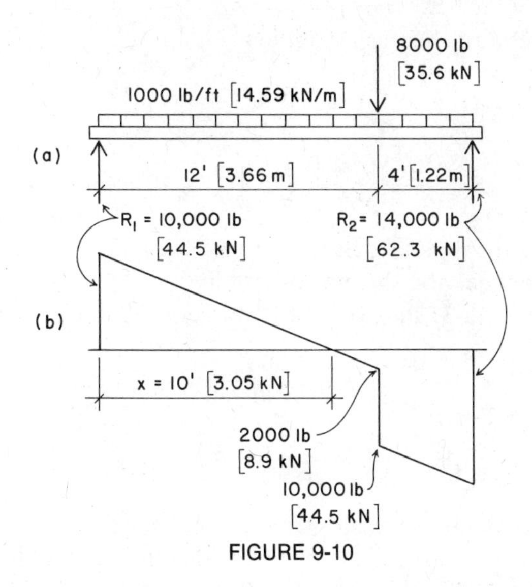

FIGURE 9-10

To compute the maximum bending moment it is necessary to determine the point along the span where the shear is zero. This may be done graphically by constructing the shear diagram to scale (Fig. 9-10*b*) and measuring the distance x from R_1. If the shear diagram is not drawn accurately to scale, x may be found by writing an expression for the shear at this point and equating it to zero. Thus

$$V_x = 10{,}000 - (1000 \times x) = 0 \quad \text{and} \quad x = 10 \text{ ft}$$

$$[V_x = 44.50 - (14.59 \times x) = 0 \quad \text{and} \quad x = 3.05 \text{ m}]$$

Then

$$M_{(x=10)} = (10{,}000 \times 10) - (1000 \times 10 \times 5) = 50{,}000 \text{ ft-lb}$$

$$\begin{bmatrix} M_{(x=3.05)} = (44.50 \times 3.05) - (14.59 \times 3.05 \times 1.525) \\ = 67.86 \text{ kN-m} \end{bmatrix}$$

With the beam supported laterally throughout its length the maximum allowable bending stress of 24 ksi [165 MPa] can be used; the required section modulus is

$$S_x = \frac{M}{F_b} = \frac{50 \times 12}{24} = 25 \text{ in.}^3$$

$$\left[S_x = \frac{67.86 \times 10^6}{165} = 411 \times 10^3 \text{ mm}^3\right]$$

Referring to Appendix B, we see that the lightest choice is a W 12 × 22. To check the shear stress we find on Appendix Table A-1 values of d = 12.31 in. [313 mm] and t_w = 0.260 in. [6.60 mm]:

$$f_v = \frac{V}{A_w} = \frac{14{,}000}{12.31 \times 0.260} = 4374 \text{ psi}$$

$$\left[f_v = \frac{62.3 \times 10^3}{313 \times 6.60} = 30.0 \text{ MPa}\right]$$

which is well within the allowable of 14,500 psi [100 MPa].

There are no convenient formulas for the computation of the exact deflection for this loading combination. However, an approximate deflection (within 10 to 15% of the true deflection) can be found by using the equivalent uniform load method. We thus find the hypothetical total uniform load that will cause the same maximum moment as the true loading:

If

$$M = \frac{WL}{8}$$

Then

$$W = \frac{8M}{L} = \frac{8 \times 50}{16} = 25 \text{ kips}$$

Using this as the actual load, we obtain the deflection factor and table load for the W 12 × 22 from Appendix D and use the procedure described in Example 1. Thus

$$\Delta = \frac{\text{(deflection factor)}}{\text{(beam depth)}} \times \frac{\text{(actual load on beam}}{\text{(load from table)}}$$

$$\Delta = \frac{(6.36)}{(12.31)} \times \frac{25}{23.8} = 0.54 \text{ in.}$$

Because this is slightly over the stated limit of 0.5 in., a more accurate deflection calculation may be justified. In the absence of that it may be desirable to increase the beam depth. A second look at the tables will show that an alternative is a W 14 × 22, which does not result in additional weight of steel. A similar computation for the approximate deflection will produce a value of 0.40 in. for this section. If the additional depth is not a problem, the 14-in.-deep beam is probably the best choice.

Example 4. An overhanging beam with a total length of 24 ft [7.32 m] projects 6 ft [1.83 m] beyond its right support. There is a uniformly distributed load of 2 kips/ft [29.18 kN/m] over its entire length; this includes the weight of the beam. The beam is braced laterally at 6 ft [1.83 m] intervals. The right support is a bearing type with a length of 6 in. [152 mm] along the beam length. Design the beam.

Solution: The beam diagram in Fig. 9-11 is drawn to show the loading and span dimensions. Note that for this beam the bottom flange is in compression for a portion of the span; thus the lateral bracing must brace the bottom of the beam in this region. The deformation curve is similar to that shown in Fig. 5-18.

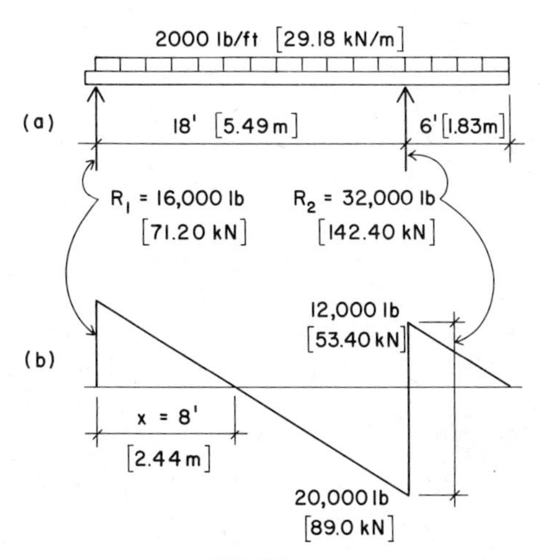

FIGURE 9-11

By computing the reactions we have

$$18R_1 = 2000 \times 24 \times 6 \quad \text{and} \quad R_1 = 16{,}000 \text{ lb}$$

$$18R_2 = 2000 \times 24 \times 12 \quad \text{and} \quad R_2 = 32{,}000 \text{ lb}$$

$$\begin{bmatrix} 5.49R_1 = 29.18 \times 7.32 \times 1.83 \quad \text{and} \quad R_1 = 71.20 \text{ kN} \\ 5.49R_2 = 29.18 \times 7.32 \times 3.66 \quad \text{and} \quad R_2 = 142.40 \text{ kN} \end{bmatrix}$$

To find the maximum bending moment the shear diagram is drawn next (Fig. 9-11*b*). We note that the shear passes through zero at the right support and at some point between the two supports. These are the points of maximum negative and maximum positive bending moments. Letting x be the distance from R_1 to the point of zero shear, we obtain

$$16{,}000 - (2000 \times x) = 0 \quad \text{and} \quad x = 8 \text{ ft}$$

$$M_{(x=8)} = (16{,}000 \times 8) - (2000 \times 8 \times 4) = 64{,}000 \text{ ft-lb}$$

$$\begin{bmatrix} 71.20 - (29.18 \times x) = 0 \quad \text{and} \quad x = 2.44 \text{ m} \\ M_{(x=2.44)} = (71.20 \times 2.44) - (29.18 \times 2.44 \times 1.22) \\ = 88.86 \text{ kN-m} \end{bmatrix}$$

which is the maximum positive bending moment.

$$M_{(x=18)} = (16{,}000 \times 18) - (2000 \times 18 \times 9) = -36{,}000 \text{ ft-lb}$$

$$\begin{bmatrix} M_{(x=5.49)} = (71.20 \times 5.49) - (29.18 \times 5.49 \times 2.74) \\ = -48.1 \text{ kN-m} \end{bmatrix}$$

which is the maximum negative bending moment.

Because the positive moment is greater, it is used in determining the required section modulus. Assuming the maximum allowable bending stress of 24 ksi [165 MPa],

$$S_x = \frac{M}{F_b} = \frac{64 \times 12}{24} = 32 \text{ in.}^3$$

$$\left[\frac{88.86 \times 10^6}{165} = 538.5 \times 10^3 \text{ mm}^3\right]$$

The lightest weight beam listed in Appendix B with a value of S_x at least that required is a W 12 × 26 for which the L_c is 6.9 ft; this indicates that the assumed allowable stress is alright.

Note on Lateral Bracing. Returning to the point raised in the first step of the solution for positive and negative bending moments in this overhanging beam, we find that the *bottom* flange of the beam needs bracing against lateral buckling where the negative moment occurs. To locate the position along the span where negative moment changes to positive we use a procedure developed in Example 2 of Art. 5-17. As in that example, the zero moment point occurs at 2 ft [0.61 m] to the left of the right support. From this point to the right end of the beam the bottom flange must be braced.

From the shear diagram (Fig. 9-11*b*) we see that the maximum vertical shear is 20 kips [89.0 kN]. In Appendix Table A-1 we find that d = 12.22 in. [310 mm] for the W 12 × 26 and that t_w = 0.230 in. [5.84 mm]. Then

$$f_v = \frac{V}{A_w} = \frac{20{,}000}{12.22 \times 0.230} = 7116 \text{ psi}$$

$$\left[f_v = \frac{89.0 \times 10^3}{310 \times 5.84} = 49.2 \text{ MPa}\right]$$

Because this value is well within the 14,500 psi [100 MPa] allowable, the beam is acceptable for shear.

For investigation of the bearing condition at the right reaction we use the procedure developed in Art. 9-6. In Appendix Table A-1 we find k = 0.875 in. [22.23 mm]. Then, for the maximum reaction allowable,

$$R = F_p \times t_w \times (N + 2k)$$

$$= 27 \times 0.230 \times (6 + 1.75) = 48.1 \text{ kips}$$

$$\left[R = \frac{186 \times 5.84 \times (152.4 + 44.46)}{10^3} = 213.8 \text{ kN}\right]$$

Because this is greater than the computed value of R_2, the beam is not critical for web crippling at the support.

Example 5. A cantilever beam extends 6 ft [1.83 m] beyond the face of its support. The loading consists of a uniformly distributed load (including the beam weight) of 500 lb/ft [7.295 kN/m] and a concentrated load of 6 kips [26.69 kN] applied at the unsupported end. The compression flange is braced laterally only at the ends of the beam. Determine the required beam size.

Solution: The beam diagram is drawn as in Fig. 9-12. The value for the vertical reaction force and the maximum shear at the support is

$$R = V = 6 + (0.5 \times 6) = 9 \text{ kips}$$

$$[R = V = 26.69 + (7.295 \times 1.83) = 40.04 \text{ kN}]$$

The maximum bending moment is at the face of the wall and negative.

$$M = (6 \times 6) + (0.5 \times 6 \times 3) = 45 \text{ kip-ft}$$

$$[M = (26.69 \times 1.83) + (7.295 \times 1.83 \times 0.915) = 61.06 \text{ kN-m}]$$

Assuming the maximum allowable bending stress of 24 ksi [165 MPa], the required section modulus is

$$S = \frac{M}{F_b} = \frac{45 \times 12}{24} = 22.5 \text{ in.}^3$$

$$\left[S = \frac{61.06 \times 10^6}{165} = 370 \times 10^3 \text{ mm}^3\right]$$

From Appendix B we find that a W 10 × 22 is the lightest beam with adequate section modulus. The table also shows that L_c = 6.1 ft and the assumed allowable stress is confirmed.

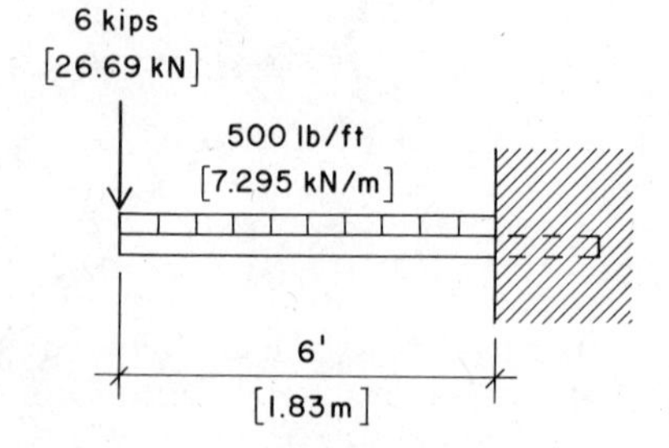

FIGURE 9-12

As noted, the maximum shear at the support is 9 kips [40.04 kN]. In Appendix Table A-1 we find that d = 10.17 in. [258 mm] and t_w = 0.240 in. [6.10 mm]. Then

$$f_v = \frac{V}{A_w} = \frac{9000}{10.17 \times 0.240} = 3687 \text{ psi}$$

$$\left[f_v = \frac{40.04 \times 10^3}{258 \times 6.10} = 25.4 \text{ MPa}\right]$$

which is considerably less than the allowable stress of 14,500 psi [100 MPa]; therefore the beam is adequate for shear.

Problem 9.11.A.* A simple beam has a span of 15 ft [4.57 m]. There is a concentrated load of 7.5 kips [33.36 kN] 3 ft [0.915 m] from the left support and a uniformly distributed load of 1.2 kips/ft [17.51 kN/m], including the beam weight, extending over the full span. The beam is laterally braced throughout its length. Design the beam.

Problem 9.11.B.* A beam whose total length is 28 ft [7.925 m] overhangs both supports a distance of 4 ft [1.22 m]. A uniformly distributed load of 1.3 kips/ft [18.97 kN/m] extends over the full length of the beam. The beam is braced laterally at 6 ft [1.83 m] intervals. Design the beam for strength in bending.

Problem 9.11.C. A beam has a total length of 25 ft [7.62 m] and overhangs the left reaction a distance of 5 ft [1.52 m]. At the end of the overhang there is a concentrated load of 6 kips [26.69 kN] and midway between the two reactions, a concentrated load of 42 kips [186.8 kN]. The beam is braced laterally at 5-ft [1.52-m] intervals. Design the beam for bending strength.

Problem 9.11.D. A cantilever beam extends 5 ft [1.52 m] beyond the face of its support. At the unsupported end there is a concentrated load of 8 kips [35.58 kN]. If the beam is supported laterally throughout its length, what is its size?

Problem 9.11.E. A cantilever beam has a length of 4 ft [1.22 m]. There is a concentrated load of 10 kips [44.5 kN] at the unsupported end and a uniformly distributed load (including the beam weight) of 1 kip/ft [14.59 kN/m] extending over the full length. Lateral support is provided at both ends. What is its size?

10

Roof and Floor Framing Systems

10-1. Layout

The arrangement of columns and the layout of beams and girders depends on a number of factors. The area of the building, the floor plan, and the occupancy requirements determine the location of columns and other points of support.

The layout of beams and girders depends entirely on the column spacing and the type of system to be used. Figure 10-1 illustrates two common methods of framing when one of the shorter span deck systems is used, such as a solid concrete slab or certain of the corrugated or ribbed steel deck systems. Third-point concentration is shown at (*a*) in Fig. 10-1 and center-point concentration at (*b*). When the area enclosed by the columns (called a *bay*) is not a square the usual custom is to run the beams

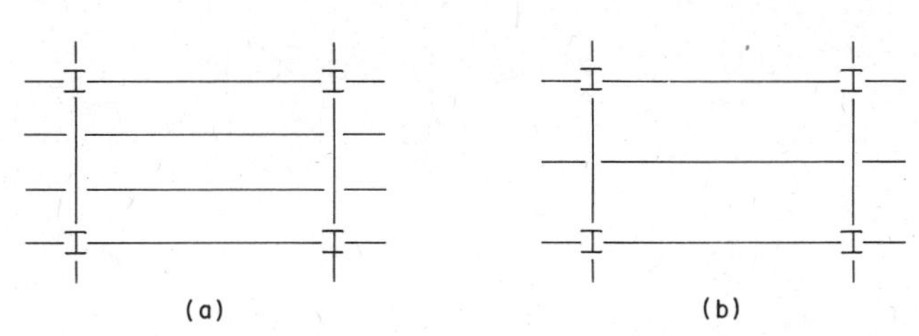

FIGURE 10-1. Framing layouts.

in the long direction, with the girders on the shorter span. This is not always done, however, because the economical length of span of the floor system adopted may be a controlling factor or it may be necessary that the beam and girder have the same depth to facilitate installation of a flush ceiling. The usual procedure is to design two or more acceptable layouts and then select the most economical.

As stated in Art. 5-3, the area of floor supported by any one beam is equal to the span length multiplied by the sum of half the distances to adjacent beams. The span lengths used in design are generally taken from center to center of supporting members, although this distance may be reduced where beams or girders frame against the flanges of large columns. The total uniform load on a floor beam is then found by multiplying this panel area by the design load in lb per sq ft; the design load for any floor is defined as the sum of the dead and live loads on that floor.

10-2. Dead Load

The dead load consists of the weight of the materials of which the building is constructed: walls, partitions, columns, framing, floors, and roofs. In the design of a beam the dead load must include an allowance for the weight of the member and its protective covering if fireproofing is required. This weight is first estimated, then checked when the design has been completed. Table 10-1, which lists the weights of many construction materials, may be used in the computation of dead loads. The weights given are approximate and the actual weights may vary somewhat from the values given in this table. Most building codes specify the weights

TABLE 10-1. Weights of Building Construction

	lb/ft²	kN/m²
Roofs		
3-ply ready roofing (roll, composition)	1	0.05
3-ply felt and gravel	5.5	0.26
5-ply felt and gravel	6.5	0.31
Shingles		
wood	2	0.10
asphalt	2–3	0.10–0.15
clay tile	9–12	0.43–0.58
concrete tile	8–12	0.38–0.58
slate, $\frac{1}{4}$ in.	10	0.48
fiber glass	2–3	0.10–0.15
aluminum	1	0.05
steel	2	0.10
Insulation		
fiber glass batts	0.5	0.025
rigid foam plastic	1.5	0.075
foamed concrete, mineral aggregate	2.5/in.	3.0/mm
Wood rafters		
2 × 6 at 24 in.	1.0	0.05
2 × 8 at 24 in.	1.4	0.07
2 × 10 at 24 in.	1.7	0.08
2 × 12 at 24 in.	2.1	0.10
Steel deck, painted		
22 ga.	1.6	0.08
20 ga.	2.0	0.10
18 ga.	2.6	0.13
Skylight		
glass with steel frame	6–10	0.29–0.48
plastic with aluminum frame	3–6	0.15–0.29
Plywood or softwood board sheathing	3.0/in.	3.6/mm
Ceilings		
Suspended steel channels	1	0.05
Lath		
steel mesh	0.5	0.025
gypsum board, $\frac{1}{2}$ in.	2	0.10
Fiber tile	1	0.05

TABLE 10-1 (*Continued*)

	lb/ft²	kN/m²
Drywall, gypsum board, $\frac{1}{2}$ in.	2.5	0.12
Plaster		
gypsum, acoustic	5	0.24
cement	8.5	0.41
Suspended lighting and air distribution systems, average	3	0.15
Floors		
Hardwood, $\frac{1}{2}$ in.	2.5	0.12
Vinyl tile, $\frac{1}{8}$ in.	1.5	0.07
Asphalt mastic	12/in.	14.6/mm
Ceramic tile		
$\frac{3}{4}$ in.	10	0.48
thin set	5	0.24
Fiberboard underlay, $\frac{5}{8}$ in.	3	0.15
Carpet and pad, average	3	0.15
Timber deck	2.5/in.	3.0/mm
Steel deck, stone concrete fill, average	35–40	1.68–1.92
Concrete deck, stone aggregate	12.5/in.	15.2/mm
Wood joists		
2 × 8 at 16 in.	2.1	0.10
2 × 10 at 16 in.	2.6	0.13
2 × 12 at 16 in.	3.2	0.16
Lightweight concrete fill	8.0/in.	9.73/mm
Walls		
2 × 4 studs at 16 in., average	2	0.10
Steel studs at 3.5 in., average	4	0.20
Lath, plaster; see Ceilings		
Gypsum drywall, $\frac{5}{8}$ in. single	2.5	0.12
Stucco, $\frac{7}{8}$ in., on wire and paper or felt	10	0.48
Windows, average, glazing + frame		
small pane, single glazing, wood or metal frame	5	0.24
large pane, single glazing, wood or metal frame	8	0.38
increase for double glazing	2–3	0.10–0.15
curtain walls, manufactured units	10–15	0.48–0.72

TABLE 10-1 (*Continued*)

	lb/ft²	kN/m²
Brick veneer		
4 in., mortar joints	40	1.92
½ in., mastic	10	0.48
Concrete block		
lightweight, unreinforced—4 in.	20	0.96
6 in.	25	1.20
8 in.	30	1.44
heavy, reinforced, grouted—6 in.	45	2.15
8 in.	60	2.87
12 in.	85	4.07

of materials to be used in computing dead loads and, of course, they must be used in actual design work. Dead loads are due to gravity and result in downward vertical forces.

10-3. Roof Loads

In addition to the dead loads they support, roofs are designed for a uniformly distributed live load which accounts for one of two conditions. The first occurs where snow accumulation is possible. Based on the weather history in a particular area, this load is specified by local codes to account for an estimated maximum total snow accumulation. Because this accumulation is less likely as the roof slope increases, there is usually some provision for this reduction. In regions of light or no snowfall history a minimum load is required to account for the general loadings that occur during construction and maintenance of the roof.

Table 10-2 lists the minimum roof live-load requirements specified by the 1982 edition of the Uniform Building Code (Ref. 2). Note the adjustments for roof slope and for the total area of roof surface supported by a structural element. The latter accounts for

TABLE 10-2. Minimum Roof Live Loads[a]

Roof slope conditions	Minimum uniformly distributed load (lb/ft²)			(kN/m²)		
	Tributary loaded area for structural member (ft²)			(m²)		
	0–200	201–600	Over 600	0–18.6	18.6–55.7	Over 55.7
1. Flat or rise less than 4 in./ft (1 : 3). Arch or dome with rise less than $\frac{1}{8}$ span.	20	16	12	0.96	0.77	0.575
2. Rise 4 in./ft (1 : 3) to less than 12 in./ft (1 : 1). Arch or dome with rise $\frac{1}{8}$ of span to less than $\frac{3}{8}$ of span.	16	14	12	0.77	0.67	0.575
3. Rise 12 in./ft (1 : 1) or greater. Arch or dome with rise $\frac{3}{8}$ of span or greater.	12	12	12	0.575	0.575	0.575
4. Awnings, except cloth covered.	5	5	5	0.24	0.24	0.24
5. Greenhouses, lath houses, and agricultural buildings.	10	10	10	0.48	0.48	0.48

[a] Adapted from the *Uniform Building Code*, 1982 ed. (Ref. 2) with permission of the publishers, International Conference of Building Officials.

the increase in probability of the lack of total surface loading as the size of the surface area increases.

Roof surfaces must also be designed for some wind loading. Both the magnitude of the wind pressure and the manner in which it must be applied vary for different regions. Any design work must use the specific requirements of the local code. For light roof construction a critical problem is sometimes that of the upward (suction) effect of the wind, which may often exceed the dead load and result in a net upward lifting force.

Although the term *flat roof* is often used, there is usually no such thing; all roofs must be designed for some water drainage. The minimum required is usually $\frac{1}{4}$ in. per ft, or a slope of approximately 50 : 1. On roof surfaces that are this close to flat a potential problem is that of *ponding,* a phenomenon in which the weight of water on the roof causes deflection; this in turn results in more water accumulation (in a pond), more deflection, and an accelerated collapse condition if the water due to precipitation continues to fall on the surface. Investigation of this condition is required for near-flat roofs. A procedure for this investigation is given in the AISC Manual (Ref. 1).

10-4. Floor Live Loads

The live load on a floor represents the probable effect created by the occupancy. It includes the weight of human occupants, furniture, equipment, stored materials, and so on. All building codes provide minimum live loads to be used in the design of buildings for various occupanices. Because there is a lack of uniformity among different building codes in the specification of live loads, the local code should always be used. Table 10-3 lists the values of live loads for floors given in the *Uniform Building Code,* 1982 ed. (Ref. 2).

Most codes require that the floors of offices, parking garages, and manufacturing buildings be designed for possible concentrated loads at any position. Another frequent requirement is that 100 psf [4.79 kN/m^2] be used as the live load for aisles, corridors, lobbies, and public spaces in buildings in which large numbers of people are likely to gather. Although expressed as a uniform load,

TABLE 10-3. Minimum Live Loads[a]

Use or occupancy		Uniform load		Concentrated load	
Description	Description	(psf)	(kN/m²)	(lb)	(kN)
Armories		150	7.2		
Assembly areas and auditoriums and balconies therewith	Fixed seating areas	50	2.4		
	Movable seating and other areas	100	4.8		
	Stages and enclosed platforms	125	6.0		
Cornices, marquees, and residential balconies		60	2.9		
Exit facilities		100	4.8		
Garages	General storage, repair	100	4.8	[b]	
	Private pleasure car storage	50	2.4	[b]	
Hospitals	Wards and rooms	40	1.9	1000	4.5
Libraries	Reading rooms	60	2.9	1000	4.5
	Stack rooms	125	6.0	1500	6.7
Manufacturing	Light	75	3.6	2000	9.0
	Heavy	125	6.0	3000	13.3
Offices		50	2.4	2000	9.0
Printing plants	Press rooms	150	7.2	2500	11.1
	Composing rooms	100	4.8	2000	9.0
Residential		40	1.9		
Rest rooms		[c]			
Reviewing stands, grandstands, and bleachers		100	4.8		
Roof decks (occupied)	Same as area served				
Schools	Classrooms	40	1.9	1000	4.5
Sidewalks and driveways	Public access	250	12.0	[b]	
Storage	Light	125	6.0		
	Heavy	250	12.0		
Stores	Retail	75	3.6	2000	9.0
	Wholesale	100	4.8	3000	13.3

[a] Adapted from the *Uniform Building Code*, 1982 ed., with permission of the publishers, International Conference of Building Officials.
[b] Wheel loads related to size of vehicles that have access to the area.
[c] Same as the area served or minimum of 50 psf.

code-required values are established large enough to cover the effect of ordinary concentrations that may occur. When buildings are to house heavy machinery, stored materials, or other contents of unusual weight, these items must be provided for individually.

When structural framing members support large areas most codes allow some reduction in the live load to be used for design. This issue was discussed in the preceding article with regard to roof loads; the reductions are given in the data in Table 10-2. The following is the method recommended in the 1982 edition of the *Uniform Building Code* (Ref. 2) for determining the amount of reduction permitted for beams, trusses, or columns that support large floor areas.

Except for floors in places of assembly (theaters, etc.) and for live loads greater than 100 psf [4.79 kN/m^2], the design live load on a member may be reduced in accordance with the formula

$$R = 0.08\,(A - 150)$$

The reduction shall not exceed 40% for horizontal members or vertical members receiving a load from one level only, 60% for other vertical members, nor R as determined by the formula

$$R = 23.1\left(1 + \frac{D}{L}\right)$$

In these formulas

R = reduction in percent
A = area of floor supported by the member
D = unit dead load per sq ft of supported area
L = unit live load per sq ft of supported area

10-5. Movable Partitions

In office buildings and certain other building types the partition layout may not be fixed but may be erected or moved from one position to another in accordance with the requirements of the occupant. To provide for this flexibility it is customary to require that an allowance of 15 to 20 psf [0.72 to 0.96 kN/m^2] be added to the design load to cover movable partitions. Because this requirement appears under dead loads in some codes and under live loads in others, care must be exercised to determine whether such a provision is mandatory.

10-6. Fireproofing for Beams

In fire-resistive construction some insulating material must be placed around the structural steel to protect it from contact with flames or excessive heat that might cause structural failure. Materials commonly used for this purpose are concrete, masonry, or lath and plaster. In addition, there are fibrous and cementitious coatings that can be sprayed directly on the surfaces of steel members. Figure 9-1*a* illustrates a method often used when a reinforced concrete slab is needed for the floor system. Regardless of the material, its weight must be given consideration in the design of the beam.

After the beam has been selected the weight of the fireproofing that projects below the slab may be computed accurately; but in the design of the beam its probable weight as well as the weight of the fireproofing must be estimated. Building codes differ in the thickness of fireproofing material required. A common specification for beams and girders calls for 2 in. of material on the flat surfaces and $1\frac{1}{2}$ in. on the edges of the flanges. These dimensions are given in Fig. 10-2. In this figure *d* and *b* are the depth of beam and width of flange in inches. For the thicknesses shown the

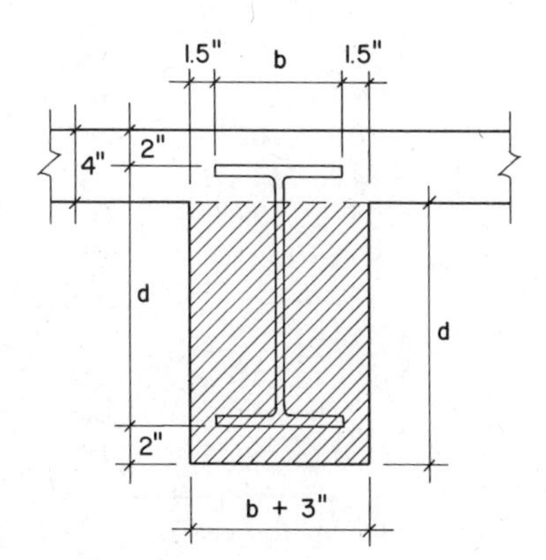

FIGURE 10-2. Fireproofing of a steel beam with concrete.

cross-sectional area of the fireproofing (drawn hatched in the figure) is $[d \times (b + 3)]$ sq in. The number of cubic feet of fireproofing *per linear foot of beam* is $[d \times (b + 3)]/144$. Taking the weight of unreinforced concrete as 144 lb per cu ft, the weight of the fireproofing per linear foot of beam becomes

$$W_{FP} = \frac{d \times (b + 3)}{144} \times 144 = [d \times (b + 3)] \text{ lb per lin ft}$$

It should be noted that this expression depends on the thickness of the fireproofing, the thickness of the structural concrete slab, and the distance from the top of the steel beam to the top of the slab. Similar equations can, of course, be established for other conditions of encasement. This procedure is sufficiently accurate to make a preliminary allowance for the weight when designing the beam. Table 10-4, based on the above formula,

TABLE 10-4. Approximate Weight of Concrete Fireproofing for Wide Flange Beams[a]

Sections	Weight	Sections	Weight
W 8 × 10 to 15	56	W 18 × 35 to 46	162
W 8 × 18, 21	66	W 18 × 50 to 71	189
W 8 × 24, 28	76	W 18 × 76 to 119	252
W 10 × 12 to 19	70	W 21 × 44 to 57	200
W 10 × 22 to 30	88	W 21 × 62 to 93	236
W 10 × 33 to 45	110	W 21 × 101 to 147	323
W 12 × 14 to 22	84	W 24 × 55, 62	240
W 12 × 26 to 35	114	W 24 × 68 to 94	288
W 12 × 40 to 50	132	W 24 × 104 to 162	381
W 12 × 53, 58	156	W 27 × 84 to 114	351
W 14 × 22, 26	112	W 27 × 146 to 178	459
W 14 × 30 to 38	137	W 30 × 99 to 132	405
W 14 × 43 to 53	154	W 30 × 173 to 211	540
W 14 × 61 to 82	182	W 33 × 118 to 152	479
W 16 × 26, 31	136	W 33 × 201 to 241	619
W 16 × 36 to 57	160	W 36 × 135 to 210	540
W 16 × 67 to 100	216	W 36 × 230 to 300	702

[a] Weight in lb/ft of beam length; does not include weight of steel beam.

gives the weight of concrete fireproofing for a number of beam sections. When the section has finally been determined the true weight of the beam and fireproofing are checked to see that an adequate weight allowance is provided.

10-7. Decks

A wide range of elements is available for the creation of the infilling surface of roofs and floors. Figure 10-3 shows four possibilities for a floor deck, all of which may be used with a steel framing system. When a plywood deck is specified it is usually

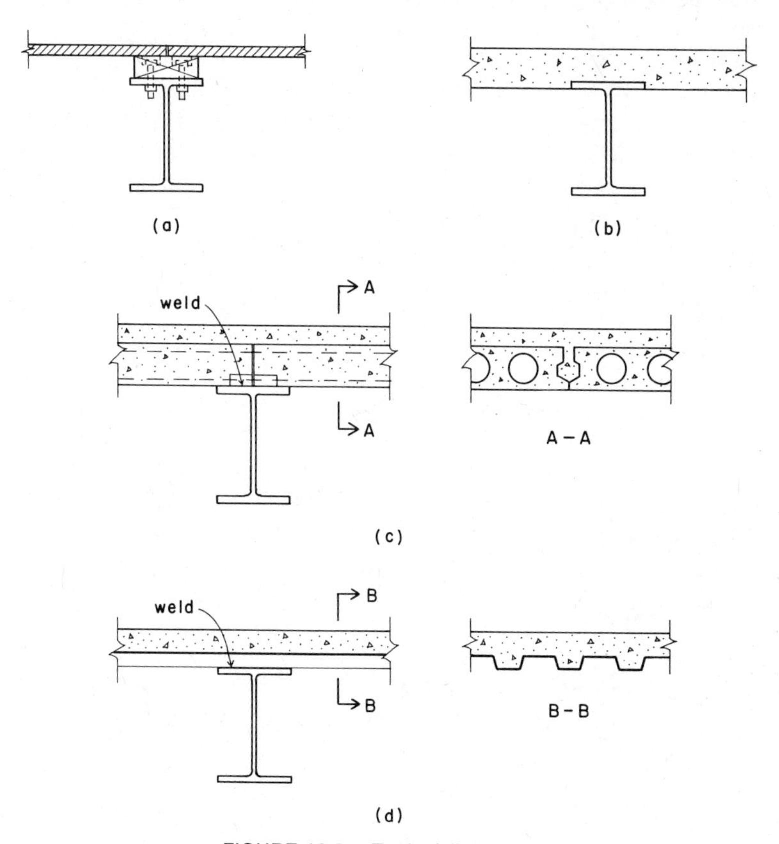

FIGURE 10-3. Typical floor decks.

nailed to a series of wood joists or trusses supported by the steel beams: in some cases, however, wood nailers are bolted to the tops of steel beams and the plywood is attached directly to the beams.

When a concrete slab is poured at the site it may be combined with poured concrete fireproofing, as shown in Fig. 10-2. When the beams are fireproofed by other means it is common to follow the detail shown in Fig. 10-3*b*. Concrete may also be used in the form of precast deck units which are welded to the steel beams through steel devices cast into the units. A concrete fill is normally required on top of precast units to provide a smooth top surface.

A decking system combined widely with steel framing is that of corrugated or fluted steel units with concrete fill (Fig. 10-3*d*). This type of system may function in one of three ways: (1) with a steel form deck where the deck serves only as a form for the concrete, which is reinforced as usual to produce a concrete structural slab. (2) with a steel structural deck in which the concrete fill is structurally inert and the deck is the structure; (3) with a composite deck in which the steel deck and the concrete interact; the steel functions in a manner similar to the reinforcing bars in an ordinary reinforced concrete slab.

Steel deck units are available from a large number of manufacturers. Specific information regarding the type of deck available, possible range of sizes, rated load-carrying capacity, and so on, should be obtained directly from the suppliers of these products who service the region in which a proposed building is to be built. The information in Table 10-5 is taken from Reference 8.

Selection of the deck for a particular situation depends on a large number of considerations. The following are some of the usual major concerns:

1. *Structural capacity*. This includes basic strength, deflection resistance, and type of loading, concentrated or uniformly distributed only.
2. *Fireproofing*. The deck itself must often have some fire rating; in addition, it may function to help fire protect the steel framing.

TABLE 10-5. Load Capacity of Steel Roof Deck[a]

Deck[b] Type	Span Condition	Weight[c] (psf)	Total (Dead & Live) Safe Load[d] for Spans Indicated in ft-in.												
			4-0	4-6	5-0	5-6	6-0	6-6	7-0	7-6	8-0	8-6	9-0	9-6	10-0
NR22	Simple	1.6	73	58	47										
NR20		2.0	91	72	58	48	40								
NR18		2.7	121	95	77	64	54	46							
NR22	Two	1.6	80	63	51	42									
NR20		2.0	96	76	61	51	43								
NR18		2.7	124	98	79	66	55	47	41						
NR22	Three or More	1.6	100	79	64	53	44								
NR20		2.0	120	95	77	63	53	45							
NR18		2.7	155	123	99	82	69	59	51	44					
IR22	Simple	1.6	86	68	55	45									
IR20		2.0	106	84	68	56	47	40							
IR18		2.7	142	112	91	75	63	54	46	40					
IR22	Two	1.6	93	74	60	49	41								
IR20		2.0	112	88	71	59	50	42							
IR18		2.7	145	115	93	77	64	55	47	41					
IR22	Three or More	1.6	117	92	75	62	52	44							
IR20		2.0	140	110	89	74	62	53	46	40					
IR18		2.7	181	143	116	96	81	69	59	52	45	40			
WR22	Simple	1.6			(89)	(70)	(56)	(46)							
WR20		2.0			(112)	(87)	(69)	(57)	(47)	(40)					
WR18		2.7			(154)	(119)	(94)	(76)	(63)	(53)	(45)				
WR22	Two	1.6			98	81	68	58	50	43					
WR20		2.0			125	103	87	74	64	55	49	43			
WR18		2.7			165	137	115	98	84	73	65	57	51	46	41
WR22	Three or More	1.6			122	101	85	72	62	54	(46)	(40)			
WR20		2.0			156	129	108	92	80	(67)	(57)	(49)	(43)		
WR18		2.7			207	171	144	122	105	(91)	(76)	(65)	(57)	(50)	(44)

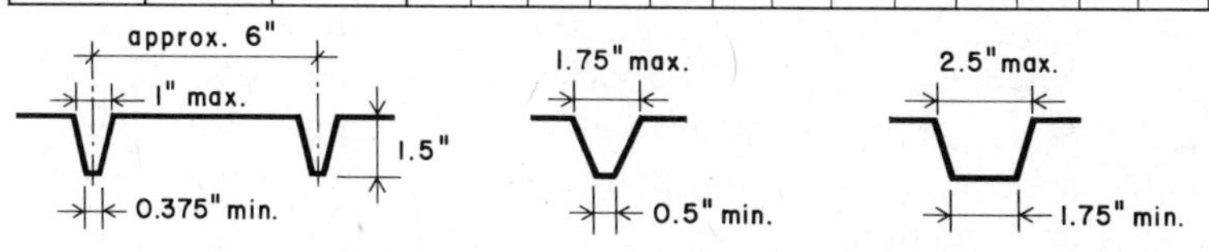

Narrow Rib Deck – NR   Intermediate Rib Deck – IR   Wide Rib Deck – WR

[a] Adapted from the *Steel Deck Institute Design Manual for Composite Decks, Form Decks, and Roof Decks,* 1981–82 issue, with permission of the publishers, the Steel Deck Institute. May not be reproduced without express permission of the Steel Deck Institute.
[b] Letters refer to rib type (see illustrations). Number indicates gage (thickness) of steel.
[c] Approximate weight with paint finish; other finishes also available.
[d] Total safe allowable load in lb/sq ft. Loads in parentheses are governed by live load deflection not in excess of 1/240 of the span, assuming a dead load of 10 psf.

3. *Diaphragm action.* Decks must often function as horizontally stiff elements in the lateral load-resisting system of the building. The decks shown in Fig. 10-3 have some capacity for this action but they differ widely in relative strength and stiffness.
4. *Incorporation of wiring.* Wiring for electrical power and communication systems is often incorporated into the floor construction for which some decks have a natural facility and some have not.
5. *Dead load.* The weight of the deck as part of the load that must be carried by the framing may be a consideration. Comparison of the data in Table 10-1 indicates a wide range of weights for the decks shown in Fig. 10-3.

For low slope roofs the decking problem is essentially the same as that for floors and many of the same types of deck may be used. However, different issues are involved in roof construction and some products are used exclusively for roofs. Examples are shown in Fig. 10-4.

Although structural capacity, fireproofing, diaphragm action, and dead load—all discussed with regard to floor decks—are also roof problems, some special concerns relate only to roofs:

1. *Insulation.* As one of the building's exterior surfaces the roof must have an unsulative character. Some roof decks have a natural thermal barrier effect; others have not and will require the addition of insulative materials.
2. *Water drainage.* Water must be drained off roof surfaces and the means may relate to the details of the roof-deck construction.
3. *Attachment for uplift.* When the construction is light the wind may exert a net upward force; an attachment of the deck by a means that develops resistance is required.
4. *Diaphragm action.* Roof decks must also help to brace the building against lateral loads due to wind or earthquakes. Many roof-deck constructions are adequate for this purpose but some have low or negligible capacity.

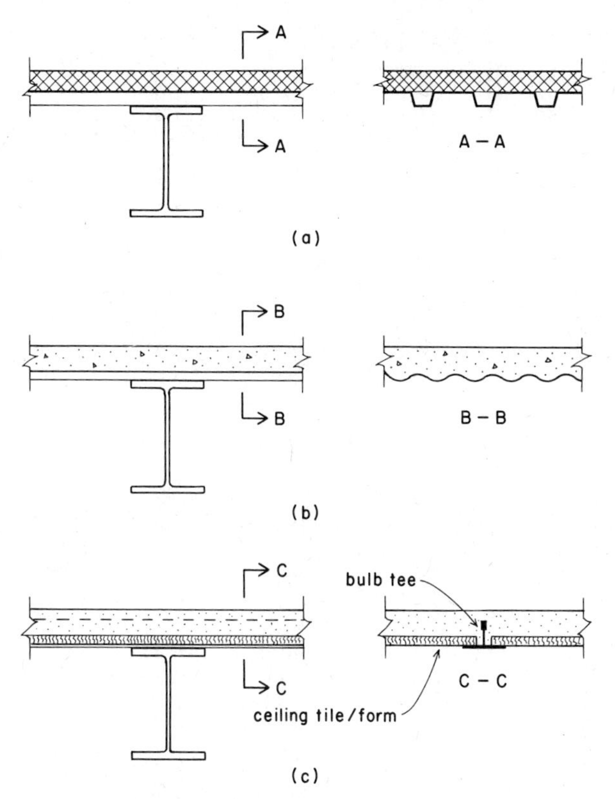

FIGURE 10-4. Typical roof decks.

10-8. Design of Typical Framing

The following example illustrates the design of beams and girders for a typical bay of floor framing:

Example. The floor plan of a building has columns spaced 20 ft [6.1 m] in one direction and 21 ft [6.40 m] in the other. The floor construction, shown in Fig. 10-5*a*, consists of a poured concrete deck, poured concrete fireproofing, a carpeted floor above, and a suspended ceiling below. The live load on the floor is 60 psf [2.87 kN/m^2] and an allowance of 15 psf [0.72 kN/m^2] is included for

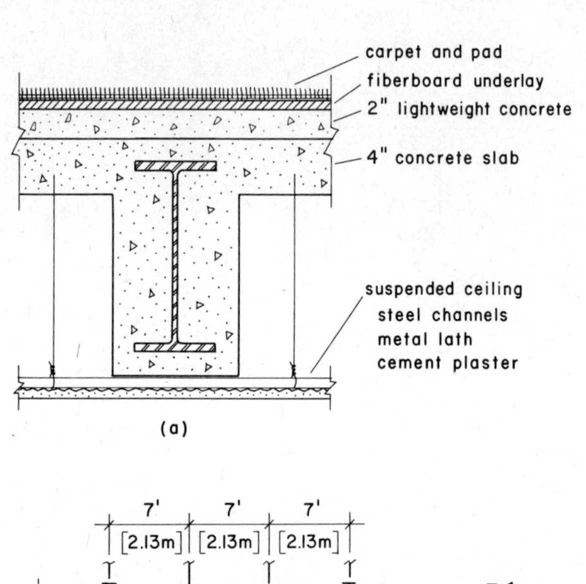
carpet and pad
fiberboard underlay
2" lightweight concrete
4" concrete slab
suspended ceiling
steel channels
metal lath
cement plaster
(a)

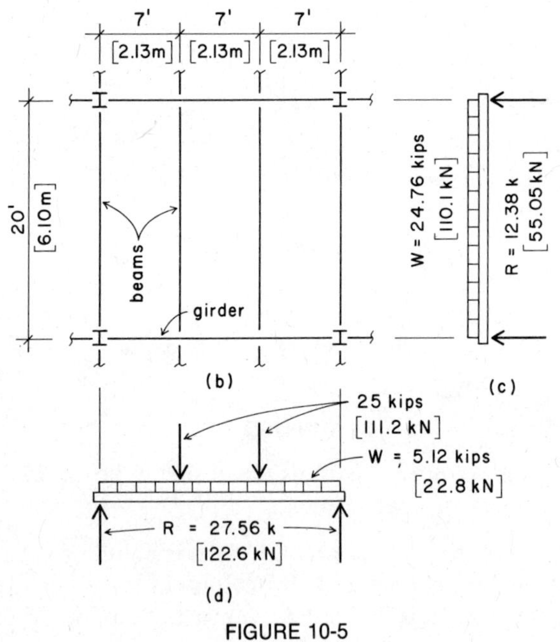
7'
7'
7'
[2.13m]
[2.13m]
[2.13m]
20'
[6.10m]
beams
girder
(b)
W = 24.76 kips
[110.1 kN]
R = 12.38 k
[55.05 kN]
(c)
25 kips
[111.2 kN]
W = 5.12 kips
[22.8 kN]
R = 27.56 k
[122.6 kN]
(d)

FIGURE 10-5

movable partitions. Design the framing of A36 steel using the layout shown in Fig. 10-5*b*.

Solution for the beam: The unit floor load that constitutes the uniformly distributed load superimposed on the beams is determined from the data in Table 10-1:

Dead load		
4 in. conc slab	=	50 psf
2 in. lt wt conc fill	=	16
carpet underlay (fiberbd)	=	3
carpet and pad	=	3
suspended ceiling	=	10
movable partitions	=	15
Total dead load	=	97 psf
Live load	=	60 psf
Total floor load:	=	157 psf [7.52 kN/m²]

The total superimposed load on one beam is the panel area times this unit load, or

$$W = 7 \times 20 \times 0.157 = 22.0 \text{ kips}$$

$$[W = 2.13 \times 6.1 \times 7.52 = 97.7 \text{ kN}]$$

The maximum bending moment due to this total uniformly distributed load is

$$M = \frac{WL}{8} = \frac{22.0 \times 20}{8} = 55.0 \text{ kip-ft}$$

$$\left[M = \frac{97.7 \times 6.10}{8} = 74.5 \text{ kN-m}\right]$$

Because the top flange of the beam is adequately supported laterally by the encasing concrete, the allowable bending stress is 24 ksi [165 MPa] and the required section modulus is

$$S = \frac{M}{F_b} = \frac{55.0 \times 12}{24} = 27.5 \text{ in.}^3$$

$$\left[S = \frac{74.5 \times 10^6}{165} = 452 \times 10^3 \text{ mm}^3\right]$$

Bearing in mind that so far we have not taken into account the weight of the beam and its fireproofing, we refer to Appendix B and look for a beam with a slightly greater section modulus than that computed. A possible consideration is a W 14 × 26, with $S_x = 35.3$ in.³ From Table 10-4 we find that the fireproofing for this section weighs 112 lb/ft, making a total weight of 26 + 112 = 138 lb/ft for the beam and fireproofing. This will add a total uniformly distributed load of

$$W = 20 \times 0.138 = 2.76 \text{ kips}$$

$$[12.28 \text{ kN}]$$

which is added to the superimposed load for a revised load of

$$W = 2.76 + 22.0 = 24.76 \text{ kips}$$

$$[110.1 \text{ kN}]$$

and results in a new moment and section modulus of

$$M = \frac{WL}{8} = \frac{24.76 \times 20}{8} = 61.9 \text{ kip-ft}$$

$$[83.9 \text{ kN-m}]$$

$$S = \frac{M}{F_b} = \frac{61.9 \times 12}{24} = 31.0 \text{ in.}^3$$

$$[508 \times 10^3 \text{ mm}^3]$$

Returning to Appendix B, we find that no section that will fulfill this requirement is lighter than the W 14 × 26, although there are alternatives if a shallower beam is desired. If we stay with our choice, it is necessary to consider the problems of shear and deflection. As discussed in Chapter 8, it is unlikely that shear will be critical in the typical uniformly loaded floor and roof beam under normal loading. If any doubt exists, however, an investigation should be made by using the procedures developed in Art. 9-5.

Although the requirements of the actual building code of jurisdiction must be used for any real design work, the following are typical deflection criteria:

Loading condition	Maximum deflection
Live load only	$L/360$
Total load, roofs	$L/180$
Total load, floors	$L/240$

Because the dead load is high in our case, we use the total load limit; thus the maximum deflection to be allowed is

$$\Delta = \frac{L}{240} = \frac{20 \times 12}{240} = 1.0 \text{ in.}$$

The actual deflection may be determined in several ways; for example, by the formula for Case 2 in Fig. 5-26 or the deflection factor from the table in Appendix D. Using the latter procedure, we find from the table in Appendix D that the load capacity of the W 14 × 26 is 28 kips and the deflection factor for the 20-ft span is 9.93. Then the actual deflection under the total load is

$$\Delta = \frac{9.93}{14} \times \frac{24.76}{28} = 0.63 \text{ in.}$$

which is less than that allowed.

If the end of the beam is connected to the girder in the usual way (Fig. 9-6), it may be necessary to investigate the condition of shear across the net area of the web. This type of investigation is discussed in Art. 12-7.

An alternative to the procedure of using the superimposed load to find a required section modulus would have been to compare the load directly with the tables in Appendix D. Scanning these tables for a beam to carry slightly more than 22 kips on a 20-ft span produces the same results as the computations that were performed. This is the simpler procedure because the beam carries only uniform loads.

Solution for the girder: The girder loading takes the form shown in Fig. 10-5*d*. The superimposed loading consits of the concentrated loads due to the end reactions of the beams. With two beams being supported at each point, this load is equal to the total load on one beam, or approximately 25 kips. The maximum bending moment due to this loading (Case 3, Fig. 5-26) is

$$M = \frac{PL}{3} = \frac{25 \times 21}{3} = 175 \text{ kip-ft}$$

$$[237 \text{ kN-m}]$$

and the required section modulus, assuming that the concrete encasement provides full lateral support and using $F_b = 24$ ksi, is

$$S = \frac{M}{F_b} = \frac{175 \times 12}{24} = 87.5 \text{ in.}^3$$

$$[1434 \times 10^3 \text{ mm}^3]$$

We now consider the section listings in Appendix B, bearing in mind that the beam weight and fireproofing has not been taken into account. One possible choice is the W 18 × 55, with S_x of 98.3 in.^{3}. From Table 10-4 the fireproofing will add 189 lb/ft to the beam weight, making a total of 244 lb/ft and producing an additional moment of

$$M = \frac{WL}{8} = \frac{(0.244 \times 21) \times 21)}{8} = \frac{5.12 \times 21}{8} = 13.4 \text{ kip-ft}$$

which, when added to the moment due to the superimposed load, produces a total moment on the beam of

$$M = 175 + 13.4 = 188.4 \text{ kip-ft } [255 \text{ kN-m}]$$

and requires a section modulus of

$$S = \frac{M}{F_b} = \frac{188.4 \times 12}{24} = 94.2 \text{ in.}^3 \ [1544 \times 10^3 \text{ mm}^3]$$

From Appendix B we find that the W 21 × 50 just satisfies this requirement with an S of 94.5. The deeper section, however, will have slightly more fireproofing concrete; therefore the margin of excess is probably not as indicated. If the deeper section is acceptable, the W 21 × 50 would be the lightest selection. Let us consider the case for deflection with the W 18 × 55. As for the beam, we consider the limit to be that under total load; thus

$$\Delta = \frac{L}{240} = \frac{21 \times 12}{240} = 1.05 \text{ in.}$$

For the actual deflection we consider the two loadings separately, compute two deflections, and add the results. From Fig. 5-26, Case 2, the deflection due to the uniformly distributed load is

$$\Delta = \frac{5WL^3}{384EI} = \frac{5 \times 5.12 \times (21 \times 12)^3}{384 \times 29{,}000 \times 890} = 0.041 \text{ in.}$$

and from Fig. 5-26, Case 3, the deflection due to the beam loads is

$$\Delta = \frac{23PL^3}{648EI} = \frac{23 \times 25 \times (21 \times 12)^3}{648 \times 29{,}000 \times 890} = 0.550 \text{ in.}$$

The total deflection is thus 0.041 + 0.550 = 0.591 in., which is considerably less than the limit of 1.05 in.

Shear stress on the gross web (depth times web thickness) will be low on the girder. Other possible concerns for shear on a net web section or for web crippling may exist if the end framing of the girder is known. In this example we finish with a consideration of bending and deflection.

It is common with framing systems of this type to provide for cambering of the beams to compensate for the dead-load deflection. This consists of cold bending the beams to produce a residual upward deflection (crowning) so that the beam top is flat under the dead load. In this example, which concerns relatively short spans, this is not really critical, but when spans are greater it is well advised.

Problem 10.8.A. The floor framing for a typical interior bay of a building is shown in Fig. 10-6; the layout is indicated in Fig. 10-6*b* and the floor construction, in Fig. 10-6*a*. Design the beam and girder for a live load of 100 psf using A36 steel. Assume that the steel deck provides full lateral support for the beams but that the girders are braced only at the column and beam intersections.

10-9 Open-Web Steel Joists

Open-web steel joists are shop-fabricated, parallel chord trusses used for direct support of roof and floor decks. (See Fig. 2-3*a* and 9-1*c*.) They are produced in a wide range of sizes with various fabrication details by a number of manufacturers. Although spe-

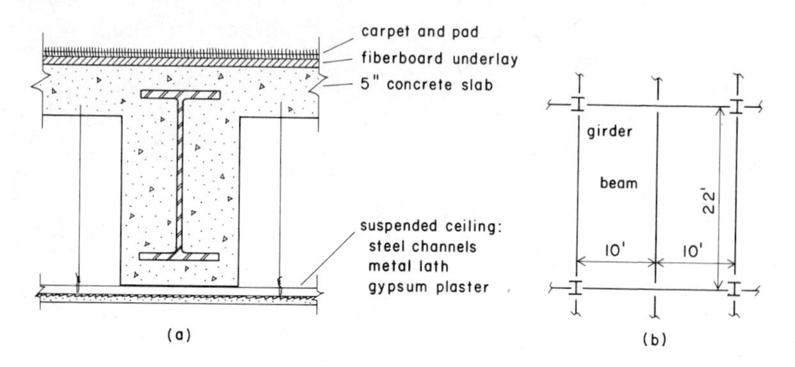

FIGURE 10-6

cific data should be obtained from individual suppliers, design and fabrication usually comply with the requirements of the standard specifications of the Steel Joist Institute (see Ref. 6).

Joists are produced in two series, designated J and H. J series joists are fabricated typically from A36 steel, whereas H series joists use a steel with a yield stress of 50 ksi [350 MPa]. In addition to the basic J and H series joists, there are two series of large joists designated LJ and LH (for longspan) and DLJ and DLH (for deep longspan).

Table 10-6 is adapted from the standard tables of the Steel Joist Institute (Ref. 6). This table lists the range of joists sizes available in the basic H series. (*Note:* A few of the heavier sizes have been omitted to shorten the table.) Joists are identified by a three-unit designation. The first number indicates the overall depth of the joist, the letter, the series, and the second number, the class of size of the members used; the higher the number, the heavier the joist.

Table 10-6 can be used to select the proper joist for a determined load and span condition. There are two entries in the table for each span; the first number represents the total load capacity of the joist and the number in parentheses identifies the load that will produce a deflection of 1/360 of the span. The following examples illustrate the use of the table data for some typical design situations:

TABLE 10-6. Load Table for Open Web Steel Joists[a]

Joist designation	8H3	10H3	10H4	12H3	12H4	12H5	12H6	14H3	14H4	14H5	14H6	14H7	16H4	16H5	16H6	16H7	16H8
Joist weight (lb/ft)	5.0	5.0	6.1	5.2	6.2	7.1	8.2	5.5	6.5	7.4	8.6	10.0	6.6	7.8	8.6	10.3	11.4
Span (ft)																	
8	600																
10	480 (460)	500	560														
12	400 (266)	417	467	467	533	600	650										
14	310 (167)	357 (270)	400 (334)	400 (393)	457	514	557	457	500	543	600	657					
16	232 (112)	302 (181)	350 (223)	350 (264	400 (345)	450 (404)	488 (480)	400 (366)	438	475	525	575	475	538	575	613	650
18		239 (127)	305 (157)	288 (185)	356 (242)	400 (284)	433 (337)	340 (257)	389 (336)	422 (393)	467	511	422 (413)	478	511	544	578
20		193 (92)	247 (114)	233 (135)	300 (177)	360 (207)	390 (246)	275 (187)	350 (245)	380 (287)	420 (342)	460 (403)	368 (301)	430 (370)	460 (437)	490	520
22				193 (101)	248 (133)	306 (155)	355 (185)	227 (141)	292 (184)	345 (215)	382 (257)	418 (302)	304 (226)	391 (278)	418 (328)	445 (395)	473 (454)
24				162 (78)	208 (102)	257 (120)	301 (142)	191 (108)	245 (142)	300 (166)	350 (198)	383 (233)	256 (174)	334 (214)	383 (253)	408 (304)	433 (350)
26								163 (85)	209 (111)	255 (131)	303 (156)	354 (183)	218 (137)	285 (169)	339 (199)	377 (239)	400 (275)
28								140 (68)	180 (89)	220 (104)	261 (125)	314 (147)	188 (110)	246 (135)	293 (159)	350 (192)	371 (220)
30													164 (89)	214 (110)	255 (129)	306 (155)	347 (179)
32													144 (74)	188 (90)	224 (107)	269 (128)	311 (148)

Note: The Steel Joist Institute publishes both Specifications and Load Tables; each of these contain standards which are to be used in conjunction with one another.

TABLE 10-6 (*Continued*)

Joist designation	18H5	18H6	18H7	18H8	18H9	18H10	20H5	20H6	20H7	20H8	20H9	20H10	22H6	22H7	22H8	22H9	22H10	22H11
Joist weight (lb/ft)	8.0	9.2	10.4	11.6	12.6	14.0	8.4	9.6	10.7	12.2	13.2	14.6	9.7	10.7	12.0	13.8	15.2	16.9
Span (ft)																		
18	500	533	578	600	621	629												
20	450	480	520	540	590	629	480	510	540	560	640	636						
22	409	436	473	491	536	600	436	464	491	509	582	636	491	509	527	609	626	648
	(356)	(420)																
24	375	400	433	450	492	550	400	425	450	467	533	583	450	467	483	558	600	648
	(274)	(324)	(388)	(444)	(484)	(546)	(335)	(382)					(446)					
26	321	369	400	415	454	508	360	392	415	431	492	538	415	431	446	515	554	623
	(216)	(255)	(305)	(349)	(380)	(429)	(263)	(300)	(365)		(476)		(351)	(426)				
28	276	326	371	386	421	471	310	345	386	400	457	500	359	400	414	479	514	579
	(173)	(204)	(244)	(280)	(305)	(344)	(211)	(240)	(292)	(352)	(381)	(431)	(281)	(341)		(468)		
30	241	284	345	360	393	440	270	301	360	373	427	467	313	373	387	447	480	540
	(140)	(166)	(199)	(227)	(248)	(280)	(171)	(195)	(238)	(286)	(310)	(350)	(228)	(277)	(343)	(381)	(428)	(487)
32	212	249	303	338	369	413	238	264	325	350	400	438	275	342	363	419	450	506
	(116)	(137)	(164)	(187)	(204)	(230)	(141)	(161)	(196)	(236)	(255)	(288)	(188)	(228)	(282)	(314)	(352)	(401)
34	187	221	269	311	347	388	210	234	288	329	376	412	243	303	341	394	424	476
	(96)	(114)	(136)	(156)	(170)	(192)	(118)	(134)	(163)	(196)	(213)	(240)	(157)	(190)	(235)	(261)	(294)	(335)
36	167	197	240	278	323	363	188	209	257	310	356	389	217	271	322	372	400	450
	(81)	(96)	(115)	(132)	(143)	(162)	(99)	(113)	(137)	(166)	(179)	(203)	(132)	(160)	(198)	(220)	(247)	(282)
38							169	187	230	278	324	364	195	243	301	353	379	426
							(84)	(96)	(117)	(141)	(153)	(172)	(112)	(136)	(169)	(187)	(210)	(240)
40							152	169	208	251	292	329	176	219	272	323	360	405
							(72)	(82)	(100)	(121)	(131)	(148)	(96)	(117)	(145)	(161)	(180)	(205)
42													159	199	247	293	330	381
													(83)	(101)	(125)	(139)	(156)	(177)
44													145	181	225	267	301	347
													(72)	(88)	(109)	(121)	(136)	(154)

Note: The Steel Joist Institute publishes both Specifications and Load Tables; each of these contain standards which are to be used in

Joist designation	24H6	24H7	24H8	24H9	24H10	24H11	26H8	26H9	26H10	26H11	28H8	28H9	28H10	28H11	30H8	30H9	30H10	30H11
Joist weight (lb/ft)	10.3	11.5	12.7	14.0	15.5	17.5	12.8	14.8	16.2	17.9	13.5	15.2	16.8	18.3	14.2	15.4	17.3	18.8
Span (ft)																		
24	467	483	500	583	625	631												
26	431	446	462	538	577	631	515	554	585	638								
28	393 (336)	414 (406)	429	500	536	586	479	514	543	593	479	514	550	600				
30	342 (273)	387 (330)	400	467 (457)	500	547	447	480	507	553	447	480	513	560	453	500	540	580
32	301 (225)	363 (272)	375 (339)	438 (376)	469 (423)	513 (482)	419 (380)	450 (445)	475	519	419	450	481	525	425	469	506	544
34	266 (188)	332 (227)	353 (283)	412 (314)	441 (353)	482 (402)	394 (317)	424 (371)	447 (417)	488 (476)	394 (370)	424	453	494	400	441	476	512
36	238 (158)	296 (191)	333 (238)	389 (264)	417 (297)	456 (339)	372 (267)	400 (312)	422 (352)	461 (401)	372 (311)	400 (364)	428 (410)	467	378 (359)	417	450	483
40	193 (115)	240 (139)	298 (174)	350 (193)	375 (217)	410 (247)	327 (194)	360 (228)	380 (256)	415 (292)	335 (227)	360 (266)	385 (299)	420 (342)	340 (262)	375 (306)	405 (345)	435 (395)
44	159 (87)	198 (105)	247 (130)	293 (145)	330 (163)	373 (186)	270 (146)	319 (171)	345 (193)	377 (220)	291 (171)	327 (200)	350 (225)	382 (257)	309 (196)	341 (230)	368 (259)	395 (297)
48	134 (67)	167 (81)	207 (100)	246 (111)	277 (125)	320 (143)	227 (112)	268 (132)	301 (148)	346 (169)	245 (131)	289 (154)	321 (173)	350 (198)	263 (151)	311 (177)	338 (200)	363 (229)
52							193 (88)	228 (104)	256 (117)	297 (133)	209 (103)	247 (121)	277 (136)	321 (156)	224 (119)	265 (139)	298 (157)	335 (180)
56											180 (83)	213 (97)	239 (109)	276 (125)	193 (95)	229 (112)	257 (126)	297 (144)
60															168 (77)	199 (91)	224 (102)	259 (117)

[a] Loads in pounds per ft of joist span; first entry represents the total joist capacity; entry in parentheses is the load that produces a maximum deflection of 1/360 of the span. Loads above the heavy line are governed by shear; loads below the dotted line are used for roofs only. Data adapted from more extensive tables published in the Standard Specifications, Load Tables, and Weight Tables for Steel Joists and Joist Girders, 1982 ed. (Ref. 6) with permission of the publishers, the Steel Joist Institute.

Note: The Steel Joist Institute publishes both Specifications and Load Tables; each of these contain standards which are to be used in conjunction with one another.

Example 1. Open-web steel joists are to be used to support a roof with a unit live load of 20 psf and a unit dead load of 15 psf (not including the weight of the joists) on a span of 40 ft. Joists are spaced at 6 ft center to center. Select the lightest joist if deflection under live load is limited to 1/360 of the span.

Solution: We first determine the load per ft on the joist:

Live load	$6 \times 20 =$	120 lb/ft
Dead load	$6 \times 15 =$	90 lb/ft
Total load		210 lb/ft

We then scan the entries in Table 10-6 for the joists that will just carry these loads, noting that the joist weight must be deducted from the entry for total capacity. The possible choices for this example are summarized in Table 10-7. Although the joist weights are all very close, the 24H7 is the lightest choice.

Example 2. Open-web steel joists are to be used for a floor with a unit live load of 75 psf and a unit dead load of 40 psf (not including the joists) on a span of 30 ft. Joists are 2 ft center to center and deflection is limited to 1/360 of the span under live load only and to 1/240 of the span under total load. Determine (a) the lightest joist possible and (b) the joist with the least depth possible.

TABLE 10-7. Possible Choices for the Roof Joist

	Load per foot for the indicated joists			
Load condition	20H8	22H8	24H7	26H8
Total capacity (from Table 10-6)	251	272	240	327
Joist weight	12.2	12.0	11.5	12.8
Net usable capacity	238.8	260	228.5	314.2
Load for $\frac{1}{360}$ deflection (from Table 10-6)	121	145	139	194

Solution: As in the preceding example, we first find the loads on the joist:

Live load	2 × 75 =	150 lb/ft
Dead load	2 × 40 =	80 lb/ft
Total load		230 lb/ft

To satisfy the deflection criteria we must find a table entry in parentheses of 150 plf (for live load only) or 240/360 × 230 = 153 plf (for total load). The possible choices obtained from scanning Table 10-6 are summarized in Table 10-8, from which we observe

for (a) the lightest joist is the 20H5

for (b) the shallowest joist is the 18H6

Note that, although an entry is given for the 16H7, Table 10-6 recommends it for roofs only for this span.

Problem 10.9.A.* Open-web steel joists are to be used for a roof with a live load of 25 psf and a dead load of 20 psf (not including joists) on a span of 48 ft. Joists are 4 ft center to center and the deflection under live load is limited to 1/360 of the span. Select the lightest joist possible.

Problem 10.9.B. Open web steel joists are to be used for a roof with a live load of 30 psf and a dead load of 18 psf (not including joists) on a span of 44 ft. Joists are 5 ft center to center and deflection under live load is limited to 1/360 of the span. Select the lightest joist.

TABLE 10-8. Possible Choices for the Floor Joist

Load condition	Load per foot for the indicated joists		
	16H7	18H6	20H5
Total capacity (from Table 10-6)	306	284	270
Joist weight	10.3	9.2	8.4
Net usable capacity	295.7	274.8	261.6
Load for $\frac{1}{360}$ deflection (from Table 10-6)	156	166	171

Problem 10.9.C.* Open-web steel joists are to be used for a floor with a live load of 50 psf and a dead load of 45 psf (not including joists) on a span of 36 ft. Joists are 2 ft center to center and deflection is limited to 1/360 of the span under live load and to 1/240 of the span under total load. Select (a) the lightest possible joist and (b) the shallowest possible joist.

Problem 10.9.D. Repeat Problem 10.9.C, except that the live load is 100 psf, the dead load is 35 psf, and the span is 26 ft.

11

Columns

11-1. Introduction

A column or strut is a compression member, the length of which is several times greater than its least lateral dimension. The term column denotes a relatively heavy vertical member, whereas the lighter vertical and inclined members, such as braces and the compression members of roof trusses, are called struts.

Under the discussion of direct stress in Art. 3-2 it was pointed out that the unit compressive stress in the short block shown in Fig. 3-1*b* could be expressed by the direct stress formula $f_a = P/A$; but it was also stated that this relationship became invalid as the ratio of the length of the compression member to its least width increased. To pursue this further consider a small block of steel 1 in. by 1 in. in cross section and 1 or 2 in. high. If the allowable compressive stress is 20 ksi, the block will safely support a load P (Fig. 11-1*a*) of 20 kips. If, however, we consider a bar of the same cross section with a length of, say, 30 to 40 in., we find that the value of P it will sustain is considerably less because of the tendency of this slenderer bar to buckle or bend (Fig. 11-1*b*). Therefore in columns the element of *slenderness* must be taken into account when determining allowable loads. A short column or block fails by crushing but long slender columns fail by stresses that result from bending.

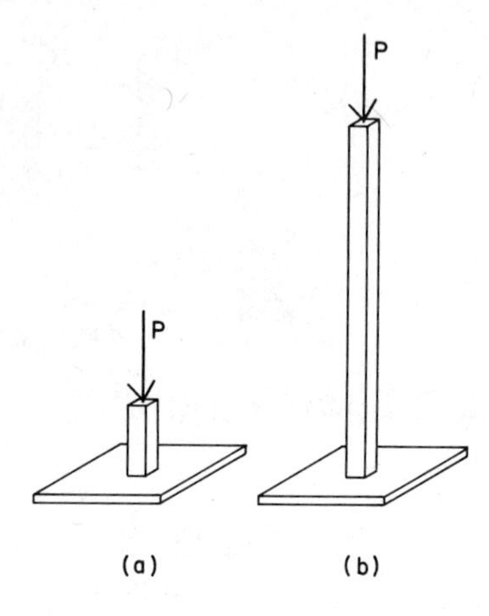

FIGURE 11-1. Effect of column slenderness.

11-2. Column Shapes

Because of the tendency to bend, the safe load on a column depends not only on the number of square inches in the cross section but also on the manner in which the material is distributed with respect to the axes of the cross section; that is, the *shape* of the column section is an important factor. An axially loaded column tends to bend in a plane perpendicular to the axis of the cross section about which the moment of inertia is least. Since column cross sections are seldom symmetrical with respect to both major axes, the ideal section would be one in which the moment of inertia for each major axis is equal. Pipe columns and structural tubing meet this condition, but their use is somewhat limited because of difficulties in making beam connections.

Of the two major axes of a standard I-beam, the moment of inertia about the axis parallel to the web is much the smaller; hence for the amount of material in the cross section I-beams are not economical shapes when used as columns or struts. In former years built-up sections such as Fig. 11-2*c* and *d* were used extensively, but wide flange sections (Fig. 11-2*a*) are now rolled in a large variety of sizes and are used universally because they require a minimum of fabrication. They are sometimes called H-columns. For excessive loads or unusual conditions plates are welded to the flanges of wide flange sections to give added

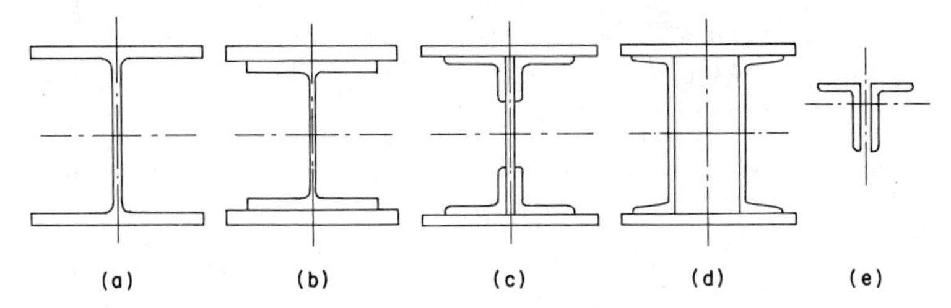

FIGURE 11-2. Typical column sections.

strength (Fig. 11-2*b*). The compression members of steel trusses are often formed of two angles, as shown in Fig. 11-2*e*.

11-3. Slenderness Ratio

In the design of timber columns the term *slenderness ratio* is defined as the unbraced length divided by the dimension of the least side, both in inches. For structural shapes such as those shown in Fig. 11-2 the least lateral dimension is not an accurate criterion; and the radius of gyration r, which relates more precisely to the stiffness of columns in general, is used in steel column design. As discussed in Art. 6-7, $r = \sqrt{I/A}$. For rolled sections the value of the radius of gyration with respect to both major axes is given in the tables of properties for designing. For built-up sections it may be necessary to compute its value. The slenderness ratio of a steel column is then l/r, where l is the effective length of the column in inches and r is the *least* radius of gyration of the cross section, also in inches. The slenderness ratio for compression members should not exceed 200.

Example. A W 10 × 49 is used as a column whose effective length is 20 ft [6.10 m]. Compute the slenderness ratio.
Solution: Reference to Appendix Table A-1 reveals that the radii of gyration for this section are $r_X = 4.35$ in. and $r_Y = 2.54$ in. Therefore the *least* radius of gyration is 2.54 in.

Because the effective length of the column is 20 ft [6.10 m], the slenderness ratio is

$$\frac{L}{r} = \frac{20 \times 12}{2.54} = 94.5 \left[\frac{6.10 \times 10^3}{64.5} = 94.6\right]$$

It should be remembered that the tendency to bend due to buckling under the compression load is in a direction perpendicular to the axis about which the radius of gyration is least.

Problem 11.3.A.* The effective length of a W 8 × 31 used as a column is 16 ft [4.88 m]. Compute the slenderness ratio.

Problem 11.3.B. What is the slenderness ratio of a column whose section is a W 12 × 65 with an effective length of 30 ft [9.14 m]?

11-4. Effective Column Length

The AISC Specification requires that, in addition to the unbraced length of a column, the condition of the ends must be given consideration. The slenderness ratio is Kl/r, where K is a factor dependent on the restraint at the ends of a column and the means

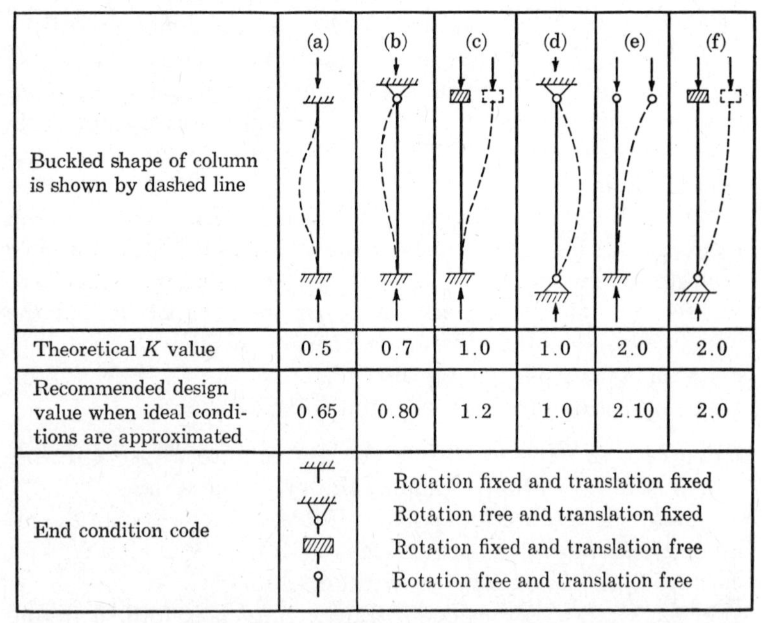

	(a)	(b)	(c)	(d)	(e)	(f)
Buckled shape of column is shown by dashed line						
Theoretical K value	0.5	0.7	1.0	1.0	2.0	2.0
Recommended design value when ideal conditions are approximated	0.65	0.80	1.2	1.0	2.10	2.0
End condition code	Rotation fixed and translation fixed Rotation free and translation fixed Rotation fixed and translation free Rotation free and translation free					

FIGURE 11-3. Determination of effective column length. Reprinted from the *Manual of Steel Construction,* 8th ed. (Ref. 1), with permission of the publishers, American Institute of Steel Construction.

available to resist lateral motion. Figure 11-3 shows diagrammatically six idealized conditions in which joint rotation and joint translation are illustrated. The term K is the ratio of the effective column length to the actual unbraced length. For average conditions in building construction the value of K is taken as 1; therefore the slenderness ratio Kl/r becomes simply l/r. (See Fig. 11-3d).

11-5. Column Formulas

The AISC Specification gives the following requirements for use in the design of compression members. The allowable unit stresses shall not exceed the following values:

On the gross section of axially loaded compression members, when Kl/r, the largest effective slenderness ratio of any unbraced segment is less than C_c,

$$F_a = \frac{[1 - (Kl/r)^2/2C_c^2]F_y}{\text{FS}} \qquad \text{(Formula 11.5.1)}$$

where

$$\text{FS} = \text{factor of safety} = \frac{5}{3} + \frac{3(Kl/r)}{8C_c} - \frac{(Kl/r)^3}{8C_c^3}$$

and

$$C_c = \sqrt{\frac{2\pi^2 E}{F_y}}$$

On the gross section of axially loaded columns, when Kl/r exceeds C_c,

$$F_a = \frac{12\pi^2 E}{23(Kl/r)^2} \qquad \text{(Formula 11.5.2)}$$

On the gross section of axially loaded bracing and secondary members, when l/r exceeds 120 (for this case, K is taken as unity),

$$F_{as} = \frac{F_a(\text{by Formula 11.5.1 or 11.5.2})}{1.6 - l/200r} \qquad \text{(Formula 11.5.3)}$$

In these formulas

F_a = the axial compression stress permitted in the absence of bending stress
K = effective length factor (see Art. 11-4)
l = actual unbraced length
r = governing radius of gyration (usually the least)
C_c = $\sqrt{2\pi^2 E/F_y}$; for A36 steel $C_c = 126.1$
F_y = the minimum yield point of the steel being used (for A36 steel $F_y = 36{,}000$)
FS = factor of safety (see above)
E = the modulus of elasticity of structural steel, 29,000 ksi
F_{as} = the axial compressive stress permitted in the absence of bending stress for bracing and other secondary members

To determine the allowable axial load that a main column will support F_a, the allowable unit stress, is computed by formula 1 or 2, and this stress is multiplied by the cross-sectional area of the column. If the column is a secondary member or is used for bracing, formula 3 gives the allowable unit stress; these allowable unit stresses are somewhat greater than those permitted for main members. Table 11-1 gives allowable stresses computed in accordance with these formulas. It should be examined carefully because it will be of great assistance. Note particularly that this table is for use with A36 steel; tables based on other grades of steel are contained in the AISC Manual.

11-6. Allowable Column Loads

The allowable axial load that a steel column will support is found by multiplying the allowable unit stress by the cross-sectional area of the column. The value of Kl/r is first determined, and by referring to Table 11-1 we can establish the allowable unit stress.

Example 1. A W 12 × 65 is used as a column with an unbraced length of 16 ft [4.88 m]. Compute the allowable load.
Solution: Referring to Appendix Table A-1, we find that A = 19.1 in.^{2} [12,323 mm^2], $r_X = 5.28$ in. [134 mm], and $r_Y = 3.02$ in. [76.7 mm]. Because the column is unbraced with respect to both

TABLE 11-1. Allowable Unit Stresses for Columns of A36 Steel (ksi)[a]

Main and Secondary Members Kl/r not over 120						Main Members Kl/r 121 to 200				Secondary Members[b] l/r 121 to 200			
$\frac{Kl}{r}$	F_a (ksi)	$\frac{Kl}{r}$	F_a (ksi)	$\frac{Kl}{r}$	F_a (ksi)	$\frac{Kl}{r}$	F_a (ksi)	$\frac{Kl}{r}$	F_a (ksi)	$\frac{l}{r}$	F_{as} (ksi)	$\frac{l}{r}$	F_{as} (ksi)
1	21.56	41	19.11	81	15.24	121	10.14	161	5.76	121	10.19	161	7.25
2	21.52	42	19.03	82	15.13	122	9.99	162	5.69	122	10.09	162	7.20
3	21.48	43	18.95	83	15.02	123	9.85	163	5.62	123	10.00	163	7.16
4	21.44	44	18.86	84	14.90	124	9.70	164	5.55	124	9.90	164	7.12
5	21.39	45	18.78	85	14.79	125	9.55	165	5.49	125	9.80	165	7.08
6	21.35	46	18.70	86	14.67	126	9.41	166	5.42	126	9.70	166	7.04
7	21.30	47	18.61	87	14.56	127	9.26	167	5.35	127	9.59	167	7.00
8	21.25	48	18.53	88	14.44	128	9.11	168	5.29	128	9.49	168	6.96
9	21.21	49	18.44	89	14.32	129	8.97	169	5.23	129	9.40	169	6.93
10	21.16	50	18.35	90	14.20	130	8.84	170	5.17	130	9.30	170	6.89
11	21.10	51	18.26	91	14.09	131	8.70	171	5.11	131	9.21	171	6.85
12	21.05	52	18.17	92	13.97	132	8.57	172	5.05	132	9.12	172	6.82
13	21.00	53	18.08	93	13.84	133	8.44	173	4.99	133	9.03	173	6.79
14	20.95	54	17.99	94	13.72	134	8.32	174	4.93	134	8.94	174	6.76
15	20.89	55	17.90	95	13.60	135	8.19	175	4.88	135	8.86	175	6.73
16	20.83	56	17.81	96	13.48	136	8.07	176	4.82	136	8.78	176	6.70
17	20.78	57	17.71	97	13.35	137	7.96	177	4.77	137	8.70	177	6.67
18	20.72	58	17.62	98	13.23	138	7.84	178	4.71	138	8.62	178	6.64
19	20.66	59	17.53	99	13.10	139	7.73	179	4.66	139	8.54	179	6.61
20	20.60	60	17.43	100	12.98	140	7.62	180	4.61	140	8.47	180	6.58
21	20.54	61	17.33	101	12.85	141	7.51	181	4.56	141	8.39	181	6.56
22	20.48	62	17.24	102	12.72	142	7.41	182	4.51	142	8.32	182	6.53
23	20.41	63	17.14	103	12.59	143	7.30	183	4.46	143	8.25	183	6.51
24	20.35	64	17.04	104	12.47	144	7.20	184	4.41	144	8.18	184	6.49
25	20.28	65	16.94	105	12.33	145	7.10	185	4.36	145	8.12	185	6.46
26	20.22	66	16.84	106	12.20	146	7.01	186	4.32	146	8.05	186	6.44
27	20.15	67	16.74	107	12.07	147	6.91	187	4.27	147	7.99	187	6.42
28	20.08	68	16.64	108	11.94	148	6.82	188	4.23	148	7.93	188	6.40
29	20.01	69	16.53	109	11.81	149	6.73	189	4.18	149	7.87	189	6.38
30	19.94	70	16.43	110	11.67	150	6.64	190	4.14	150	7.81	190	6.36
31	19.87	71	16.33	111	11.54	151	6.55	191	4.09	151	7.75	191	6.35
32	19.80	72	16.22	112	11.40	152	6.46	192	4.05	152	7.69	192	6.33
33	19.73	73	16.12	113	11.26	153	6.38	193	4.01	153	7.64	193	6.31
34	19.65	74	16.01	114	11.13	154	6.30	194	3.97	154	7.59	194	6.30
35	19.58	75	15.90	115	10.99	155	6.22	195	3.93	155	7.53	195	6.28
36	19.50	76	15.79	116	10.85	156	6.14	196	3.89	156	7.48	196	6.27
37	19.42	77	15.69	117	10.71	157	6.06	197	3.85	157	7.43	197	6.26
38	19.35	78	15.58	118	10.57	158	5.98	198	3.81	158	7.39	198	6.24
39	19.27	79	15.47	119	10.43	159	5.91	199	3.77	159	7.34	199	6.23
40	19.19	80	15.36	120	10.28	160	5.83	200	3.73	160	7.29	200	6.22

[a] Reprinted from the *Manual of Steel Construction,* 8th ed, with permission of the publishers, American Institute of Steel Construction.
[b] K is taken as 1.0 for secondary members.

axes, the least radius of gyration is used to determine the slenderness ratio. Also, with no qualifying conditions given, $K = 1.0$. The slenderness ratio is then

$$\frac{KL}{r} = \frac{1 \times 16 \times 12}{3.02} = 63.6$$

$$\left[\frac{4.88 \times 10^3}{76.7} = 63.6\right]$$

In design work it is usually considered acceptable to round the slenderness ratio off in front of the decimal point because a typical lack of accuracy in the design data does not warrant greater precision. Therefore we consider the slenderness ratio to be 64; the allowable stress given in Table 11-1 is $F_a = 17.04$ ksi [117.5 MPa]. The allowable load on the column is then

$$P = A \times F_a = 19.1 \times 17.04 = 325.5 \text{ kips}$$

$$\left[P = \frac{12{,}323 \times 117.5}{10^3} = 1448 \text{ kN}\right]$$

Example 2. A built-up column of A36 steel consists of a W 14 × 311 core section with two 18 × 1 in. [457 × 25 mm] cover plates (Fig. 11-4). If K is 1 and the unbraced height is 20 ft [6.096 m], compute the axial load that this combined section can support.

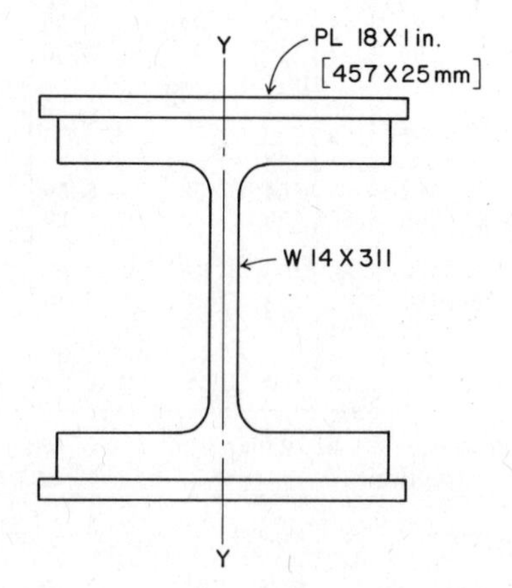

FIGURE 11-4. The built-up section.

Solution: The first step in this problem is to find the properties of the combined section. We need the total area, the moment of inertia about the Y-Y axis, and the value for r_y. From Appendix Table A-1, for the W 14 × 311, we find $A = 91.4$ in.^{2} [58,971 mm^2] and $I_y = 1610$ in.^{2} [670.1 × 10^6 mm^4]. The total area is thus

$$A = 91.4 + 2 \times (18 \times 1) = 91.4 + 36 = 127.4 \text{ in.}^2$$

$$[A = 58{,}971 + 2 \times (457 \times 25) = 1821 \text{ mm}^2]$$

Reference to Table 6-1 shows that the moment of inertia of one plate about the Y-Y axis is

$$I = \frac{bd^3}{12} = \frac{1 \times (18)^3}{12} = 486 \text{ in.}^4$$

$$[198.8 \times 10^6 \text{ mm}^4]$$

and the total I for the section is thus

$$I = 1610 + (2 \times 486) = 1610 + 972 = 2582 \text{ in.}^4$$

$$[1067.7 \times 10^6 \text{ mm}^4]$$

Referring to Art. 6-7,

$$r = \sqrt{\frac{I}{A}} = \sqrt{\frac{2582}{127.4}} = 4.50 \text{ in.}$$

$$\left[\sqrt{\frac{1067.7 \times 10^6}{1821}} = 114.2 \text{ mm}\right]$$

Then

$$\frac{KL}{r_y} = \frac{1 \times 20 \times 12}{4.50} = 53.33$$

$$\left[\frac{6.096 \times 10^3}{114.2} = 53.38\right]$$

Rounding the slenderness ratio off to 53, we find from Table 11-1 that $F_a = 18.08$ ksi [124.7 MPa]. The allowable load on the section is

$$P = A \times F_a = 127.4 \times 18.08 = 2303 \text{ kips}$$

$$\left[P = \frac{81{,}821 \times 124.7}{10^3} = 10{,}203 \text{ kN}\right]$$

Note: In the following problems assume A36 steel and a value of $K = 1$.

Problem 11.6.A.* Compute the allowable axial load on a W 10 × 88 column with an unbraced height of 15 ft [4.57 m].

Problem 11.6.B. A W 12 × 65 used as a column has an unbraced height of 22 ft [6.71 m]. Compute the allowable axial load.

Problem 11.6.C.* A built-up column section is composed of a W 14 × 342 with two 22 × 2 [559 × 50 mm] cover plates. If the unbraced length is 18 ft [5.49 m], compute the allowable axial load.

11-7. Design of Steel Columns

In practice, the design of steel columns is accomplished largely by the use of safe load tables; if these tables are not available, design is carried out by the trial method. Data include the load and length of the column. The designer selects a trial cross section on the basis of his experience and judgment and, by means of a column formula, computes the allowable load that it can support. If this load is less than the actual load the column will be required to support, the trial section is too small and another section is tested in a similar manner.

Table 11-2 lists allowable loads on a number of column sections. It has been compiled from more extensive tables in the AISC Manual and the loads are computed in accordance with the formulas in Art. 11-5. Note particularly that these allowable loads are for main members of A36 steel. Loads for Kl/r ratios between 120 and 200 are for main members. The significance of the bending factors, given at the extreme right of the table, is considered in Art. 11-11.

To illustrate the use of Table 11-2, refer to Example 1 of Art. 11-6. This problem asked for the allowable load that could be supported by a W 12 × 65 of A36 steel with an unbraced height of 16 ft. Referring to Table 11-2, we see at a glance that the allowable axial load is 326 kips, which agrees closely with the value found by computation.

Although the designer may select the proper column section by merely referring to the safe load tables, it is well to understand the application of the formulas by means of which the tables have

TABLE 11-2. Allowable Column Loads for Selected W Shapes[a]

	Effective length (KL) in feet										Bending factor	
Shape	8	9	10	11	12	14	16	18	20	22	B_x	B_y
M 4 × 13	48	42	35	29	24	18					0.727	2.228
W 4 × 13	52	46	39	33	28	20	16				0.701	2.016
W 5 × 16	74	69	64	58	52	40	31	24	20		0.550	1.560
M 5 × 18.9	85	78	71	64	56	42	32	25			0.576	1.768
W 5 × 19	88	82	76	70	63	48	37	29	24		0.543	1.526
W 6 × 9	33	28	23	19	16	12					0.482	2.414
W 6 × 12	44	38	31	26	22	16					0.486	2.367
W 6 × 16	62	54	46	38	32	23	18				0.465	2.155
W 6 × 15	75	71	67	62	58	48	38	30	24	20	0.456	1.424
M 6 × 20	98	92	87	81	74	61	47	37	30	25	0.453	1.510
W 6 × 20	100	95	90	85	79	67	54	42	34	28	0.438	1.331
W 6 × 25	126	120	114	107	100	85	69	54	44	36	0.440	1.308
W 8 × 24	124	118	113	107	101	88	74	59	48	39	0.339	1.258
W 8 × 28	144	138	132	125	118	103	87	69	56	46	0.340	1.244
W 8 × 31	170	165	160	154	149	137	124	110	95	80	0.332	0.985
W 8 × 35	191	186	180	174	168	155	141	125	109	91	0.330	0.972
W 8 × 40	218	212	205	199	192	127	160	143	124	104	0.330	0.959
W 8 × 48	263	256	249	241	233	215	196	176	154	131	0.326	0.940
W 8 × 58	320	312	303	293	283	263	240	216	190	162	0.329	0.934
W 8 × 67	370	360	350	339	328	304	279	251	221	190	0.326	0.921
W 10 × 33	179	173	167	161	155	142	127	112	95	78	0.277	1.055
W 10 × 39	213	206	200	193	186	170	154	136	116	97	0.273	1.018
W 10 × 45	247	240	232	224	216	199	180	160	138	115	0.271	1.000
W 10 × 49	279	273	268	262	256	242	228	213	197	180	0.264	0.770
W 10 × 54	306	300	294	288	281	267	251	235	217	199	0.263	0.767
W 10 × 60	341	335	328	321	313	297	280	262	243	222	0.264	0.765
W 10 × 68	388	381	373	365	357	339	320	299	278	255	0.264	0.758
W 10 × 77	439	431	422	413	404	384	362	339	315	289	0.263	0.751
W 10 × 88	504	495	485	475	464	442	417	392	364	335	0.263	0.744
W 10 × 100	573	562	551	540	428	503	476	446	416	383	0.263	0.735
W 10 × 112	642	631	619	606	593	565	535	503	469	433	0.261	0.726
W 12 × 40	217	210	203	196	188	172	154	135	114	94	0.227	1.073
W 12 × 45	243	235	228	220	211	193	173	152	129	106	0.227	1.065
W 12 × 50	271	263	254	246	236	216	195	171	146	121	0.227	1.058
W 12 × 53	301	295	288	282	275	260	244	227	209	189	0.221	0.813
W 12 × 58	329	322	315	308	301	285	268	249	230	209	0.218	0.794
W 12 × 65	378	373	367	361	354	341	326	311	294	277	0.217	0.656
W 12 × 72	418	412	406	399	392	377	361	344	326	308	0.217	0.651
W 12 × 79	460	453	446	439	431	415	398	379	360	339	0.217	0.648
W 12 × 87	508	501	493	485	477	459	440	420	398	376	0.217	0.645
W 12 × 96	560	552	544	535	526	506	486	464	440	416	0.215	0.635
W 12 × 106	620	611	602	593	583	561	539	514	489	462	0.215	0.633
W 12 × 120	702	692	660	636	611	584	555	525	493	460	0.217	0.630
W 12 × 136	795	772	747	721	693	662	630	597	561	524	0.215	0.621
W 12 × 152	891	866	839	810	778	745	710	673	633	592	0.214	0.614
W 12 × 170	998	970	940	908	873	837	798	757	714	668	0.213	0.608

TABLE 11-2 (*Continued*)

Shape	Effective length (KL) in feet										Bending factor	
	8	10	12	14	16	18	20	22	24	26	B_x	B_y
W 12 × 190	1115	1084	1051	1016	978	937	894	849	802	752	0.212	0.600
W 12 × 210	1236	1202	1166	1127	1086	1042	995	946	894	840	0.212	0.594
W 12 × 230	1355	1319	1280	1238	1193	1145	1095	1041	985	927	0.211	0.589
W 12 × 252	1484	1445	1403	1358	1309	1258	1203	1146	1085	1022	0.210	0.583
W 12 × 279	1642	1600	1554	1505	1452	1396	1337	1275	1209	1141	0.208	0.573
W 12 × 305	1799	1753	1704	1651	1594	1534	1471	1404	1333	1260	0.206	0.564
W 12 × 336	1986	1937	1884	1827	1766	1701	1632	1560	1484	1404	0.205	0.558
W 14 × 43	230	215	199	181	161	140	117	96	81	69	0.201	1.115
W 14 × 48	258	242	224	204	182	159	133	110	93	79	0.201	1.102
W 14 × 53	286	268	248	226	202	177	149	123	104	88	0.201	1.091
W 14 × 61	345	330	314	297	278	258	237	214	190	165	0.194	0.833
W 14 × 68	385	369	351	332	311	289	266	241	214	186	0.194	0.826
W 14 × 74	421	403	384	363	341	317	292	265	236	206	0.195	0.820
W 14 × 82	465	446	425	402	377	351	323	293	261	227	0.196	0.823
W 14 × 90	536	524	511	497	482	466	449	432	413	394	0.185	0.531
W 14 × 99	589	575	561	546	529	512	494	475	454	433	0.185	0.527
W 14 × 109	647	633	618	601	583	564	544	523	501	478	0.185	0.523
W 14 × 120	714	699	682	663	644	623	601	578	554	528	0.186	0.523
W 14 × 132	786	768	750	730	708	686	662	637	610	583	0.186	0.521
W 14 × 145	869	851	832	812	790	767	743	718	691	663	0.184	0.489
W 14 × 159	950	931	911	889	865	840	814	786	758	727	0.184	0.485
W 14 × 176	1054	1034	1011	987	961	933	904	874	842	809	0.184	0.484
W 14 × 193	1157	1134	1110	1083	1055	1025	994	961	927	891	0.183	0.477
W 14 × 211	1263	1239	1212	1183	1153	1121	1087	1051	1014	975	0.183	0.477
W 14 × 233	1396	1370	1340	1309	1276	1241	1204	1165	1124	1081	0.183	0.472
W 14 × 257	1542	1513	1481	1447	1410	1372	1331	1289	1244	1198	0.182	0.470
W 14 × 283	1700	1668	1634	1597	1557	1515	1471	1425	1377	1326	0.181	0.465
W 14 × 311	1867	1832	1794	1754	1711	1666	1618	1568	1515	1460	0.181	0.459
W 14 × 342		2022	1985	1941	1894	1845	1793	1738	1681	1621	0.181	0.457
W 14 × 370		2181	2144	2097	2047	1995	1939	1881	1820	1756	0.180	0.452
W 14 × 398		2356	2304	2255	2202	2146	2087	2025	1961	1893	0.178	0.447
W 14 × 426		2515	2464	2411	2356	2296	2234	2169	2100	2029	0.177	0.442
W 14 × 455		2694	2644	2589	2430	2467	2401	2332	2260	2184	0.177	0.441
W 14 × 500		2952	2905	2845	2781	2714	2642	2568	2490	2409	0.175	0.434
W 14 × 550		3272	3206	3142	3073	3000	2923	2842	2758	2670	0.174	0.429
W 14 × 605		3591	3529	3459	3384	3306	3223	3136	3045	2951	0.171	0.421
W 14 × 665		3974	3892	3817	3737	3652	3563	3469	3372	3270	0.170	0.415
W 14 × 730		4355	4277	4196	4100	4019	3923	3823	3718	3609	0.168	0.408

[a] Loads in kips for shapes of steel with yield stress of 36 ksi [250 MPa]. Adapted from data in the *Manual of Steel Construction,* 8th ed. (Ref. 1), with permission of the publisher, American Institute of Steel Construction.

been computed. To that end the *design procedure* is outlined below. When the design load and length have been ascertained, the following steps are taken:

Step 1: Assume a trial section and note from the table of properties the cross-sectional area and the least radius of gyration.

Step 2: Compute the slenderness ratio Kl/r, l being the unsupported length of the column in inches. For the value of K see Art. 10-4.

Step 3: Compute F_a, the allowable unit stress, by using a column formula or Table 11-1.

Step 4: Multiply F_a found in step 3 by the area of the column cross section. This gives the allowable load *on the trial column section*.

Step 5: Compare the allowable load found in step 4 with the design load. If the allowable load on the trial section is less than the design load (or if it is so much greater that it makes use of the section uneconomical), try another section and test it in the same manner. The reader should note that, except for assuming a trial section, these operations were carried out in Example 1 of Art. 11-6.

Problem 11.7.A.* Using Table 11-2, select a column section to support an axial load of 148 kips [658 kN] if the unbraced height is 12 ft [3.66 m]. A36 steel is to be used and K is assumed to be 1.

Problem 11.7.B. Same data as in Problem 11.7.A except that the load is 258 kips [1148 kN] and the unbraced height is 15 ft [4.57 m].

Problem 11.7.C. Same data as in Problem 11.7.A except that the load is 355 kips [1579 kN] and the unbraced height is 20 ft [6.10 m].

11-8. Steel Pipe Columns

Round steel pipe columns are frequently installed in both steel and wood framed buildings. In routine work they are designed for simple axial load by the use of safe load tables.

Table 11-3 gives allowable axial loads for Standard weight steel pipe columns with a yield point of 36 ksi [250 MPa]. The outside diameters at the head of the table are *nominal* dimensions that

TABLE 11-3. Allowable Column Loads, Standard Steel Pipe of A36 Steel (kips)[a]

Nominal Dia.		12	10	8	6	5	4	$3\frac{1}{2}$	3
Wall Thickness		0.375	0.365	0.322	0.280	0.258	0.237	0.226	0.216
Weight per Foot		49.56	40.48	28.55	18.97	14.62	10.79	9.11	7.58
F_y		36 ksi							
Effective length in feet KL with respect to radius of gyration	0	315	257	181	121	93	68	58	48
	6	303	246	171	110	83	59	48	38
	7	301	243	168	108	81	57	46	36
	8	299	241	166	106	78	54	44	34
	9	296	238	163	103	76	52	41	31
	10	293	235	161	101	73	49	38	28
	11	291	232	158	98	71	46	35	25
	12	288	229	155	95	68	43	32	22
	13	285	226	152	92	65	40	29	19
	14	282	223	149	89	61	36	25	16
	15	278	220	145	86	58	33	22	14
	16	275	216	142	82	55	29	19	12
	17	272	213	138	79	51	26	17	11
	18	268	209	135	75	47	23	15	10
	19	265	205	131	71	43	21	14	9
	20	261	201	127	67	39	19	12	
	22	254	193	119	59	32	15	10	
	24	246	185	111	51	27	13		
	25	242	180	106	47	25	12		
	26	238	176	102	43	23			
	28	229	167	93	37	20			
	30	220	158	83	32	17			
	31	216	152	78	30	16			
	32	211	148	73	29				
	34	201	137	65	25				
	36	192	127	58	23				
	37	186	120	55	21				
	38	181	115	52					
	40	171	104	47					
Properties									
Area A (in.^{2})		14.6	11.9	8.40	5.58	4.30	3.17	2.68	2.23
I (in.^{4})		279	161	72.5	28.1	15.2	7.23	4.79	3.02
r (in.)		4.38	3.67	2.94	2.25	1.88	1.51	1.34	1.16
B Bending factor		0.333	0.398	0.500	0.657	0.789	0.987	1.12	1.29
* a		41.7	23.9	10.8	4.21	2.26	1.08	0.717	0.447

* Tabulated values of a must be multiplied by 10^6.
Note: Heavy line indicates Kl/r of 200.

[a] Reprinted from the *Manual of Steel Construction,* 8th ed., with permission of the publishers, American Institute of Steel Construction.

designate the pipe sizes. True outside diameters are slightly larger and can be found from the tables in Appendix A. The AISC Manual (Ref. 1) contains additional tables that list allowable loads for the two heavier weight groups of steel pipe: extra strong and double-extra strong.

Example. Using Table 11-3, select a steel pipe column to carry a load of 41 kips [182 kN] if the unbraced height is 12 ft [3.58 m]. Verify the value in the table by computing the allowable axial load.

Solution: Entering Table 11-3 with an effective length of 12 ft, we find that a load of 43 kips can be supported by a 4-in. column. From Appendix Table A-1 we find that this section has $A = 3.17$ in.^{2} [2045 mm^2] and $r = 1.51$ in. [38.35 mm]. The slenderness ratio, with $K = 1$, is

$$\frac{KL}{r} = \frac{1 \times 12 \times 12}{1.51} = 95.4 \quad \text{say } 95$$

$$\left[\frac{KL}{r} = \frac{3658}{38.35} = 95.4\right]$$

Referring to Table 11-1, we find that the allowable unit stress F_a for a slenderness ratio of 95 is 13.60 ksi [93.8 MPa]. Thus the allowable axial load is

$$P = A \times F_a = 3.17 \times 13.60 = 43.1 \text{ kips}$$

$$\left[P = \frac{2045 \times 93.8}{10^3} = 192 \text{ kN}\right]$$

which agrees with the table value of 43 kips.

11-9. Structural Tubing Columns

Steel columns are fabricated from structural tubing in both square and rectangular shapes. Square tubing is available in sizes of 2 to 16 in. and rectangular sizes range from 3 × 2 to 20 × 12 in. Sections are produced with various wall thicknesses, thus allowing a considerable range of structural capacities. Although round

TABLE 11-4. Allowable Column Loads in Kips for Square Structural Steel Tubing with F_y = 46 ksi[a]

Nominal Size		4 x 4					3 x 3		
Thickness		1/2	3/8	5/16	1/4	3/16	5/16	1/4	3/16
Wt./ft.		21.63	17.27	14.83	12.21	9.42	10.58	8.81	6.87
F_y		46 ksi							
Effective length in feet KL with respect to radius of gyration	0	176	140	120	99	76	86	71	56
	2	168	134	115	95	73	80	67	53
	3	162	130	112	92	71	77	64	50
	4	156	126	108	89	69	73	61	48
	5	150	121	104	86	67	68	57	45
	6	143	115	100	83	64	63	53	42
	7	135	110	95	79	61	57	49	39
	8	126	103	90	75	58	51	44	35
	9	117	97	84	70	55	44	38	31
	10	108	89	78	65	51	37	33	27
	11	98	82	72	60	47	31	27	22
	12	87	74	65	55	43	26	23	19
	13	75	65	58	49	39	22	19	16
	14	65	57	51	43	35	19	17	14
	15	57	49	44	38	30	16	15	12
	16	50	43	39	33	27	14	13	11
	17	44	38	34	29	24	13	11	9
	18	39	34	31	26	21		10	8
	19	35	31	28	24	19			
	20	32	28	25	21	17			
	21	29	25	23	19	16			
	22	26	23	21	18	14			
	23	24	21	19	16	13			
	24		19	17	15	12			
	25				14	11			
Properties									
A (in.^{2})		6.36	5.08	4.36	3.59	2.77	3.11	2.59	2.02
I (in.^{4})		12.3	10.7	9.58	8.22	6.59	3.58	3.16	2.60
r (in.)		1.39	1.45	1.48	1.51	1.54	1.07	1.10	1.13
B Bending factor		1.04	0.949	0.910	0.874	0.840	1.30	1.23	1.17
*a		1.83	1.59	1.43	1.22	0.983	0.533	0.470	0.387

* Tabulated values of a must be multiplied by 10^6.

Note: Heavy line indicates Kl/r of 200.

[a] Reprinted from the *Manual of Steel Construction,* 8th ed., with permission of the publishers, American Institute of Steel Construction.

pipe is specified by a nominal outside dimension, tubing is specified by its actual outside dimensions.

The AISC Manual (Ref. 1) contains safe load tables for square and rectangular tubing based on $F_y = 46$ ksi [317 MPa]. Table 11-4 is a reproduction of one of these tables—for 3 and 4 in. square tubing—and is presented to illustrate the form of the tables.

Example. Using the data given for the example on steel pipe columns in Art. 11-8 (P = 41 kips and height = 12 ft), select a square structural tubing column from Table 11-4. Compare the two solutions.

Solution: Entering Table 11-4 with an effective length of 12 ft, we find that a load of 43 kips can be supported by a square section 4 × 4 × 3/16. With data from Appendix Table A-1, the properties of the two columns are compared in Table 11-5.

Both pipe and tubing may be available in various steel strengths. We have used the properties in these examples because they appear in the AISC Manual. The choice between round pipe or rectangular tubing for a column is usually made for reasons other than simple structural efficiency. Free-standing columns are often round but when built into wall construction the rectangular shapes are often preferred.

Problem 11.9.A. A structural tubing column TS 4 × 4 × 3/8 of steel with $F_y = 46$ ksi [317 MPa] is used on an effective length of 12 ft [3.66 m]. Compute the allowable axial load and compare it with the value in Table 11-4.

Problem 11.9.B.* Refer to Table 11-4 and select the lightest weight square tubing column with an effective length of 10 ft [3.05 m] that will support an axial load of 64 kips [285 kN].

TABLE 11-5. Comparison of the Round Pipe and Square Tube

Properties	Pipe 4-in. std	TS 4 × 4 × 0.1875
Area in in.^{2}	3.17	2.77
Weight in lb/ft	10.79	9.42
Outside dimension in in.	4.5	4.0
Axial load capacity, 12-ft length in kips	43	43

TABLE 11-6. Allowable Axial Compression for Double Angle Struts[a]

Size (in.)			8 × 6			6 × 4				5 × 3½			5 × 3		
Thickness (in.)			$\frac{3}{4}$	$\frac{1}{2}$		$\frac{5}{8}$	$\frac{1}{2}$	$\frac{3}{8}$		$\frac{1}{2}$	$\frac{3}{8}$		$\frac{1}{2}$	$\frac{3}{8}$	$\frac{5}{16}$
Weight (lb/ft)			67.6	46.0		40.0	32.4	24.6		27.2	20.8		25.6	19.6	16.4
Effective length (KL) with respect to indicated axis	X–X axis	0	430	266	0	253	205	142	0	173	129	0	162	121	94
		10	370	231	8	214	174	122	4	159	119	4	149	112	88
		12	353	222	10	200	163	115	6	150	113	6	141	106	83
		14	334	211	12	185	151	107	8	139	105	8	130	98	77
		16	315	200	14	168	137	99	10	126	96	10	119	90	71
		20	271	175	16	150	123	89	12	113	86	12	106	81	64
		24	222	148	20	110	90	69	14	97	75	14	92	70	57
		28	168	117	24	76	62	48	16	81	63	16	76	59	49
		32	129	90	28	56	46	36	20	52	40	20	49	38	32
		36	102	71											
	Y–Y axis	0	430	266	0	253	205	142	0	173	129	0	162	121	94
		10	368	229	6	222	179	125	4	158	118	4	145	108	85
		12	351	219	8	207	167	117	6	148	110	6	132	99	78
		14	332	207	10	190	153	108	8	136	101	8	118	88	69
		16	311	195	12	171	137	97	10	122	91	10	101	75	60
		20	266	169	14	151	120	86	12	107	79	12	82	61	49
		24	216	139	16	129	102	74	14	90	67	14	62	46	38
		28	162	106	20	85	66	49	16	72	53	16	47	35	29
		32	124	81	24	59	46	34	20	46	34	20	30	22	19
		36	98	64											

[a] Loads in kips; yield stress of 36 ksi [250 MPa]; long legs back-to-back with $\frac{3}{8}$ in. separation. Adapted from data in the *Manual of Steel Construction,* 8th ed. (Ref. 1), with permission of the publishers, American Institute of Steel Construction.

11-10. Double-Angle Struts

Two angle sections separated by the thickness of a connection plate at each end and fastened together at intervals by fillers and welds or bolts are commonly used as compression members in roof trusses. (See Fig. 11-2*e*.) These members, whether or not in a vertical position, are called struts; their size is determined in accordance with the requirements and formulas for columns in Art. 10-5. To ensure that the angles act as a unit the intermittent connections are made at intervals such that the slenderness ratio l/r of either angle between fasteners does not exceed the governing slenderness ratio of the built-up member. The least radius of

TABLE 11-6 (*Continued*)

	4 × 3				3½ × 2½				3 × 2				2½ × 2		
	½	⅜	5/16		⅜	5/16	¼		⅜	5/16	¼		⅜	5/16	¼
	22.2	17.0	14.4		14.4	12.2	9.8		11.8	10.0	8.2		10.6	9.0	7.2
0	140	107	90	0	91	77	60	0	75	63	51	0	67	57	46
2	134	103	86	2	86	73	57	2	70	59	48	2	61	52	42
4	126	96	81	4	80	67	53	3	67	57	46	3	58	49	40
6	115	88	74	6	71	60	48	4	63	54	44	4	53	45	37
8	102	78	66	8	61	52	41	6	55	46	38	5	48	41	34
10	88	67	57	10	50	42	34	8	44	38	31	6	42	36	30
12	71	55	47	12	37	31	26	10	32	27	23	8	30	26	21
14	54	42	36	14	27	23	19	12	22	19	16	10	19	16	14
16	41	32	27	16	21	18	15	14	16	14	12	12	13	11	9
18	33	25	22	18	16	14	12								
20	26	20	17												
0	140	107	90	0	91	77	60	0	75	63	51	0	67	57	46
2	135	103	86	2	87	73	57	2	70	59	48	2	63	53	43
4	127	97	81	4	80	67	53	3	67	56	46	3	60	51	41
6	117	89	74	6	72	60	47	4	63	53	43	4	57	48	39
8	105	80	67	8	62	52	41	6	54	45	36	6	49	41	33
10	92	70	58	10	50	42	33	8	43	36	28	8	40	34	27
12	77	58	48	12	37	31	25	10	30	25	20	10	30	24	19
14	61	45	37	14	28	23	18	12	21	17	14	12	21	17	13
16	47	35	29	16	21	17	14	14	15	13	10	14	15	12	10
18	37	27	23	18	17	14	11								
20	30	22	18												

gyration r is used in computing the slenderness ratio of each angle.

The AISC Manual contains safe load tables for struts of two angles with ⅜-in. separation back-to-back. Three series are given: equal-leg angles, unequal-leg angles with short legs back-to-back, and unequal-leg angles with long legs back-to-back. Table 11-6 has been abstracted from the latter series and lists allowable loads with respect to the X–X and Y–Y axes. The smaller (least) radius of gyration gives the smaller allowable load and, unless the member is braced with respect to the weaker axis, this is the tabular load to be used. The usual practice is to assume K equal to 1.0. The following example shows how the loads in the table are computed.

Example. Two $5 \times 3\frac{1}{2} \times \frac{1}{2}$ angle sections spaced with their long legs $\frac{3}{8}$ in. back-to-back are used as a compression member. If the member is A36 steel and has an effective length of 10 ft, compute the allowable axial load.

Solution: From Appendix A we find that the area of the two-angle member is 8.0 in. and that the radii of gyration are $r_x = 1.58$ in. and $r_y = 1.49$ in. Using the smaller r, the slenderness ratio is

$$\frac{Kl}{r} = 1 \times \frac{10 \times 12}{1.49} = 80.5 \quad \text{say } 81$$

Referring to Table 11-1, we find that $F_a = 15.24$ ksi, making the allowable load

$$P = A \times F_a = 8.0 \times 15.24 = 121.9 \text{ kips}$$

This value is, of course, readily verified by entering Table 11-6 under "*Y–Y Axis*" with an effective length of 10 ft and then proceeding horizontally to the column of loads for the $5 \times 3\frac{1}{2} \times \frac{1}{2}$ angle.

The design of double-angle members for the compression elements in trusses is considered in Chapter 14.

When designing double angles or structural tees as compression members without the help of safe load tables consideration must be given to the possibility that it may be necessary to reduce the allowable stress when these members have thin parts. This condition is indicated by the presence of a value for Q_s in the tabulated properties in Appendix A. When a value is given for Q_s the safe axial load, as calculated normally, must be multiplied by this value for the true allowable load. Load values given in the safe load tables in the AISC Manual have incorporated this requirement.

Problem 11.10.A.* A double-angle compression member 8 ft [2.44 m] long is composed of two angles 4 × 3 × 3/8 in. with the long legs 3/8 in. back-to-back. If the member is fabricated from A36 steel, compute the allowable concentric load.

Problem 11.10.B. Using Table 11-6, select a double-angle compression member that will support an axial load of 50 kips [222 kN] if the effective length is 10 ft [3.05 m].

11-11. Eccentrically Loaded Columns

The columns previously discussed have been axially or concentrically loaded. It frequently happens, however, that in addition to the axial load the column may be subjected to bending stresses that result from eccentric loads. Figure 11-5 shows a column with concentric and eccentric loads. The design of eccentrically loaded columns is accomplished by testing trial sections. As an aid to design it is convenient to convert the axial and eccentric loads into a single equivalent axial load. Having done this, the safe load tables may be used to select the trial section.

The *bending factors* B_x and B_y are listed on the right-hand side of Table 11-2. The bending factor is the area of the cross section divided by its section modulus. Because there are two major section moduli, there are two bending factors: B_x for the X–X axis

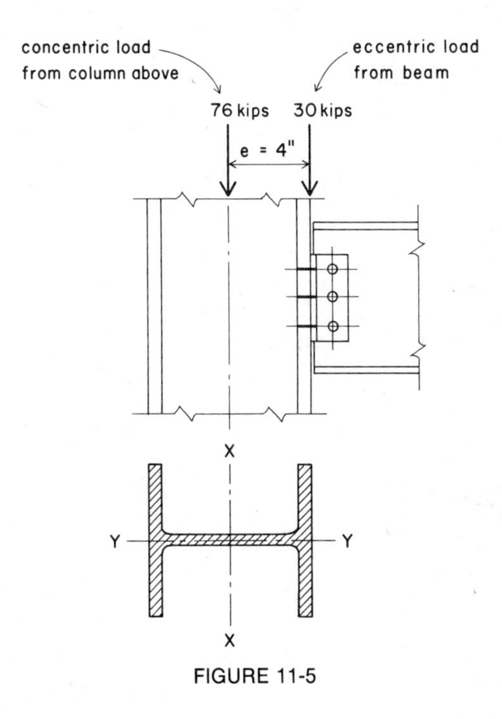

FIGURE 11-5

and B_y for the Y–Y axis; for example, the area of a W 10 × 49 is 14.4 sq in.; the section moduli with respect to the X–X and Y–Y axes are 54.6 and 18.7 in.^{3}, respectively. Then

$$B_x = \frac{A}{S_x} = \frac{14.4}{54.6} = 0.264$$

and

$$B_y = \frac{A}{S_y} = \frac{14.4}{18.7} = 0.770$$

Note that these are the values given in Table 11-2.

Bending factors are used to convert the effect of eccentricity to an equivalent axial load. To accomplish this the *bending moment* resulting from the eccentric load is multiplied by the appropriate bending factor. Then *the total equivalent axial load* (P') *is equal to the sum of the axial load and the eccentric load plus the product of the bending moment due to the eccentric load and the appropriate bending factor.*

The trial section used in designing a column subjected to both axial and eccentric loads may be established by first finding an *approximate* equivalent axial load. This procedure is illustrated in the following example:

Example. An 8-in. wide flange column of A36 steel with an unsupported height of 13 ft supports an axial load of 76 kips and a load of 30 kips applied 4 in. from the X–X axis. The arrangement is shown in Fig. 11-5. Determine the trial column section.

Solution: (1) The bending moment produced by the eccentric load is

$$M = 30 \times 4 = 120 \text{ kip-in.}$$

However, only the general dimensions of the section are known at this point; therefore we do not know the exact value of the bending factor.

(2) Referring to the 8-in. column sections in Table 11-2, tentatively select a bending factor of 0.333. This may be revised later. Then the bending moment multiplied by the bending factor is 120 × 0.333 = 40 kips, which is an equivalent axial load for the eccentric load.

(3) Now, in accordance with the principle stated in Art. 10-12, the approximate total equivalent load on the column is

$$P' = 76 + 30 + 40 = 146 \text{ kips}$$

Referring again to Table 11-2, we find that a W 8 × 35 will carry 162 kips with an effective length of 13 ft and a W 8 × 31 is listed for 143 kips.

It is necessary, of course, to verify the accuracy of our estimate for the bending factor. We assume a value of 0.333 and note from the tables that the W 8 × 35 has $B_x = 0.330$ and the W 8 × 31 has $B_x = 0.332$. These are sufficiently close to our estimate to make more work unnecessary. Had we been off by more than a few percent it would have been necessary to compute a new value for P' from the true B_x values to verify our selections.

Because the foregoing procedure gives results that are approximate on the safe side, the W 8 × 35 could be the accepted section. However, if it is desired to determine the lightest weight section that can be used, the W 8 × 31 should be investigated more precisely for compliance with the AISC Specification requirements for the design of columns with combined loading. This is not a simple procedure, and the diversity of factors involved makes it inadvisable to include treatment of these complex requirements in a book of this scope. Reference to the AISC Manual is recommended for readers who wish to study the complete specification requirements that cover combined axial compression and bending in columns.

The selection of a trial section by the equivalent axial load method is always conservative—increasingly so as the ratio of the eccentric load to the axial load and the column slenderness ratio increase. Nevertheless, for many situations that arise in routine practice the trial section determined by this method may be taken as the accepted section.

In conventional building construction it is assumed that the effect of an eccentric load disappears at each story height. Consequently in the above example the design load that the W 8 × 35 transmits to the column in the story below or to a base plate is not 146 kips but 106 kips, plus the column weight. In this example only one eccentric load was given. If, in addition, there is an eccentric load about the $Y–Y$ axis, its equivalent axial load plus the magnitude of the load is added to the 146 kips to determine the total approximate equivalent axial load on the column.

11-12. Column Splices

In the construction of steel frames of buildings it is common practice to use columns of two-story lengths. This results in a greater cross-sectional area in the upper story than the load requires, but the cost of the excess material is offset by the saving in fabricating costs for the extra splice. When columns of two-story lengths are used the load on the lower story length determines the required cross section of the column.

In order not to conflict with the beam and girder connections, column splices are made 2 ft or more above the floor level. In general, splices are made with plates $\frac{3}{8}$ or $\frac{1}{2}$ thick that are bolted, welded, or riveted to the flanges of the columns, as indicated in Fig. 11-6 and Fig. 13-9. The splice plates are not designed to resist compressive stresses; their function is to hold the column sections in position. Because the upper column transmits its load directly to the column below, the surfaces in contact should be milled to provide full bearing areas. When the upper column has a smaller width than the supporting column filler plates are used (Fig. 11-6*b*). If the difference in width is so great that a full bear-

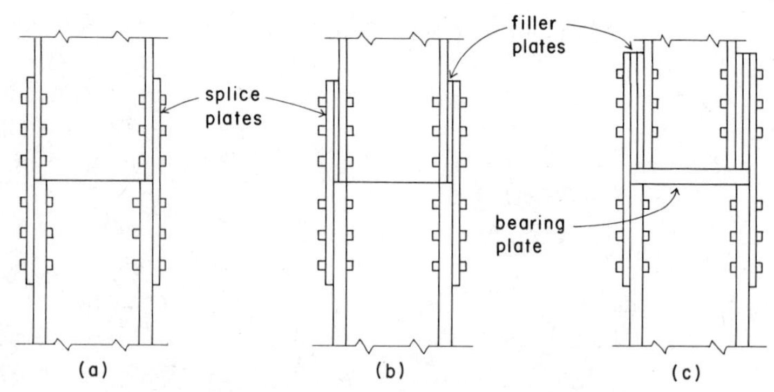

FIGURE 11-6. Typical bolted column splices.

ing area between the columns is not achieved, a horizontal plate is used as in Fig. 11-6*c*.

11-13. Column Base Plates

Steel columns generally transfer their loads to the supporting ground by means of reinforced concrete footings. As the allowable compressive strength of concrete is considerably less than the actual unit stresses in the steel column, it is necessary to provide a steel base plate under the column to spread the load sufficiently to prevent over stressing of the concrete. The typical arrangement is shown in Fig. 11-7.

Rolled steel bearing plates used for column bases may be obtained in a great variety of sizes. The lengths and widths have dimensions usually in even inches, and for the sizes commonly used in building construction the plates may be obtained in $\frac{1}{8}$-in. increments of thickness. To distribute the column load uniformly over the base plate it is important that the column and plate be in absolute contact. Rolled steel bearing plates more than 2 in. thick but not more than 4 in. thick may be straightened by pressing or planing. Material more than 4 in. thick must be planed on the

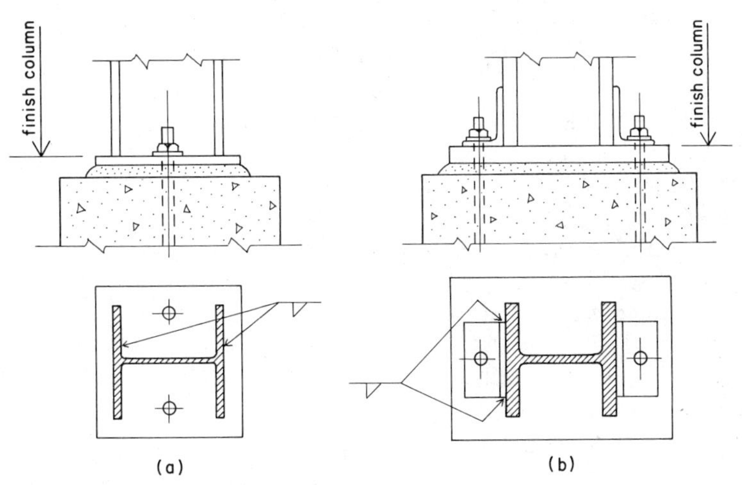

FIGURE 11-7. Typical column base details.

upper surface. The under surface need not be planed because a full bearing contact is obtained by using a layer of cement grout on the concrete surface.

Steel columns are usually secured to the foundation by steel bolts embedded in the concrete that pass through the base plate and are secured to angles bolted or welded to the column flange. See Fig. 11-7*b*. For light columns the angles are often omitted and the base plate is secured to the column by fillet welds (Fig. 11-7*a*).

The first step in the design of a base plate is to determine its area. This is accomplished by use of the basic formula

$$A = \frac{P}{F_p}$$

where $A = B \times N =$ the area of the base plate in sq in. (see Fig. 11-8*a*),

$P =$ the column load in pounds,

$F_p =$ the allowable bearing value of the concrete in psi; the AISC Specification gives this stress as $0.25\, f'_c$ when the entire area of the concrete support is covered and $0.375\, f'_c$ when only one third of the area is covered; a concrete commonly used has 3000 psi for the value of f'_c; hence $F_p = 750$ or 1125 psi.

The column load is assumed to be uniformly distributed over the dotted rectangle shown in Fig. 11-8*a*. In addition, the base plate is assumed to exert a uniform pressure on the concrete foundation.

After the minimum required area of the base plate has been found B and N are established so that the projections m and n are approximately equal.

The final step is to determine m and n in inches and to use the larger value in computing the thickness of the plate according to the formula

$$t = \sqrt{\frac{3f_p m^2}{F_b}} \quad \text{or} \quad t = \sqrt{\frac{3f_p n^2}{F_b}}$$

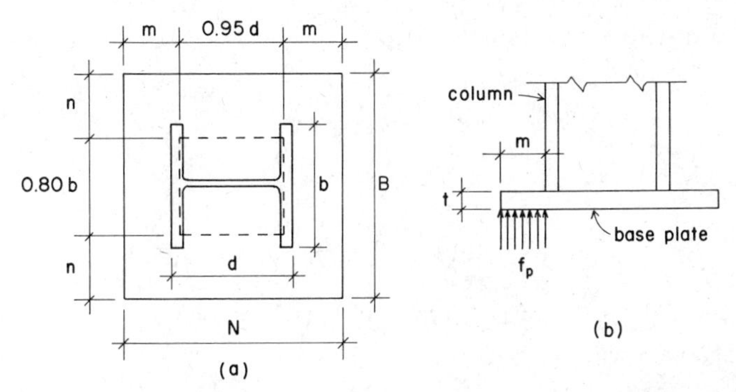

FIGURE 11-8. Reference dimensions for design of column base plates.

where t = the thickness of the bearing plate, in inches,
f_p = the *actual* bearing pressure on the foundation, in psi,
F_b = the allowable bending stress in the base plate, in psi; the AISC Specification gives the value of F_b as 0.75 F_y, F_y being the yield point of the steel plate; thus for A36 steel F_y = 36,000 psi and $F_b = 0.75 \times 36{,}000$, or F_b = 27,000 psi.

The foregoing formulas are readily derived. Refer to Fig. 11-8*b* and consider a strip of base plate 1 in. wide and t in. thick projecting m in. beyond the face of the column flange. Because the upward pressure on the steel strip is f_p pounds per sq in., the strip is actually a cantilever bending upward. Hence the bending moment at the face of the column is

$$M = f_p m \times \frac{m}{2} = \frac{f_p m^2}{2}$$

For this rectangular strip

$$\frac{I}{c} = \frac{bd^2}{6} \qquad \text{(Art. 6-6)}$$

and because b = 1 in. and $d = t$ in.

$$\frac{I}{c} = \frac{1 \times t^2}{6} = \frac{t^2}{6}$$

As the strip is in flexure,

$$\frac{M}{f} = \frac{I}{c} \qquad \text{(Art. 7-1)}$$

and if $f = F_b$

$$\frac{f_p m^2}{2} \times \frac{1}{F_b} = \frac{t^2}{6} \quad \text{and} \quad t^2 = \frac{6 f_p m^2}{2F_b} \quad \text{or} \quad t = \sqrt{\frac{3 f_p m^2}{F_b}}$$

If A36 steel is used for the plate, $F_b = 27{,}000$ psi and the two thickness formulas become

$$t = \sqrt{\frac{f_p m^2}{9000}} \quad \text{and} \quad t = \sqrt{\frac{f_p n^2}{9000}}$$

Review Problems

Note: In the following problems assume that the columns and base plates are A36 steel and that K, the effective length factor, is 1.0.

Problem 11.13A. A W 10 × 60 is used as a column with an effective length of 16 ft [4.8 m]. What safe load will it support?

Problem 11.13B. A W 6 × 25 is used as a column with an effective length of 12.5 ft [3.8 m]. Compute its allowable concentric load.

Problem 11.13C. The effective length of a W 8 × 40 used as a column is 22 ft [6.6 m]. Compute its allowable axial load.

Problem 11.13D.* What is the lightest weight wide-flange section that can be used to support an axial compression load of 250 kips [1112 kN] if the effective length is 18 ft [5.4 m]?

Problem 11.13E. An S 10 × 25.4 is used as a column with an effective length of 8 ft [2.4 m]. Compute its allowable axial load.

Problem 11.13F. A W 14 × 283 has 20 × 1½ in. [500 × 38 mm] plates welded to its flanges. Determine the allowable axial load when the combined section is used as a column with an effective length of 16 ft [4.8 m].

Problem 11.13G. Two 12 × ½ in. [300 × 13 mm] are welded to the flanges of an S 12 × 31.8 to constitute a built-up member for use as a column. If its effective length is 13 ft [3.9 m], compute the allowable concentric load.

Problem 11.13H.* A pair of angles, 6 × 4 × ½, is placed with the long legs back-to-back and separated by ⅜ in. for use as a compression member. If the effective length is 12 ft [3.6 m], compute the allowable axial load.

Problem 11.13I. An 8-in. column of A36 steel has an axial load of 90 kips [400 kN] and a load of 30 kips [133 kN] applied 4 in. [100 mm] from the *X–X* axis. The arrangement is similar to that shown in Fig. 11-5. The unsupported length of the column is 13 ft [3.9 m] and the column end conditions are shown in Fig. 11-3. Is a W 8 × 35 acceptable for this loading?

Problem 11.13J.* Design a column base plate for a W 8 × 31 column that is supported on concrete for which the allowable bearing capacity is 750 psi [5000 kPa]. The load on the column is 178 kips [792 kN].

Problem 11.13K. Design a column base plate for the W 8 × 31 in Problem 11.13J if the bearing pressure allowed on the concrete is 1125 psi [7500 kPa].

12

Bolted Connections

12-1. General

Elements of structural steel are often connected by mating flat parts with common holes and inserting a pin-type device to hold them together. In times past the pin device was a rivet; today it is usually a bolt. A great number of types and sizes of bolt are available, as are many connections in which they are used. The material in this chapter deals with a few of the common bolting methods used in building structures.

12-2. Bearing-type Shear Connections

The diagrams in Fig. 12-1 show a simple connection between two steel bars that functions to transfer a tension force from one bar to another. Although this is a tension-transfer connection, it is also referred to as a shear connection because of the manner in which the connecting device (the bolt) works in the connection. (See Fig. 12-1*b*.) If the bolt tension (due to tightening of the nut) is relatively low, the bolt serves primarily as a pin in the matched holes, bearing against the sides of the holes as shown at Fig. 12-

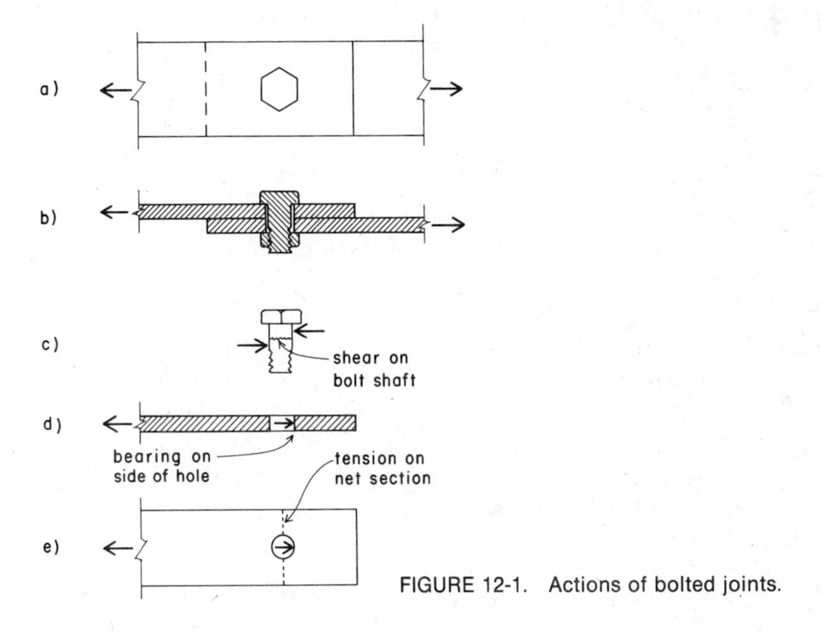

FIGURE 12-1. Actions of bolted joints.

1*c*. In addition to these functions, the bars develop tension stress that will be a maximum at the section through the bolt holes.

In the connection shown in Fig. 12-1 the failure of the bolt involves a slicing (shear) failure that is developed as a shear stress on the bolt cross section. The resistance of the bolt can be expressed as an allowable shear stress F_v times the area of the bolt cross section, or

$$R = F_v \times A$$

With the size of the bolt and the grade of steel known it is a simple matter to establish this limit. In some types of connections it may be necessary to slice the same bolt more than once to separate the connected parts. This is the case in the connection shown in Fig. 12-2, in which it may be observed that the bolt must be sliced twice to make the joint fail. When the bolt develops shear on only one section (Fig. 12-1) it is said to be in *single shear*; when it develops shear on two sections (Fig. 12-2) it is said to be in *double shear*.

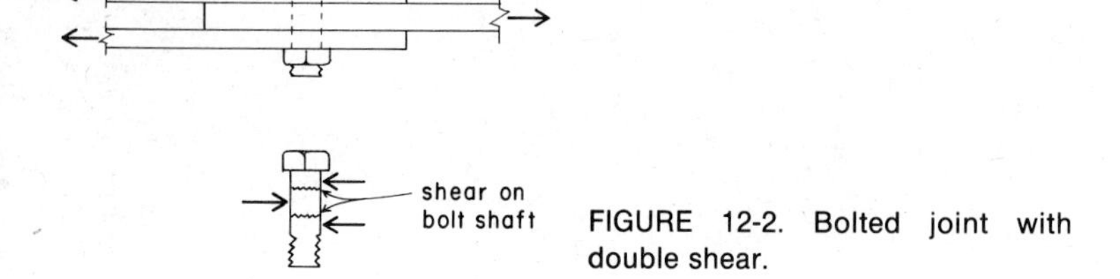

FIGURE 12-2. Bolted joint with double shear.

When the bolt diameter is large or the bolt is made of strong steel the connected parts must be sufficiently thick if they are to develop the full capacity of the bolts. The maximum bearing stress permitted for this situation by the AISC Specification is $F_p = 1.5F_u$, and F_u is the ultimate tensile strength of the steel in the part in which the hole occurs.

12-3. Friction-type Shear Connections

When the nut on a bolt is sufficiently secure the parts being bolted together will be squeezed so tightly that the connection will develop a friction stress between the two connected parts. If this friction is of sufficient magnitude, the bearing action described in the preceding article will not occur because the parts must slip somewhat to permit the bearing stress to develop. This type of joint is preferred in some cases when joint tightness or potential loosening is a problem.

High strength bolts are those that function primarily in the friction-resistive manner just described. They are tightened to a carefully controlled degree so that a determined magnitude of the clamping effect for the development of friction can be considered reliable. If the friction resistance is depended on entirely, the bolt is called a friction-type fastener. If friction is used but the ultimate resistance is in bearing, it is called a bearing-type fastener. For the bearing-type fastener there are two different specifications for loading; one when the bolt threads are excluded from the shear failure planes, the other when they are not.

12-4. Tension Connections

When tension members have reduced cross sections two stress investigations must be considered. This is the case for members with holes for bolts or for bolts or rods with cut threads. For the member with a hole (Fig. 12-1*d*) the allowable tension stress at the reduced cross section through the hole is $0.50F_u$, where F_u is the ultimate tensile strength of the steel. The total resistance at this reduced section (also called the net section) must be compared with the resistance at other, unreduced, sections at which the allowable stress is $0.60\ F_y$.

For threaded steel rods the maximum allowable tension stress at the threads is $0.33\ F_u$. For steel bolts the allowable stress is specified as a value based on the type of bolt. The load capacity of various types and sizes of bolt is given in Table 12-1.

When tension elements consist of W, M, S, and tee shapes the tension connection is usually not made in a manner that results in the attachment of all the parts of the section (e.g., both flanges plus the web for a W). In such cases the AISC Specification requires the determination of a reduced effective net area, A_e, that consists of

$$A_e = C_t A_n$$

where A_n = the actual net area of the member,
C_t = a reduction coefficient.

Unless a larger coefficient can be justified by tests, the following values are specified:

1. For W, M, or S shapes with flange widths not less than $\frac{2}{3}$ the depth and structural tees cut from such shapes, when the connection is to the flanges and has at least three fasteners per line in the direction of stress $-C_t = 0.90$.
2. For W, M, or S shapes not meeting the above conditions and for tees cut from such shapes, provided the connection has not fewer than three fasteners per line in the direction of stress $-C_t = 0.85$.
3. All members with connections that have only two fasteners per line in the direction of stress $-C_t = 0.75$.

Angles used as tension members are often connected by only one leg. In a conservative design the effective net area is only that of the connected leg, less the reduction caused by bolt holes. Rivet and bolt holes are punched larger in diameter than the nominal diameter of the fastener. The punching damages a small amount of the steel around the perimeter of the hole; consequently the diameter of the hole to be deducted in determining the net section is $\frac{1}{16}$ in. greater than the nominal diameter of the rivet.

When only one hole is involved, as in Fig. 12-1, or in a similar connection with a single row of fasteners along the line of stress the net area of the cross section of one of the plates is found by multiplying the plate thickness by its net width (width of member minus diameter of hole).

When holes are staggered in two rows along the line of stress (Fig. 12-3) the net section is determined somewhat differently. The AISC Specification reads:

> In the case of a chain of holes extending across a part in any diagonal or zigzag line, the net width of the part shall be obtained by deducting from the gross width the sum of the diameters of all the holes in the chain and adding, for each gage space in the chain, the quantity $s^2/4g$, where
>
> s = longitudinal spacing (pitch) in inches of any two successive holes
>
> and
>
> g = transverse spacing (gage) in inches for the same two holes
>
> The critical net section of the part is obtained from that chain which gives the least net width.

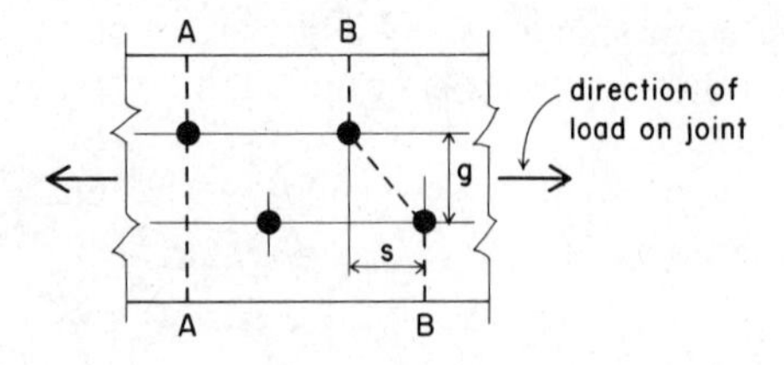

FIGURE 12-3

The AISC Specification also provides that in no case shall the net section through a hole be considered as more than 85% of the corresponding gross section.

12-5. Structural Bolts

Bolts used for the connection of structural steel members come in two types. Bolts designated A307 and called *unfinished* have the lowest load capacity of the structural bolts. The nuts for these bolts are tightened just enough to secure a snug fit of the attached parts; because of this, plus the oversizing of the holes, there is some movement in the development of full resistance. These bolts are generally not used for major connections, especially when joint movement or loosening under vibration or repeated loading may be a problem.

Bolts designated A325 or A490 are called *high-strength bolts*. The nuts of these bolts are tightened to produce a considerable tension force which results in a high degree of friction resistance between the attached parts. High-strength bolts are further designated as F, N, or X. The F designation denotes bolts for which the limiting resistance is that of friction. The N designation denotes bolts that function ultimately in bearing and shear but for which the threads are not excluded from the bolt shear planes. The X designation denotes bolts that function like the N bolts but for which the threads are excluded from the shear planes.

When bolts are loaded in tension their capacities are based on the development of the ultimate resistance in tension stress at the reduced section through the threads. When loaded in shear bolt capacities are based on the development of shear stress in the bolt shaft. The shear capacity of a single bolt is further designated as *S* for single shear (Fig. 12-1) or *D* for double shear (Fig. 12-2). The capacities of structural bolts in both tension and shear are given in Table 12-1. The size range given in the table—$\frac{5}{8}$ in. to $1\frac{1}{2}$ in.—is that listed in the AISC Manual (Ref. 1). However, the most commonly used sizes for structural steel framing are $\frac{3}{4}$ and $\frac{7}{8}$ in.

Bolts are ordinarily installed with a washer under both head and nut. Some manufactured high-strength bolts have specially formed heads or nuts that in effect have self-forming washers,

TABLE 12-1. Capacity of Structural Bolts in Kips[a]

ASTM designation	Connection type[b]	Loading condition[c]	5/8	3/4	7/8	1	1-1/8	1-1/4	1-3/8	1-1/2
			Area, based on nominal diameter (in.^{2})							
			0.3068	0.4418	0.6013	0.7854	0.9940	1.227	1.485	1.767
A307		S	3.1	4.4	6.0	7.9	9.9	12.3	14.8	17.7
		D	6.1	8.8	12.0	15.7	19.9	24.5	29.7	35.3
		T	6.1	8.8	12.0	15.7	19.9	24.5	29.7	35.3
A325	F	S	5.4	7.7	10.5	13.7	17.4	21.5	26.0	30.9
		D	10.7	15.5	21.0	27.5	34.8	42.9	52.0	61.8
	N	S	6.4	9.3	12.6	16.5	20.9	25.8	31.2	37.1
		D	12.9	18.6	25.3	33.0	41.7	51.5	62.4	74.2
	X	S	9.2	13.3	18.0	23.6	29.8	36.8	44.5	53.0
		D	18.4	26.5	36.1	47.1	59.6	73.6	89.1	106.0
	All	T	13.5	19.4	26.5	34.6	43.7	54.0	65.3	77.7
A490	F	S	6.7	9.7	13.2	17.3	21.9	27.0	32.7	38.9
		D	13.5	19.4	26.5	34.6	43.7	54.0	65.3	77.7
	N	S	8.6	12.4	16.8	22.0	27.8	34.4	41.6	49.5
		D	17.2	24.7	33.7	44.0	55.7	68.7	83.2	99.0
	X	S	12.3	17.7	24.1	31.4	39.8	49.1	59.4	70.7
		D	24.5	35.3	48.1	62.8	79.5	98.2	119.0	141.0
	All	T	16.6	23.9	32.5	42.4	53.7	66.3	80.2	95.4

Nominal diameter (in.) applies to the columns 5/8 through 1-1/2.

[a] Reproduced from data in the 8th edition of the *Manual of Steel Construction* (Ref. 1), with permission of the publishers, American Institute of Steel Construction.
[b] F = friction; N = bearing, threads not excluded; X = bearing, threads excluded.
[c] S = single shear; D = double shear; T = tension.

eliminating the need for a separate, loose washer. When a washer is used it is sometimes the limiting dimensional factor in detailing for bolt placement in tight locations, such as close to the fillet (inside radius) of angles or other rolled shapes.

For a given diameter of bolt there is a minimum thickness required for the bolted parts in order to develop the full shear capacity of the bolt. This thickness is based in the bearing stress between the bolt and the side of the hole, which is limited to a maximum of $F_p = 1.5F_u$. The stress limit may be established by either the bolt steel or the steel of the bolted parts.

Steel rods are sometimes threaded for use as anchor bolts or tie rods. When loaded in tension their capacities are usually limited by the stress on the reduced section at the threads. Tie rods are sometimes made with *upset ends,* which consist of larger diameter portions at the ends. When these enlarged ends are threaded the net section at the thread is the same as the gross section in the remainder of the rods; the result is no loss of capacity for the rod.

12-6. Layout of Bolted Connections

Design of bolted connections generally involves a number of considerations in the dimensioned layout of the bolt-hole patterns for the attached structural members. Although we cannot develop all the points necessary for the production of structural steel construction and fabrication details, the material in this article presents basic factors that often must be included in the structural calculations.

Figure 12-4 shows the layout of a bolt pattern with bolts placed in two parallel rows. Two basic dimensions for this layout are limited by the size (nominal diameter) of the bolt. The first is the center-to-center spacing of the bolts, usually called the *pitch.* The AISC Specification limits this dimension to an absolute minimum of $2\frac{2}{3}$ times the bolt diameter. The preferred minimum, however, which is used in this book, is three times the diameter.

The second critical layout dimension is the *edge distance,* which is the distance from the centerline of the bolt to the nearest edge. There is also a specified limit for this as a function of bolt size. This dimension may also be limited by edge tearing, which is discussed in Art. 12-7.

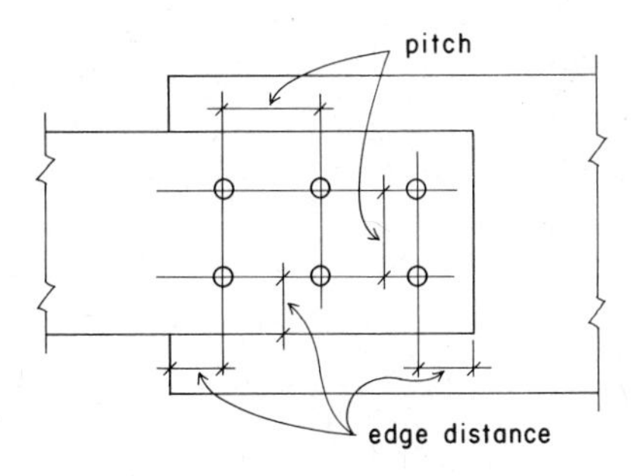

FIGURE 12-4. Pitch and edge distances for bolts.

TABLE 12-2. Pitch and Edge Distances for Bolts[a]

Rivet or bolt diameter d (in.)	Minimum edge distance for punched, reamed, or drilled holes (in.): At sheared edges	At rolled edges of plates, shapes or bars, or gas cut edges[b]	Pitch center-to center (in.): Minimum $2\text{-}\frac{2}{3}d$	Recommended $3d$
$\frac{5}{8}$	1.125	0.875	1.67	1.875
$\frac{3}{4}$	1.25	1	2	2.25
$\frac{7}{8}$	1.5[c]	1.125	2.33	2.625
1	1.75[c]	1.25	2.67	3

[a] Reproduced from data in the 8th edition of the *Manual of Steel Construction* (Ref. 1), with permission of the publishers, American Institute of Steel Construction.
[b] May be reduced $\frac{1}{8}$ in. when the hole is at a point where stress does not exceed 25% of the maximum allowed in the connected element.
[c] May be $1\frac{1}{4}$ in. at the ends of beam connection angles.

Table 12-2 gives the recommended limits for pitch and edge distance for the bolt sizes used in ordinary steel construction.

In some cases bolts are staggered in parallel rows (Fig. 12-5). In this case the diagonal distance, labeled m in the illustration, must also be considered. For staggered bolts the spacing in the direction of the rows is usually referred to as the pitch; the spacing of the rows is called the gage. The reason for staggering the

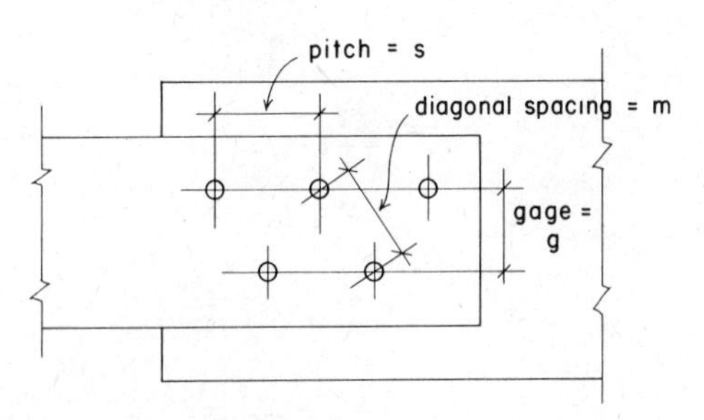

FIGURE 12-5. Standard reference dimensions for layout of bolted joints.

TABLE 12-3. Minimum Pitch to Maintain Three Diameters Center-to-Center of Holes[a]

Diameter of rivet	m	Distance, g (in.)								
		1	$1\frac{1}{4}$	$1\frac{1}{2}$	$1\frac{3}{4}$	2	$2\frac{1}{4}$	$2\frac{1}{2}$	$2\frac{3}{4}$	3
$\frac{5}{8}$	$1\frac{7}{8}$	$1\frac{5}{8}$	$1\frac{3}{8}$	$1\frac{1}{8}$	$\frac{5}{8}$	0				
$\frac{3}{4}$	$2\frac{1}{4}$	2	$1\frac{7}{8}$	$1\frac{5}{8}$	$1\frac{3}{8}$	1	0			
$\frac{7}{8}$	$2\frac{5}{8}$	$2\frac{1}{2}$	$2\frac{3}{8}$	$2\frac{1}{8}$	2	$1\frac{3}{4}$	$1\frac{3}{8}$	$\frac{3}{4}$	0	
1	3	$2\frac{7}{8}$	$2\frac{3}{4}$	$2\frac{5}{8}$	$2\frac{1}{2}$	$2\frac{1}{4}$	2	$1\frac{3}{8}$	$1\frac{1}{8}$	0

[a] Reproduced from data in the 8th edition of the *Manual of Steel Construction* (Ref. 1) with permission of the publishers, American Institute of Steel Construction. (See Fig. 12-5.)

bolts is that sometimes the rows must be spaced closer (gage spacing) than the minimum spacing required for the bolts selected. Table 12-3 gives the pitch required for a given gage spacing to keep the diagonal spacing (m) within the recommended diameter limit.

Location of bolt lines is often related to the size and type of structural members being attached. This is especially true of bolts placed in the legs of angles or in the flanges of W, M, S, C, and structural tee shapes. Figure 12-6 shows the placement of bolts in the legs of angles. When a single row is placed in a leg, its recommended location is at the distance labeled g from the back of the angle. When two rows are used the first row is placed at the distance g_1 and the second row is spaced a distance g_2 from the first. Table 12-4 gives the recommended values for these distances.

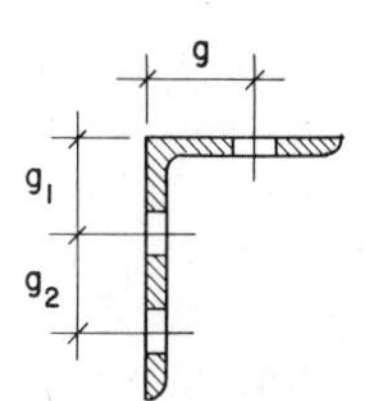

FIGURE 12-6. Gage distances for angles.

TABLE 12-4. Usual Gage Dimensions for Angles (in.)[a]

Gage dimension	Width of angle leg								
	8	7	6	5	4	$3\frac{1}{2}$	3	$2\frac{1}{2}$	2
g	$4\frac{1}{2}$	4	$3\frac{1}{2}$	3	$2\frac{1}{2}$	2	$1\frac{3}{4}$	$1\frac{3}{8}$	$1\frac{1}{8}$
g_1	3	$2\frac{1}{2}$	$2\frac{1}{4}$	2					
g_2	3	3	$2\frac{1}{2}$	$1\frac{3}{4}$					

[a] Reproduced from data in the 8th edition of the *Manual of Steel Construction* (Ref. 1), with permission of the publishers, American Institute of Steel Construction.

When placed at the recommended locations in rolled shapes bolts will end up a certain distance from the edge of the part. Based on the recommended edge distance for rolled edges given in Table 12-2, it is thus possible to determine the maximum size of bolt that can be accommodated. For S and C shapes this maximum fastener size is given in the tables of properties in Appendix A. For angles the maximum fastener may be limited by the edge distance, especially when two rows are used: however, other factors may in some cases be more critical. The distance from the center of the bolts to the inside fillet of the angle may limit the use of a large washer where one is required. Another consideration may be the stress on the net section of the angle, especially if the member load is taken entirely by the attached leg. These problems are given some discussion in the design of the truss in Chapter 14.

Sections 1.16.4 and 1.16.5 of the AISC Specification provide additional criteria for minimum spacing and edge distances for fasteners as a function of the load per fastener and the thickness and ultimate stress capacity of the connected parts. Some of these issues are illustrated in the example problem in Article 12-8.

12-7. Tearing

One possible form of failure in a bolted connection is that of tearing out the edge of one of the attached members. The diagrams in Fig. 12-7 show this potentiality in a connection between two plates. The failure in this case involves a combination of

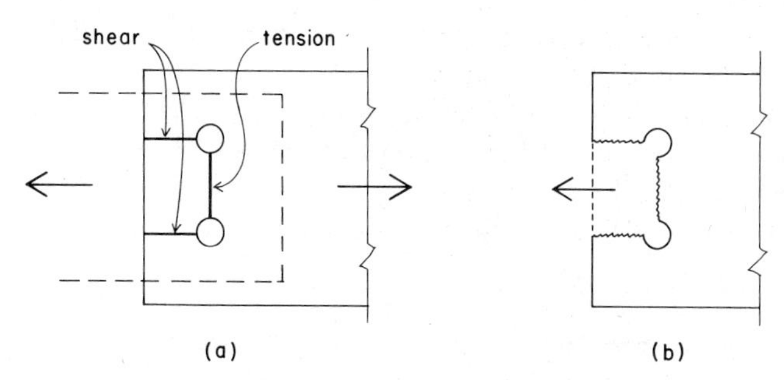

FIGURE 12-7. Tearing in a bolted tension joint.

shear and tension to produce the torn out form shown. The total tearing force is computed as the sum required to cause both forms of failure. The allowable stress on the net tension area is specified as $0.50F_u$, where F_u is the maximum tensile strength of the steel. The allowable stress on the shear areas is specified as $0.30F_u$. With the edge distance, hole spacing, and diameter of the holes known, the net widths for tension and shear are determined and multiplied by the thickness of the part in which the tearing occurs. These areas are then multiplied by the appropriate stresses to find the total tearing force that can be resisted. If this force is greater than the connection design load, the tearing problem is not critical.

Another case of potential tearing is shown in Fig. 12-8. This is the common situation for the end framing of a beam in which

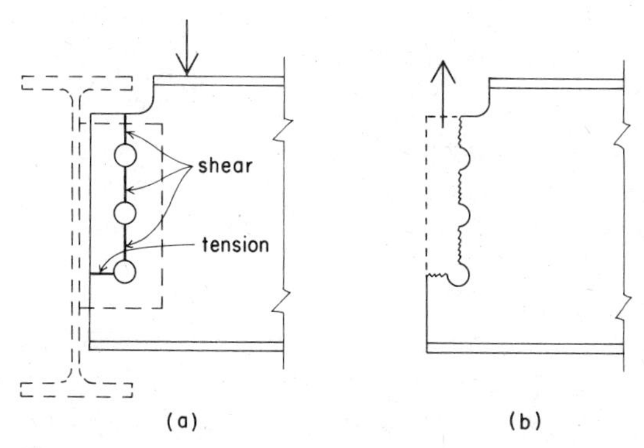

FIGURE 12-8. Tearing in a bolted beam connection.

support is provided by another beam whose top is aligned with that of the supported beam. The end portion of the top flange of the supported beam must be cut back to allow the beam web to extend to the side of the supporting beam. With the use of a bolted connection the tearing condition shown is developed. This problem is discussed further in Art. 12-10.

12-8. Design of a Bolted Tension Connection

The issues raised in several of the preceding articles are illustrated in the following design example. Before proceeding with the problem data, we should consider some of the general requirements for this joint.

If friction-type bolts are used, the surfaces of the connected parts must be cleaned and made reasonably true. If high-strength bolts are used, the determination to exclude threads from the shear failure planes must be established.

The AISC Specification has a number of general requirements for connections:

Need for a minimum of two bolts per connection.

Need for a minimum connection capacity of 6 kips.

Need for the connection to develop at least 50% of the full potential capacity of the member (for trusses only).

Although a part of the design problem may be the selection of the type of fastener or the required strength of steel for the attached parts, we provide this as given data in the example problem.

Example. The connection shown in Fig. 12-9 consists of a pair of narrow plates that transfer a load of 100 kips [445 kN] in tension to a single 10-in. [254-mm] wide plate. The plates are A36 steel with F_u = 58 ksi [400 MPa] and are attached with $\frac{3}{4}$-in. A325F bolts placed in two rows. Determine the number of bolts required, the width and thickness of the narrow plates, the thickness of the wide plate, and the layout of the bolts.

Solution: From Table 12-1 we find the double shear (D) capacity for one bolt is 15.5 kips [69 kN]. The required number of bolts is thus

$$n = \frac{\text{connection load}}{\text{bolt capacity}} = \frac{100}{15.5} = 6.45$$

$$\left[n = \frac{445}{69} = 6.45\right]$$

and the minimum number for a symmetrical connection is 8.

With eight bolts used the load on one bolt is

$$P = \frac{100}{8} = 12.5 \text{ kips}$$

$$\left[\frac{445}{8} = 55.6 \text{ kN}\right]$$

According to Table 12-2 the $\frac{3}{4}$-in. bolts require a minimum edge distance of 1.25 in. (at a sheared edge) and a recommended pitch of 2.25 in. The minimum width for the narrow plates is therefore (see Fig. 12-9)

$$w = b + 2(a)$$

$$w = 2.25 + 2(1.25) = 4.75 \text{ in. [120 mm]}$$

With no other constraining conditions given we arbitrarily select a width of 6 in. [152.4 mm] for the narrow plates. Checking first for the requirement of a maximum tension stress of 0.60 F_y on the gross area, we find

$$F_t = 0.60F_y = 0.60\ (36) = 21.6 \text{ ksi [149 MPa]}$$

$$A_{\text{req}} = \frac{100}{21.6} = 4.63 \text{ in.}^2$$

$$\left[\frac{445 \times 10^3}{149} = 2987 \text{ mm}^2\right]$$

and the required thickness with the width selected is

$$t = \frac{4.63}{2(6)} = 0.386 \text{ in.}$$

$$\left[\frac{2987}{2(152.4)} = 9.80 \text{ mm}\right]$$

We therefore select a minimum thickness of $\frac{7}{16}$ in. (0.4375) [10 mm]. The next step is to check the stress condition on the net section through the holes, for which the allowable stress is $0.50F_u$. For the computations we assume a hole diameter $\frac{1}{16}$ in. [1.59 mm] larger than the bolt. Thus

$$\text{hole size} = 0.8125 \text{ in. } [20.64 \text{ mm}]$$

$$\text{net width} = 2\{6 - (2 \times 0.8125)\} = 8.75 \text{ in.}$$

$$[2\{152.4 - (2 \times 20.64)\} = 222.2 \text{ mm}]$$

and the stress on the net section of the two plates is

$$f_t = \frac{100}{0.4375 \times 8.75} = 26.12 \text{ ksi}$$

$$\left[\frac{445 \times 10^3}{10 \times 222.2} = 200 \text{ MPa}\right]$$

This computed stress is compared with the specified allowable stress of

$$F_t = 0.50F_u = 0.50 \times 58 = 29 \text{ ksi } [200 \text{ MPa}]$$

Bearing stress is computed by dividing the load on a single bolt by the product of the bolt diameter and the plate thickness. Thus

$$f_p = \frac{12.5}{2 \times 0.75 \times 0.4375} = 19.05 \text{ ksi}$$

$$\left[f_p = \frac{55.6 \times 10^3}{2 \times 19.05 \times 10} = 144 \text{ MPa}\right]$$

This is compared with the allowable stress of

$$F_p = 1.5F_u = 1.5 \times 58 = 87 \text{ ksi } [600 \text{ MPa}]$$

For the middle plate the procedure is essentially the same, except that in this case the plate width is given. As before, on the basis of stress on the unreduced section, we determine that the total area required is 4.63 in.^{2} [2987 mm^2]. Thus the thickness required is

$$t = \frac{4.63}{10} = 0.463 \text{ in.}$$

$$\left[\frac{2987}{254} = 11.76 \text{ mm}\right]$$

We therefore select a minimum thickness of $\frac{1}{2}$ in. (0.50) [12 mm]. We then proceed as before to check the stress on the net width. The net width through the two holes is

$$w = 10 - (2 \times 0.8125) = 8.375 \text{ in.}$$

$$[w = 254 - (2 \times 20.64) = 212.7 \text{ mm}]$$

and the tension stress on this net cross section is

$$f_t = \frac{100}{8.375 \times 0.5} = 23.88 \text{ ksi}$$

$$\left[f_t = \frac{4.45 \times 10^3}{12 \times 212.7} = 174 \text{ MPa}\right]$$

which is less than the allowable stress of 29 ksi [200 MPa] determined previously.

The computed bearing stress on the wide plate is

$$f_p = \frac{12.5}{0.75 \times 0.50} = 33.3 \text{ ksi}$$

$$\left[\frac{55.6 \times 10^3}{19.05 \times 12} = 243 \text{ MPa}\right]$$

which is considerably less than the allowable determined before: F_p = 87 ksi [600 MPa].

In addition to the layout restrictions given in Art. 12-6, the AISC Specification requires that the minimum spacing in the direction of the load be

$$\frac{2P}{F_u t} + \frac{d}{2} \qquad \text{(dimension } d \text{ in Fig. 12-9)}$$

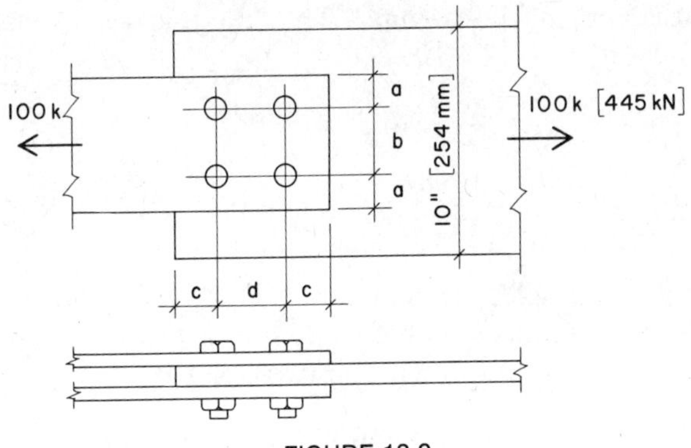

FIGURE 12-9

and that the minimum edge distance in the direction of the load be

$$\frac{2P}{F_u t} \qquad \text{(dimension } c \text{ in Fig. 12-9)}$$

where P = the force transmitted by one fastener to the critical connected part,
F_u = the specified minimum (ultimate) tensile strength of the connected part,
t = the thickness of the critical connected part.

For our case

$$\frac{2P}{F_u t} = \frac{2 \times 12.5}{58 \times 0.5} = 0.862 \text{ in.}$$

which is considerably less than the specified edge distance listed in Table 12-2 for a $\frac{3}{4}$-in. bolt at a sheared edge: 1.25 in.

For the spacing

$$\frac{2P}{F_u t} + \frac{d}{2} = 0.862 + 0.375 = 1.237 \text{ in.}$$

which is also not critical.

A final problem that must be considered is the potential of tearing out the two bolts at the end of the plates. Because the combined thickness of the two outer plates is greater than that of the middle plate, the critical case in this connection is that of the middle plate. Figure 12-10 shows the condition for the tearing, which involves tension on the section labeled 1 and shear on the two sections labeled 2.

For the tension section

$$w_{(net)} = 3 - 0.8125 = 2.1875 \text{ in. } [55.6 \text{ mm}]$$

$$F_t = 0.50F_u = 29 \text{ ksi } [200 \text{ MPa}]$$

For the shear sections

$$w_{(net)} = 2\left(1.25 - \frac{0.8125}{2}\right) = 1.6875 \text{ in. } [43.0 \text{ mm}]$$

$$F_v = 0.30F_u = 17.4 \text{ ksi } [120 \text{ MPa}]$$

The total resistance to tearing is

$$T = (2.1875 \times 0.5 \times 29) + (1.6875 \times 0.5 \times 17.4) = 46.4 \text{ kips}$$

$$\left[T = \left(\frac{55.6 \times 12 \times 200}{10^3}\right) + \left(\frac{43.0 \times 12 \times 120}{10^3}\right) = 195 \text{ kN}\right]$$

Because this is greater than the combined load of 25 kips [111.2 kN] on the two bolts, the problem is not critical.

Connections that transfer compression between the joined parts are essentially the same with regard to the bolt stresses and bearing on the parts. Stress on the net section is less likely to be critical because the compression members will usually be de-

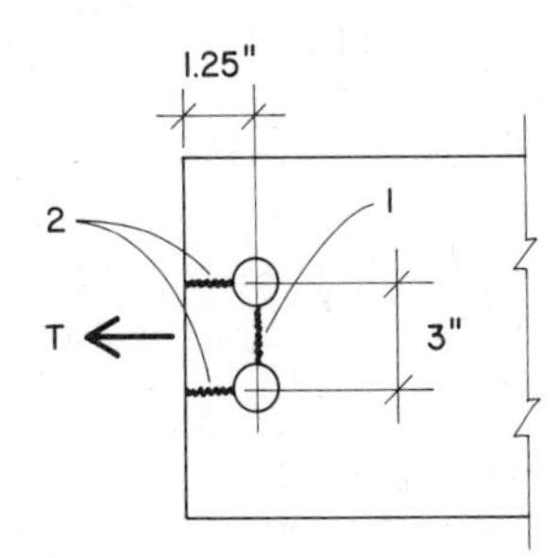

FIGURE 12-10

signed for column action, with a considerably reduced value for the allowable compression stress.

Additional examples of connection design are given in Art. 12-10 and Chapter 13.

Problem 12.8.A.* A bolted connection of the general form shown in Fig. 12-9 is to be used to transmit a tension force of 200 kips [890 kN] by using $\frac{7}{8}$-in. A490N bolts and plates of A36 steel. The outer plates are to be 8 in. [200 mm] wide and the center plate is to be 12 in. [300 mm] wide. Find the required thicknesses of the plates and the number of bolts needed if the bolts are placed in two rows. Sketch the bolt layout with the necessary dimensions.

Problem 12.8.B. Design a connection for the data in Problem 12.8.A, except that the bolts are 1 in. A325N, the outside plates are 9 in. wide, and the bolts are placed in three rows.

12-9. Framing Connections

The joining of structural steel members in a structural system generates a wide variety of situations, depending on the form of the connected parts, the type of connecting device used, and the nature and magnitude of the forces that must be transferred between the members. Figure 12-10 shows a number of common connections that are used to join steel columns and beams consisting of rolled shapes.

In the joint shown in Fig. 12-11*a* a steel beam is connected to a supporting column by the simple means of resting it on top of a steel plate that is welded to the top of the column. The bolts in this case carry no computed loads if the force transfer is limited to that of the vertical end reaction of the beam. The only computed stress condition that is likely to be of concern in this situation is that of crippling the beam web (Art. 9-6). This is a situation in which the use of unfinished bolts is indicated.

The remaining details in Fig. 12-11 illustrate situations in which the beam end reactions are transferred to the supports by attachment to the beam web. This is, in general, an appropriate form of force transfer because the vertical shear at the end of the beam is resisted primarily by the beam web. The most common form of connection is that which uses a pair of angles (Fig. 12-11*b*). The two most frequent examples of this type of connection

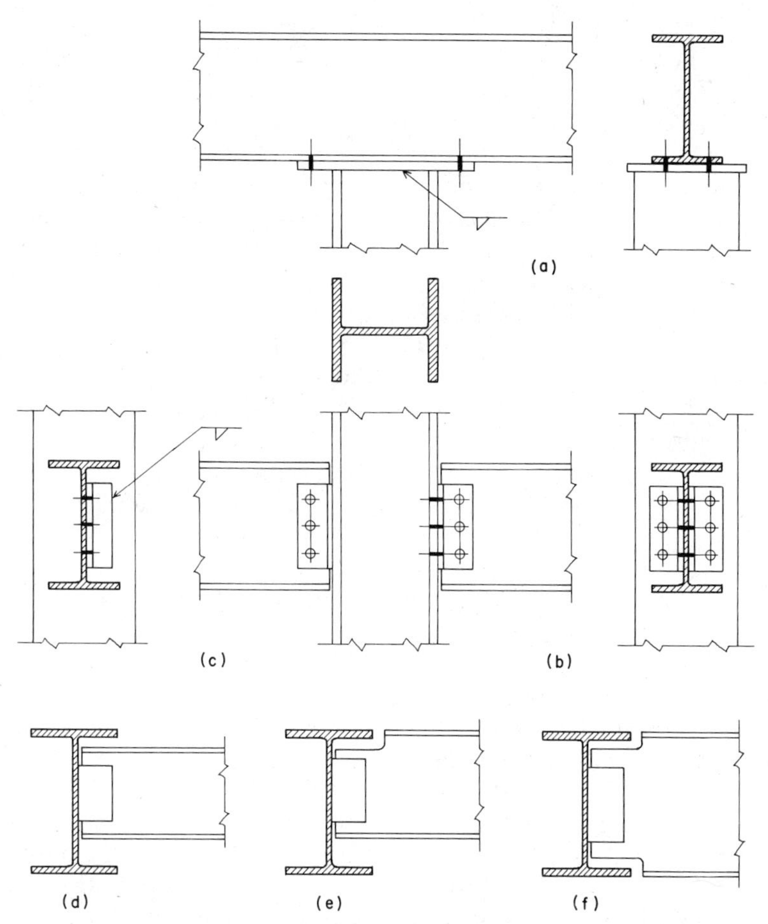

FIGURE 12-11. Typical bolted framing connections for structural steel.

are the joining of a steel beam to the side of a column (Fig. 12-11*b*) or to the side of another beam (Fig. 12-11*d*). A beam may also be joined to the web of a W shape column in this manner if the column depth provides enough space for the angles.

An alternative to this type of connection is shown in Fig. 12-11*c*, where a single plate is welded to the side of a column and the beam web is bolted to one side of the plate. This is generally acceptable only when the magnitude of the load on the beam is low because the one-sided connection experience some torsion.

When the two intersecting beams must have their tops at the same level, the supported beam must have its top flange cut back, as shown at Fig. 12-11*e*. This is to be avoided, if possible, because it represents an additional cost in the fabrication and also reduces the shear capacity of the beam. Even worse is the situation in which the two beams have the same depth and which requires cutting both flanges of the supported beam. When these conditions produce critical shear in the beam web it will be necessary to reinforce the beam end. The problem of tearing the beam web in these situations is discussed in the next article.

Alignment of the tops of beams is usually done to simplify the installation of decks on top of the framing. When steel deck is used it may be possible to adopt some form of the detail shown in Fig. 12-12, which permits the beam tops to be offset by the depth of the deck ribs. Unless the flange of the supporting beam is quite thick, it will probably provide sufficient space to permit the connection shown, which does not require cutting the flange of the supported beam.

Figure 12-13 shows additional framing details that may be used in special situations. The technique described in Fig. 12-13*a* is sometimes used when the supported beam is shallow. The vertical load in this case is transferred through the seat angle, which may be bolted or welded to the carrying beam. The connection to the web of the supported beam merely provides additional resistance to roll-over, or torsional rotation, on the part of the beam. Another reason for favoring this detail is the possibility that the seat angle may be welded in the shop and the web connection

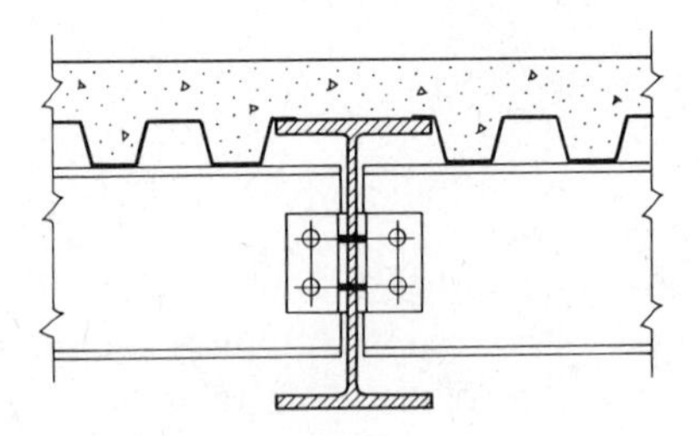

FIGURE 12-12

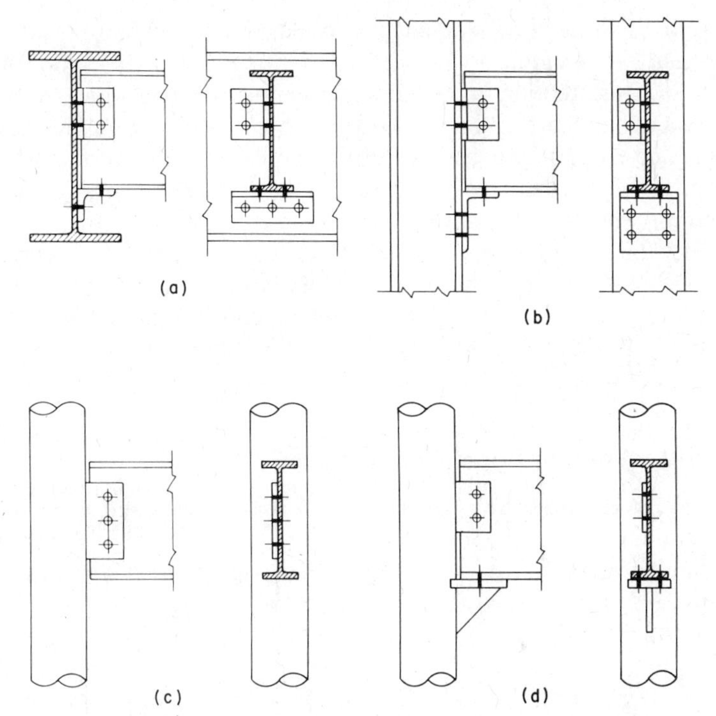

FIGURE 12-13. Special bolted framing connections.

made with small unfinished bolts in the field, which greatly simplifies the field work.

Fig. 12-13*b* shows the use of a similar connection for joining a beam and column. For heavy beam loads the seat angle may be braced with a stiffening plate. Another variation of this detail involves the use of two plates rather than the angle which may be used if more than four bolts are required for attachment to the column.

Figures 12-13*c* and *d* show connections commonly used when pipe or tube columns carry the beams. Because the one-sided connection in Fig. 12-13*c* produces some torsion in the beam, the seat connection is favored when the beam load is high.

Framing connections quite commonly involve the use of welding and bolting in a single connection, as illustrated in the figures. In general, welding is favored for fabrication in the shop and bolting, for erection in the field. If this practice is recognized, the connections must be developed with a view to the overall fabrication and erection process and some decision made regarding what is to be done where. With the best of designs, however, the contractor who is awarded the job may have some of his own ideas about these procedures and may suggest alterations in the details.

Development of connection details is particularly critical for structures in which a great number of connections occur. The truss is one such structure. Some of the problems of truss connections are discussed in Chapter 14.

12-10. Framed Beam Connections

The connection shown in Fig. 12-11*b* is the type used most frequently in the development of structures that consist of I-shaped beams and H-shaped columns. This device is referred to as a *framed beam connection,* for which there are several design considerations:

1. *Type of fastening*. This may be accomplished with rivets or with any of the several types of structural bolt. The angles may also be welded in place, as described in Chapter 13. The most common practice is to weld the angles to the beam web in the fabricating shop and to bolt them to the supports in the field.
2. *Number of fasteners*. This refers to the number of bolts used on the beam web; there are twice this number in the outstanding legs of the angles. The capacities are matched, however, because the web bolts are in double shear, the others in single shear.
3. *Size of the angles*. This depends on the size of the fasteners, the magnitude of the loads, and the size of the support, if it is a column with a particular limiting dimension. Two sizes used frequently are 4×3 in. and $5 \times 3\frac{1}{2}$ in. Thickness of the angle legs is usually based on the size and type of the fastener.

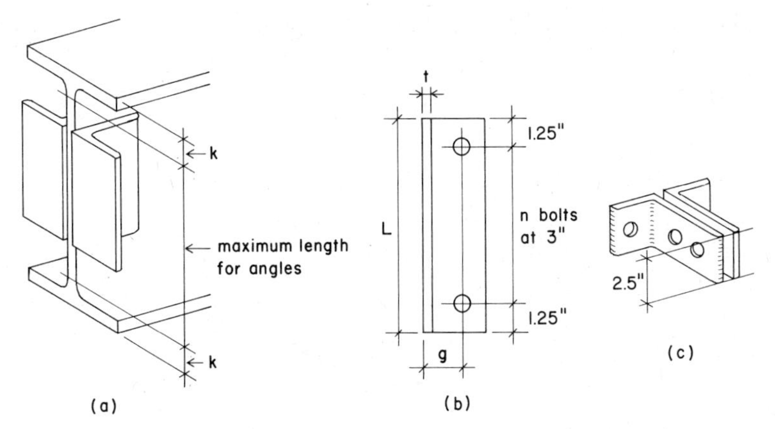

FIGURE 12-14. Framed beam connections using steel angles.

4. *Length of the angles*. This is primarily a function of the size of the fasteners. As shown in Fig. 12-14, typical dimensions are an end distance of 1.25 in. and a pitch of 3 in. In special situations, however, smaller dimensions may be used with bolts of 1 in. or smaller diameter.

The AISC Manual (Ref. 1) provides considerable information to assist in the design of this type of connection in both the bolted and welded versions. A sample for bolted connections that use A325F bolts and angles of A36 steel is given in Table 12-5. The angle lengths in the table are based on the standard dimensions, as shown in Fig. 12-14. For a given beam shape the maximum size of connection (designated by the number of bolts) is limited by the dimension of the flat portion of the beam web. By referring to the tables in Appendix A we can determine this dimension for any beam designation.

Although there is no specified limit for the minimum size of a framed connection to be used with a beam, the general rule is to choose one with an angle length of at least one-half the beam depth. This is intended in the most part to ensure some rotational stability for the beam end.

The one-bolt connection with an angle length of only 2.5 in.

TABLE 12-5. Framed Beam Connections with A325F Bolts and A36 Angles[a]

No. of bolts n	Angle length L (in.) (Fig. 12-14)	Total shear capacity of bolts (kips) — Bolt diameter, d (in.) $\frac{3}{4}$ / Usual angle thickness, t (in.) $\frac{1}{4}$	$\frac{7}{8}$ / $\frac{5}{16}$	1 / $\frac{1}{2}$	Use with the following rolled shapes
10	$29\frac{1}{2}$	155	210	275	W 36
9	$26\frac{1}{2}$	139	189	247	W 36, 33
8	$23\frac{1}{2}$	124	168	220	W 36, 33, 30
7	$20\frac{1}{2}$	108	147	192	W 36, 33, 30, 27, 24, S 24
6	$17\frac{1}{2}$	92.8	126	165	W 36, 33, 30, 27, 24, 21, S 24
5	$14\frac{1}{2}$	77.3	105	137	W 30, 27, 24, 21, 18, S 24, 20, 18, C 18
4	$11\frac{1}{2}$	61.9	84.2	110	W 24, 21, 18, 16, S24, 20, 18, 15, C 18, 15
3	$8\frac{1}{2}$	46.4	61.9[b]	82.5	W 18, 16, 14, 12, 10, S 18, 15, 12, 10, C 18, 15, 12, 10
2	$5\frac{1}{2}$	30.9	39.4[b]	55.0	W 12, 10, 8, S12, 10, 8, C 12, 10, 9, 8
1	$2\frac{1}{2}$	15.4	21.0	27.5	W 6, 5, M 6, 5, C 7, 6, 5

[a] Adapted from data in the AISC Manual, 8th ed. (Ref. 1), with permission of the publishers, the American Institute of Steel Construction.
[b] Limited by shear on the angles.

(Fig. 12-14*c*) is the shortest. This special connection has double-gage spacing of bolts in the beam web to ensure its stability.

The following example illustrates the general design procedure for a framed beam connection. In practice, this process can be shortened because experience permits judgments that will eventually make some of the steps unnecessary. Other design aids in the AISC Manual (Ref. 1) will shorten the work for some computations.

Example. A beam consists of a W 27 × 94 of A36 steel with F_u of 58 ksi [400 MPa] that is needed to develop an end reaction of 80 kips [356 kN]. Design a standard framed beam connection with A325F bolts and angles of A36 steel.

Solution: A scan of Table 12-5 reveals that the range of possible connections for a W 27 is $n = 5$ to $n = 7$. For the required load possible choices are

$$n = 6, \tfrac{3}{4} \text{ in. bolts, angle } t = \tfrac{1}{4} \text{ in., load} = 92.8 \text{ kips [413 kN]}$$

$$n = 5, \tfrac{7}{8} \text{ in. bolts, angle } t = \tfrac{5}{16} \text{ in., load} = 105 \text{ kips [467 kN]}$$

Bolt size is ordinarily established for a series of framing rather than for each element. Having no other criterion, we make an arbitrary choice of the connection with $\frac{7}{8}$-in. bolts.

The bolt capacity in double shear is the primary consideration in the development of data in Table 12-5. We must make a separate investigation of the bearing on the beam web because it is not incorporated in the table data. It is actually seldom a problem except in heavily loaded beams, but the following procedure should be used:

From Table A.1 the thickness of the beam web is 0.490 in. [12.5 mm]. The total bearing capacity of the five bolts is

$$V = n \times (\text{bolt diameter}) \times (\text{web } t) \times 1.5F_u$$

$$= 5 \times 0.875 \times 0.490 \times 87 = 186.5 \text{ kips}$$

$$\left[V = 5 \times 22.2 \times 12.5 \times \frac{600}{10^3} = 832 \text{ kN}\right]$$

which is considerably in excess of the required load of 80 kips [356 kN].

Another concern in the typical situation is that for the shear stress through the net section of the web, reduced by the chain of bolt holes. If the connection is made as shown in Fig. 12-11*b* or *d*, this section is determined as the full web width (beam depth), less the sum of the hole diameters, times the web thickness, and the allowable stress is specified as $0.40F_y$. From Table A.1 for the W 27, $d = 26.92$ in. [684 mm]. The net shear width through the bolt holes is thus

$$w = 26.92 - (5 \times 0.9375) = 22.23 \text{ in.}$$

$$[w = 684 - (5 \times 23.8) = 565 \text{ mm}]$$

and the computed stress due to the load is

$$f_v = \frac{80}{22.23 \times 0.490} = 7.34 \text{ ksi}$$

$$\left[f_v = \frac{356 \times 10^3}{565 \times 12.5} = 50.4 \text{ MPa}\right]$$

which is less than the allowable of $0.40 \times 36 = 14.4$ ksi [100 MPa].

If the top flange of the beam is cut back to form the type of connection shown in Fig. 12-11*e*, a critical condition that must be investigated is that of tearing out the end portion of the beam web, as discussed in Art. 12-7. This is also called *block shear,* which refers to the form of the failed portion (Fig. 12-15). If the angles are placed with the edge distances shown in Fig. 12-15, this failure block will have the dimensions of 14 by 2.25 in. [356 by 57 mm]. The tearing force V is resisted by a combination of tension stress on the section labeled 1 and shearing stress on the combined sections labeled 2. The allowable stresses for this situation are $0.30F_u$ for shear and $0.50F_u$ for tension. Next find the net

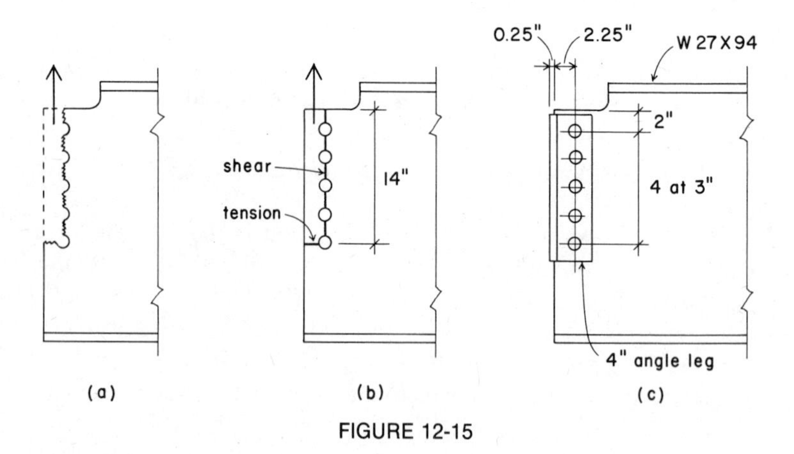

FIGURE 12-15

widths of the sections, multiply them by the web thickness to obtain the areas, and multiply by the allowable stresses to obtain the total resisting forces.

For the tension resistance

$$w = 2.25 - \frac{0.9375}{2} = 1.78 \text{ in. } [45.1 \text{ mm}]$$

For the shear resistance

$$w = 14 - (4\tfrac{1}{2} \times 0.9375) = 9.78 \text{ in. } [249 \text{ mm}]$$

For the total resisting force

$$\begin{aligned} V &= (\text{tension } w \times t_w) \times (0.50F_u) + (\text{shear } w \times t_w) \times (0.30F_u) \\ &= (1.78 \times 0.49 \times 29) + (9.78 \times 0.49 \times 17.4) \\ &= 25.3 + 83.4 = 108.7 \text{ kips} \end{aligned}$$

$$\left[\begin{aligned} V &= \left(45.1 \times 12.5 \times \frac{200}{10^3}\right) + \left(249 \times 12.5 \times \frac{120}{10^3}\right) \\ &= 113 + 374 = 487 \text{ kN} \end{aligned}\right]$$

Because this potential total resistance exceeds the load required, the tearing is not critical.

Problem 12.10.A.* A W 30 × 108 of A36 steel with $F_u = 58$ ksi [400 MPa] is required to develop an end reaction of 120 kips [534 kN]. Determine (a) the possible choices for a framed beam connection with A325F bolts and A36 angles; (b) the total resistance with the smallest bolts to bearing and shear if the beam flange is not cut back; (c) the total resistance to tearing if the connection is installed as shown in Fig. 12-15.

Problem 12.10.B. Proceed as in the preceding problem, except that the beam is a W 24 × 76 and the reaction is 100 kips [445 kN].

Problem 12.10.C. Proceed as in Problem 12.10.A, except that the beam is a W 16 × 40 and the reaction is 50 kips [222 kN].

12-11. Conventional and Moment Connections

All of the beam-to-girder and beam-to-column connections discussed in this chapter come under the category of "simple" or free-end connections; that is, insofar as gravity loading is concerned, the ends of the beams and girders are connected for shear only and are free to rotate under gravity load. We shall call connections of this nature *conventional connections;* they are used in Type 2 of the three types of steel construction recognized in the AISC Specification.

Type I construction, commonly designated as continuous or rigid frame, assumes that beam-to-column connections possess sufficient rigidity to prevent rotation of the beam ends as the member deflects under its load. This means that the connection must transmit some bending moment between beam and column. Consequently it is called a moment-resisting connection or simply a *moment connection.* Type 3 construction, called partially restrained or semi-rigid framing, assumes that the connections possess a dependable and known moment-resisting capacity of a degree between the rigidity of Type 1 and the flexibility of Type 2.

Although Type 1 continuous framing can be achieved by the proper design of bolted or riveted connections, it is accomplished much more effectively in welded construction. This aspect of welded connections is considered briefly in Chapter 13. A fully continuous frame of Type 1 construction is statically indeterminate and its analysis and design are beyond the scope of this book. Moment-resisting connections are used in multistory steel frame buildings to provide lateral stability against the effects of wind and earthquake forces.

In general, the design methods and procedures treated in this volume are applicable to Type 2 construction and follow the provisions of Section 1.12.1 of the AISC Specification, which states that "Beams, girders, and trusses shall ordinarily be designed on the basis of simple spans whose effective length is equal to the distance between centers of gravity of the members to which they deliver their end reactions.

13

Welded Connections

13-1. General

One of the distinguishing characteristics of welded construction is the facility with which one member may be attached directly to another without the use of additional plates or angles, which are necessary in bolted and riveted connections. A welded connection requires no holes for fasteners; therefore the gross rather than the net section may be considered when determining the effective cross-sectional area of members in tension.

As noted in Art. 12-11, moment-resisting connections are readily achieved by welding; consequently welded connections are customary in Type 1 construction in order to develop continuity in the framing. Welding may also be used in Type 2 construction, but care must be exercised in design to ensure that a rigid connection is not provided where free-end conditions have been assumed in the design of the framing.

Welding is often used in combination with bolting in "shop-welded and field-bolted construction." Here connection angles with holes in the outstanding legs may be welded to a beam in the fabricating shop and then bolted to a girder or column in the field.

13-2. Electric Arc Welding

Although there are many welding processes, electric arc welding is the one generally used in steel building construction. In this type of welding an electric arc is formed between an electrode and the two pieces of metal that are to be joined. The intense heat melts a small portion of the members to be joined as well as the end of the electrode or metallic wire. The term *penetration* is used to indicate the depth from the original surface of the base metal to the point at which fusion ceases. The globules of melted metal from the electrode flow into the molten seat and, when cool, are united with the members that are to be welded together. *Partial penetration* is the failure of the weld metal and base metal to fuse at the root of a weld. It may result from a number of items, and such incomplete fusion produces welds that are inferior to those of full penetration.

13-3. Welded Joints

When two members are to be joined, the ends may or may not be grooved in preparation for welding. In general, there are three classification of joints: *butt joints, tee joints,* and *lap joints*. The selection of the type of weld to use depends on the magnitude of the load requirement, the manner in which it is applied, and the cost of preparation and welding. Several joints are shown in Fig. 13-1. The type of joint and preparation permit a number of variations. In addition, welding may be done from one or both sides. The scope of this book prevents a detailed discussion of the many joints and their uses and limitations.

The weld most commonly used for structural steel in building construction is the *fillet weld*. It is approximately triangular in cross section and is formed between the two intersecting surfaces of the joined members. See Fig. 13-2*a* and *b*. The *size* of a fillet weld is the leg length of the largest inscribed isosceles right triangle, *AB* or *BC*. (See Fig. 13-2*a*.) The *root* of the weld is the point at the bottom of the weld, point *B* in Fig. 13-2*a*. The *throat* of a fillet weld is the distance from the root to the hypotenuse of the largest isosceles right triangle that can be inscribed within the

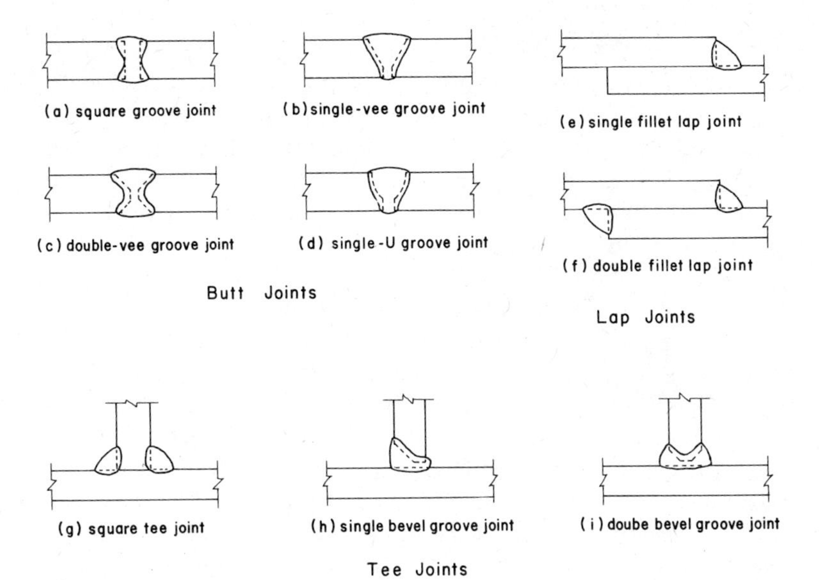

FIGURE 13-1. Typical welded joints.

weld cross section, distance *BD* in Fig. 13-2*a*. The exposed surface of a weld is not the plane surface indicated in Fig. 13-2*a* but is usually somewhat convex, as shown in Fig. 13-2*b*. Therefore the actual throat may be greater than that shown in Fig. 13-2*a*. This additional material is called *reinforcement*. It is not included in determining the strength of a weld.

A single-vee groove weld between two members of unequal thickness is shown in Fig. 13-2*c*. The *size* of a butt weld is the thickness of the thinner part joined, with no allowance made for the weld reinforcement.

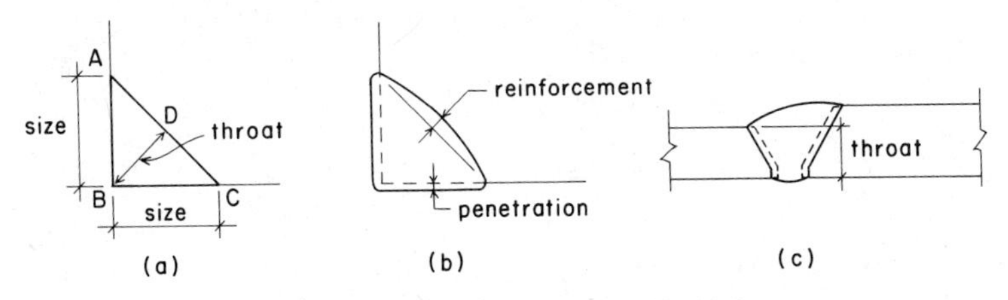

FIGURE 13-2. Properties of welded joints.

13-4. Stresses in Welds

If the dimension (size) of *AB* in Fig. 13-2*a* is 1 unit in length, $(AD)^2 + (BD)^2 = 1^2$. Because *AD* and *BD* are equal, $2(BD)^2 = 1^2$ and $BD = \sqrt{0.5}$ or 0.707. Therefore the throat of a fillet weld is equal to the *size* of the weld multiplied by 0.707. As an example, consider a $\frac{1}{2}$-in. fillet weld. This would be a weld with dimensions *AB* or *BC* equal to $\frac{1}{2}$ in. In accordance with the above, the throat would be 0.5×0.707 or 0.3535 in. Then, if the allowable unit shearing stress on the throat is 21 ksi, the allowable working strength of a $\frac{1}{2}$-in. fillet weld is $0.3535 \times 21 = 7.42$ kips *per lin in. of weld.* If the allowable unit stress is 18 ksi, the allowable working strength is $0.3535 \times 18 = 6.36$ kips *per lin in. of weld.*

The permissible unit stresses used in the preceding paragraph are for welds made with E 70 XX and E 60 XX type electrodes on A36 Steel. Particular attention is called to the fact that *the stress in a fillet weld is considered as shear on the throat, regardless of the direction of the applied load.* Neither plug nor slot welds shall be assigned any values in resistances other than shear. The allowable working strengths of fillet welds of various sizes are given in Table 13-1 with values rounded to $\frac{1}{10}$ of a kip.

TABLE 13-1. Allowable Working Strength of Fillet Welds

	Allowable load (kips/in.)		Allowable load (kN/mm)		
Size of weld (in.)	E 60 XX electrodes $F_{vw} = 18$ (ksi)	E 70 XX electrodes $F_{vw} = 21$ (ksi)	E 60 XX electrodes $F_{vw} = 124$ (MPa)	E 70 XX electrodes $F_{vw} = 145$ (MPa)	Size of weld (mm)
$\frac{3}{16}$	2.4	2.8	0.42	0.49	4.76
$\frac{1}{4}$	3.2	3.7	0.56	0.65	6.35
$\frac{5}{16}$	4.0	4.6	0.70	0.81	7.94
$\frac{3}{8}$	4.8	5.6	0.84	0.98	9.52
$\frac{1}{2}$	6.4	7.4	1.12	1.30	12.7
$\frac{5}{8}$	8.0	9.3	1.40	1.63	15.9
$\frac{3}{4}$	9.5	11.1	1.66	1.94	19.1

The stresses allowed for the metal of the connected parts (known as the *base metal*) apply to complete penetration groove welds stressed in tension and compression parallel to the axis of the weld and in tension normal to the effective throat. They apply also to complete or partial penetration groove welds stressed in compression normal to the effective throat and in shear on the effective throat. Consequently allowable stresses for butt welds are the same as for the base metal.

The relation between the weld size and the maximum thickness of material in joints connected only by fillet welds is shown in Table 13-2. The maximum size of a fillet weld applied to a square edge of a plate or section $\frac{1}{4}$ in. or more in thickness should be $\frac{1}{16}$ in. less than the nominal thickness of the edge. Along edges of material less than $\frac{1}{4}$ in. thick the maximum size may be equal to the thickness of the material.

The effective area of butt and fillet welds is considered to be the effective length of the weld multiplied by the effective throat thickness. The minimum effect length of a fillet weld should not be less than four times the weld size. For starting and stopping the arc approximately $\frac{1}{4}$ in. should be added to the design length of fillet welds.

Figure 13-3*a* represents two plates connected by fillet welds. The welds marked *A* are longitudinal; *B* indicates a transverse weld. If a load is applied in the direction shown by the arrow, the stress distribution in the longitudinal weld is not uniform and the stress in the transverse weld is approximately 30% higher per unit of length.

TABLE 13-2. Relation Between Material Thickness and Minimum Size of Fillet Welds

Material thickness of the thicker part joined		Minimum size of fillet weld	
(in.)	(mm)	(in.)	(mm)
To $\frac{1}{4}$ inclusive	To 6.35 inclusive	$\frac{1}{8}$	3.18
Over $\frac{1}{4}$ to $\frac{1}{2}$	Over 6.35 to 12.7	$\frac{3}{16}$	4.76
Over $\frac{1}{2}$ to $\frac{3}{4}$	Over 12.7 to 19.1	$\frac{1}{4}$	6.35
Over $\frac{3}{4}$	Over 19.1	$\frac{5}{16}$	7.94

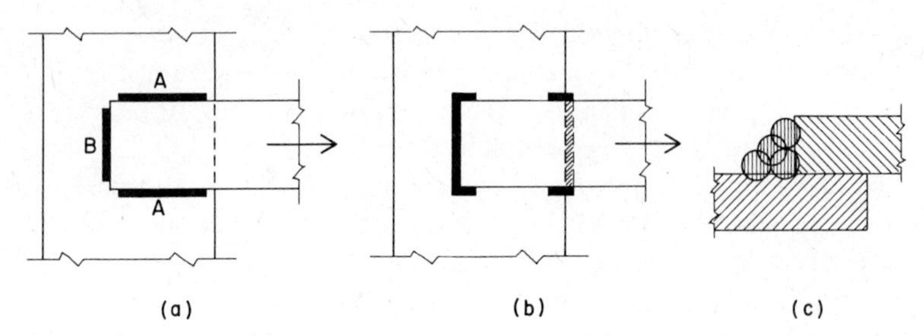

FIGURE 13-3. Welding of lapped plates.

Added strength is given to a transverse fillet weld that terminates at the end of a member, as shown in Fig. 13-3*b*, if the weld is returned around the corner for a distance not less than twice the weld size. These end returns, sometimes called *boxing,* afford considerable resistance to the tendency of tearing action on the weld.

The $\frac{1}{4}$-in. fillet weld is considered to be the minimum practical size, and a $\frac{5}{16}$-in. weld is probably the most economical size that can be obtained by one pass of the electrode. A small-size continuous weld is generally more economical than a larger discontinuous weld if both are made in one pass. Some specifications limit the single-pass fillet weld to $\frac{5}{16}$ in. Large-size fillet welds require two or more passes (multipass welds) of the electrode, as shown in Fig. 13-3*c*.

Example. A W 12 × 26 of A36 steel is to be welded to the face of a steel column with E 70 XX electrodes. (See Fig. 13-4*a*.) With respect to the upper flange only, compute the strength of fillet and butt welds. Assume that the beam is to be welded to produce continuous action as in Type 1 construction (Art. 12-11) so the upper flange will be in tension at the column. (See Fig. 13-7*a*.)
Solution: First assume that the left end of the beam is in contact with the column and that a $\frac{3}{8}$ in. [9.52 mm] fillet weld is placed across the upper face of the beam flange as shown in Fig. 13-4*b* and *c*. Appendix Table A.1 shows that the flange of the beam is 6.49 in. [165 mm] wide and 0.38 in. [9.65 mm] thick. From Table

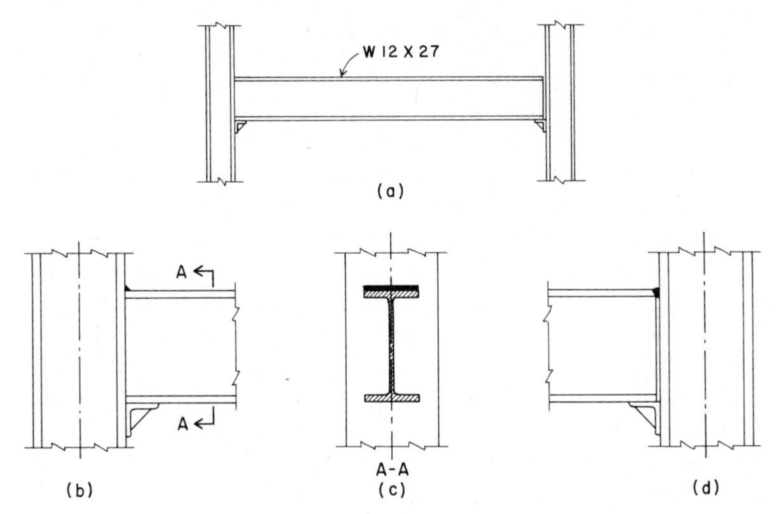

FIGURE 13-4. Welding of a restrained beam.

13.1 the strength of a $\frac{3}{8}$-in. fillet weld is given as 5.6 kips per lin. in. [0.98 kN/mm], making the total resistance of the weld $5.6 \times 6.49 = 36.3$ kips [$0.98 \times 165 = 162$ kN]. Note that this weld resists tensile force but, as stated earlier, the strength of the weld is determined by the shear at the throat of the weld.

Next, instead of the fillet weld, suppose that the upper flange is beveled and that a butt weld is used, as shown in Fig. 13-4*d*. The area of the weld resisting tension is the flange width multiplied by the flange thickness, or $6.49 \times 0.38 = 2.47$ in.^{2} [$165 \times 9.65 = 1592$ mm^2]. Assuming that the maximum allowable bending stress for the beam is 24 ksi [165 MPa], the allowable strength of the butt weld in tension is $24 \times 2.47 = 59.3$ kips [$165 \times 1592/10^3 = 263$ kN].

It is clear, therefore, that if the full bending capacity of the beam must be developed at this connection the butt weld (or possibly a larger fillet weld) must be used.

Problem 13.4.A. A W 14 × 43 of A36 steel is to be welded with E70XX electrodes to a column, as in the preceding example. Determine the capacity of (a) a $\frac{1}{2}$-in. fillet weld and (b) a full butt weld.

13-5. Design of Welded Joints

The most economical choice of weld to use for a given condition depends on several factors. It should be borne in mind that members to be connected by welding must be firmly clamped or held rigidly in position during the welding process. When riveting a beam to a column it is necessary to provide a seat angle as a support to keep the beam in position for riveting the connecting angles. The seat angle is not considered as adding strength to the connection. Similarly, seat angles are commonly used with welded connections. The designer must have in mind the actual conditions during erection and must provide for economy and ease in working the welds. Seat angles or similar members used to facilitate erection are *shop-welded* before the material is sent to the site. The welding done during erection is called *field welding*. In preparing welding details the designer indicates shop or field welds on the drawings. Conventional welding symbols are used to identify the type, size, and position of the various welds. Only engineers or architects experienced in the design of welded connections should design or supervise welded construction. It is apparent that a wide variety of connections is possible; experience is the best aid in determining the most economical and practical connection.

The following examples illustrate the basic principles on which welded connections are designed:

Example 1. A bar of A36 steel, $3 \times \frac{7}{16}$ in. [76.2 × 11 mm] in cross section, is to be welded with E 70 XX electrodes to the back of a channel so that the full tensile strength of the bar may be developed. What is the size of the weld? (See Fig. 13-5.)

Solution: The area of the bar is $3 \times 0.4375 = 1.313$ in.^{2} [$76.2 \times 11 = 838.2$ mm^2]. Because the allowable unit tensile stress of the steel is 22 ksi [Table 4-2), the tensile strength of the bar is $F_t \times A = 22 \times 1.313 = 28.9$ kips. [$152 \times 838.2/10^3 = 127$ kN]. The weld must be of ample dimensions to resist a force of this magnitude.

A $\frac{3}{8}$-in. [9.52 mm] fillet weld will be used. Table 13-1 gives the allowable working strength as 5.6 kips per in. [0.98 kN per mm]. Hence the required length of weld to develop the strength of the bar is $28.9 \div 5.6 = 5.16$ in. [$127 \div 0.98 = 130$ mm]. The position

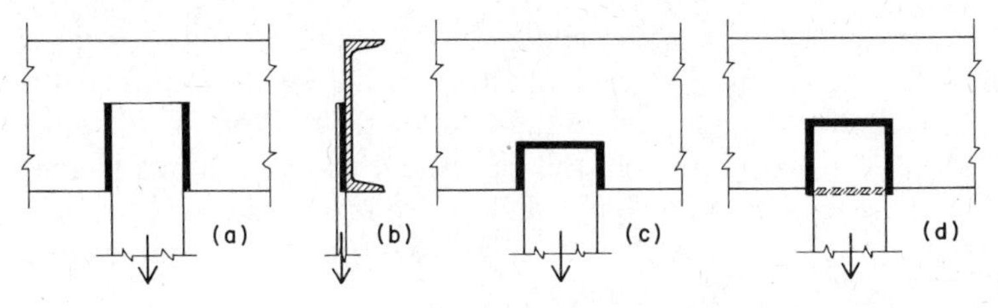

FIGURE 13-5.

of the weld with respect to the bar has several options, three of which are shown in Fig. 13-5*a, c,* and *d.*

Example 2. A $3\frac{1}{2} \times 3\frac{1}{2} \times \frac{5}{16}$ in. [89 × 89 × 7.94 mm] angle of A36 steel subjected to a tensile load is to be connected to a plate by fillet welds, using E 70 XX electrodes. What should the dimensions of the welds be to develop the full tensile strength of the angle?

Solution: We shall use a $\frac{1}{4}$-in. fillet weld which has an allowable working strength of 3.7 kips per in. [0.65 kN per mm] (Table 13-1). From Appendix Table A-1 the cross-sectional area of the angle is 2.09 in.^{2} [1348 mm^2]. By using the allowable tension stress of 22 ksi [152 MPa] for A36 steel (Table 4-2) the tensile strength of the angle is 22 × 2.09 = 46 kips [152 × 1348/10^3 = 205 kN]. Therefore the required total length of weld to develop the full strength of the angle is 46 ÷ 3.7 = 12.4 in. [205 ÷ 0.65 = 315 mm].

An angle is an unsymmetrical cross section and the welds marked L_1 and L_2 in Fig. 13-6 are made unequal in length so that their individual resistance will be proportioned in accordance to

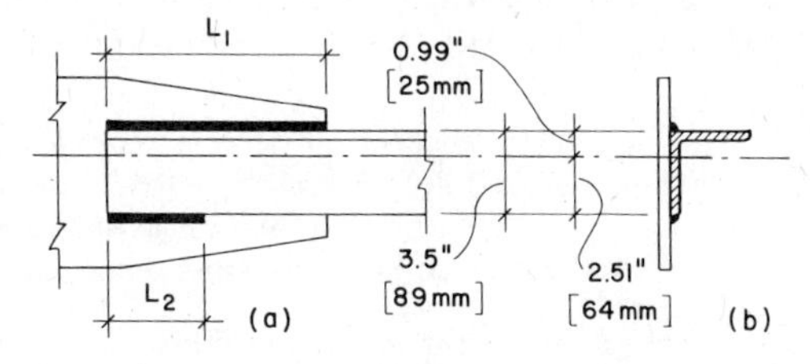

FIGURE 13-6.

the distributed area of the angle. From Appendix Table A-1 we find that the centroid of the angle section is 0.99 in. [25 mm] from the back of the angle; hence the two welds are 0.99 in. [25 mm] and 2.51 in. [64 mm] from the centroidal axis, as shown in Fig. 13-6. The lengths of welds L_1 and L_2 are made inversely proportional to their distances from the axis, but the sum of their lengths is 12.4 in. [315 mm]. Therefore

$$L_1 = \frac{2.51}{3.5} \times 12.4 = 8.9 \text{ in.}$$

$$\left[\frac{64}{89} \times 315 = 227 \text{ mm}\right]$$

and

$$L_2 = \frac{0.99}{3.5} \times 12.4 = 3.5 \text{ in.}$$

$$\left[\frac{25}{89} \times 315 = 88 \text{ mm}\right]$$

These are the design lengths required; and, as noted earlier, each weld would actually be made $\frac{1}{4}$-in. [6.5 mm] longer than its computed length.

When angle shapes are used as tension members and connected by fastening only one leg, it is questionable to assume a stress distribution of equal magnitude on the entire cross section. Some designers therefore prefer to ignore the stress in the unconnected leg and to limit the capacity of the member in tension to the force obtained by multiplying the allowable stress by the area of the connected leg only. If this is done, it is logical to use welds of equal length on each side of the leg, as in Example 1.

Problem 13.5.A.* A 4 × 4 × $\frac{1}{2}$ in. angle of A36 steel is to be welded to a plate with E 70 XX electrodes to develop the full tensile strength of the angle. Using $\frac{3}{8}$-in. fillet welds, compute the design lengths L_1 and L_2, as shown in Fig. 13-6, assuming the development of tension on the entire cross section of the angle.

Problem 13.5.B. Redesign the welded connection in Problem 13.5.A assuming that the tension force is developed only by the connected leg of the angle.

13-6. Beams with Continuous Action

As noted earlier, one of the advantages of welding is that beams having continuous action at the supports (Type 1 construction) are readily provided for. The usual bolted or riveted connections of Type 2 construction are assumed to offer no rigidity at the supports and the bending moment throughout the length of the beam is positive. By the use of welding, however, a beam may be connected at its supports in such a manner that the beam is *fixed* or *restrained* and a negative bending moment results (Art. 5-12). For the same span and loading the maximum bending moment for a continuous beam is smaller than for a simple beam and a lighter beam section is required.

When beams are rigidly connected by means of moment-resisting connections the fibers in the upper flange *at the supports* are in tension and the lower flange is in compression. This is shown diagrammatically in Fig. 13-7*a*. Therefore, in designing the welds for beams that have continuous action, we must provide for tension and compression at the supports in the upper and lower flanges, respectively. A wide variety of welds is possible. The following example illustrates the principles by which they are designed:

Example. A W 12 × 40 framing into column flanges at its ends is to be connected by welding to provide continuous action. For erection of the beam it is necessary that its length be slightly shorter than the distance between the flanges of the columns to which it will be welded. Seat angles are shop-welded to the columns, and the beam is supported on them during field welding of the connection. For this example consider that the left end of the beam is tightly held against the column flange; this leaves a short space between the right end of the beam and the column on the right, as shown in Fig. 13-7*b*. Because of this difference in end conditions, each weld must receive individual consideration. As a means of identification, the different welds are referred to as *A*, *B*, *C*, and *D*, as indicated in Fig. 13-7*b*.

The negative bending moment in the beam at the supports is 1150 kip-in. [130 kN-m] and E 70 XX electrodes are used. Design the welds.

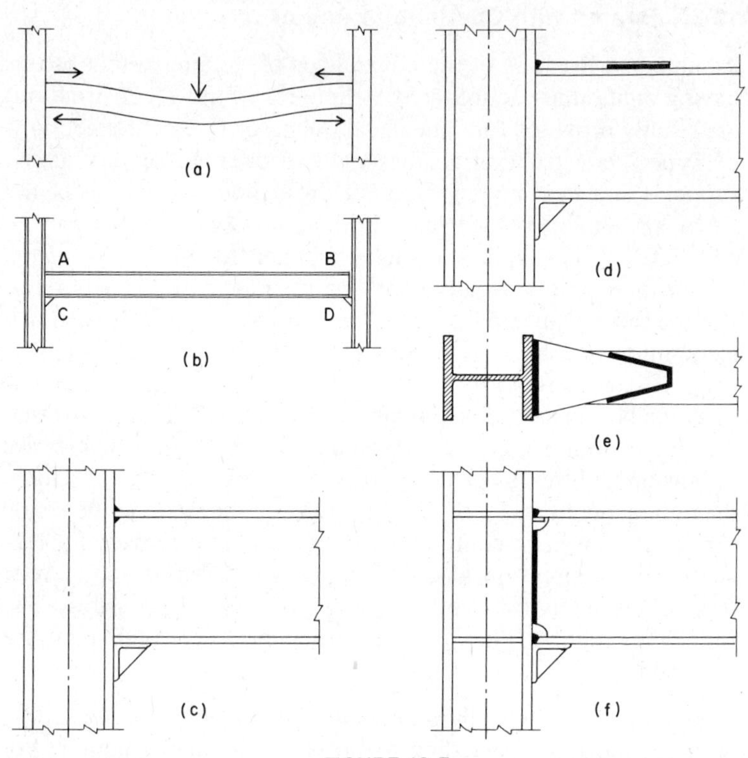

FIGURE 13-7.

Solution: The tensile and compressive forces in the beam flanges constitute a *mechanical couple,* which consists of two equal parallel forces, opposite in direction but not having the same line of action. The moment of a couple is the magnitude of one of the forces multiplied by the perpendicular distance between their lines of action. Therefore, if the negative bending moment is 1150 kip-in. [130 kN-m] and the distance between the two flange forces is approximately 12 in. [305 mm], the forces are each 1150 ÷ 12 = 96 kips [130 ÷ 0.305 = 426 kN].

From Appendix Table A-1 we find that the beam has a flange width of 8.005 in. [203 mm] and a flange thickness of 0.515 in. [13.08 mm]. Suppose we run a $\frac{3}{8}$-in. [9.52-mm] fillet weld across the upper flangc for weld A. This weld has an allowable working load of 5.6 kips per in. [0.98 kN per mm] (Table 13-1), and, because the flange is 8 in. [203 mm] wide, its resistance is 8 × 5.6 = 44.8 kips [0.98 × 203 = 199 kN]. If $\frac{3}{8}$-in. fillet welds are to be used at this joint, the total length of the weld must be 96 ÷ 5.6 = 17.2 in. [426 ÷ 0.98 = 435 mm]. However, the flange is only 8 in. wide, and welds on the upper and under surfaces of the flange, as shown in Fig. 13-7*c*, do not provide sufficient length. A solution for this condition would be to investigate larger welds, but the weld at the underside of the flange requires overhead welding which should be avoided whenever possible.

An alternate detail for the joint at *A* is shown in Fig. 13-7*d* and *e*. This joint has a tapered plate which is welded to the column face with a large butt weld and to the beam flange with fillet welds on the edges of the plate. This joint could also be used at *B* because it does not require the end of the beam to be placed against the column face.

At joints *C* and *D* the force to be transferred from the beam flange to the column face is one of compression. If the beam end is placed flush against the column face, this transfer may be achieved by direct bearing, with no welding required. However, fillet welds probably should be used on the upper surface of the lower beam flange at the column face or at the edges of the beam flange on top of the seat angle to ensure that the beam will not slip with respect to the column. This connection cannot be used at *D*, however, because the beam is not in contact with the column.

The best connection for use in developing a moment close to the limit for the beam is probably that shown in Fig. 13-7*f*. In this joint the beam flanges are cut to permit the placing of a full penetration groove weld, which will develop the capacity of the flange fully in both tension and compression. In the illustration the beam web is shown welded to the column face to develop the vertical shear force, or end reaction, of the beam. The seat angle in this case is used strictly as a temporary erection device to hold the beam in place while the welds are being made. If desired, it could be removed after the welds are completed.

The forces from the beam flanges in connections like those shown in Fig. 13-7 exert considerable bending on the column flanges, making reinforcing plates (Fig. 13-7*f*) necessary. Although generally desirable, this reinforcement may not be required when the beam flanges are considerably narrower than the column or the column flanges are quite thick.

13-7. Plug and Slot Welds

One method of connecting two overlapping plates uses welds in holes made in one of the two plates. (See Fig. 13-8.) Plug and slot welds are those in which the entire area of the hole or slot receives weld metal. The maximum and minimum diameters of plug and slot welds and the maximum length of slot welds are shown in Fig. 13-8. If the plate containing the hole is not more than $\frac{5}{8}$ in. thick, the hole should be filled with weld metal. If the plate is more than $\frac{5}{8}$ in. thick, the weld metal should be at least one-half the thickness of the material but not less than $\frac{5}{8}$ in.

The stress in a plug or slot weld is considered to be shear on the area of the weld at the plane of contact of the two plates being connected. The allowable unit shearing stress, when E 70 XX electrodes are used, is 21 ksi [145 MPa.]

A somewhat similar weld consists of a continuous fillet weld at

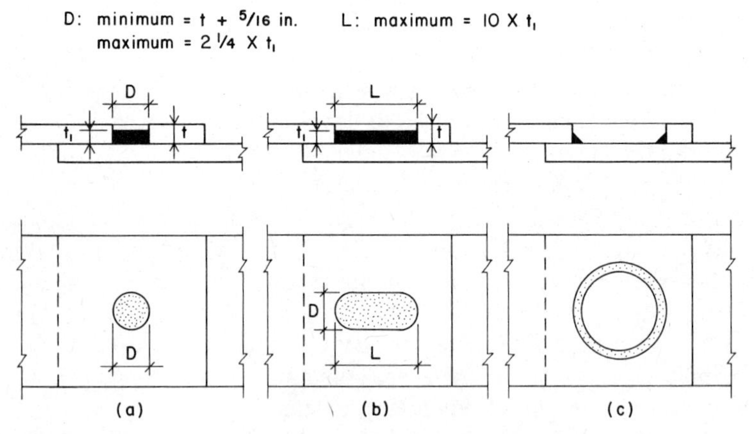

FIGURE 13-8. Welds in holes; a) plug weld, b) slot weld, c) fillet weld in large hole.

the circumference of a hole, as shown in Fig. 13-8*c*. This is not a plug or slot weld and is subject to the usual requirements for fillet welds discussed in Art. 13-1.

13-8. Miscellaneous Welded Connections

Part 4 of the AISC Manual contains a series of tables that pertain to the design of welded connections. The tables cover free-ended as well as moment-resisting connections. In addition, suggested framing details are shown for various situations.

A few common connections are shown in Fig. 13-9. As an aid to erection, certain parts are welded together in the shop before being sent to the site. Connection angles may be shop-welded to beams and the angles field-welded or field-bolted to girders or columns. The beam connection in Fig. 13-9*a* shows a beam supported on a seat that has been shop-welded to the column. A small connection plate is shop-welded to the lower flange of the beam and the plate is bolted to the beam seat. After the beams have been erected and the frame plumbed the beams are field-welded to the seat angles. This type of connection provides no degree of continuity in the form of moment transfer between the beam and column.

The connections shown in Fig. 13-9*b* and *c* are designed to develop some moment transfer between the beam and its supporting column. Auxiliary plates are used to make the connection at the upper flanges.

Beam seats shop-welded to columns are shown in Fig. 13-9*d*, *e*, and *f*. A short length of angle welded to the column with no stiffeners is shown in Fig. 13-9*d*. Stiffeners consisting of triangular-shaped plates are welded to the legs of the angles shown in Fig. 13-9*e* and add materially to the strength of the seat. Another method of forming a seat, using a short piece of structural tee, is shown in Fig. 13-9*f*.

Various types of column splice are shown in Fig. 13-9*g*, *h*, and *i*. The auxiliary plates and angles are shop-welded to the columns and provide for bolted connections in the field before the permanent welds are made. Welded connections for column base plates are shown in Fig. 11-7.

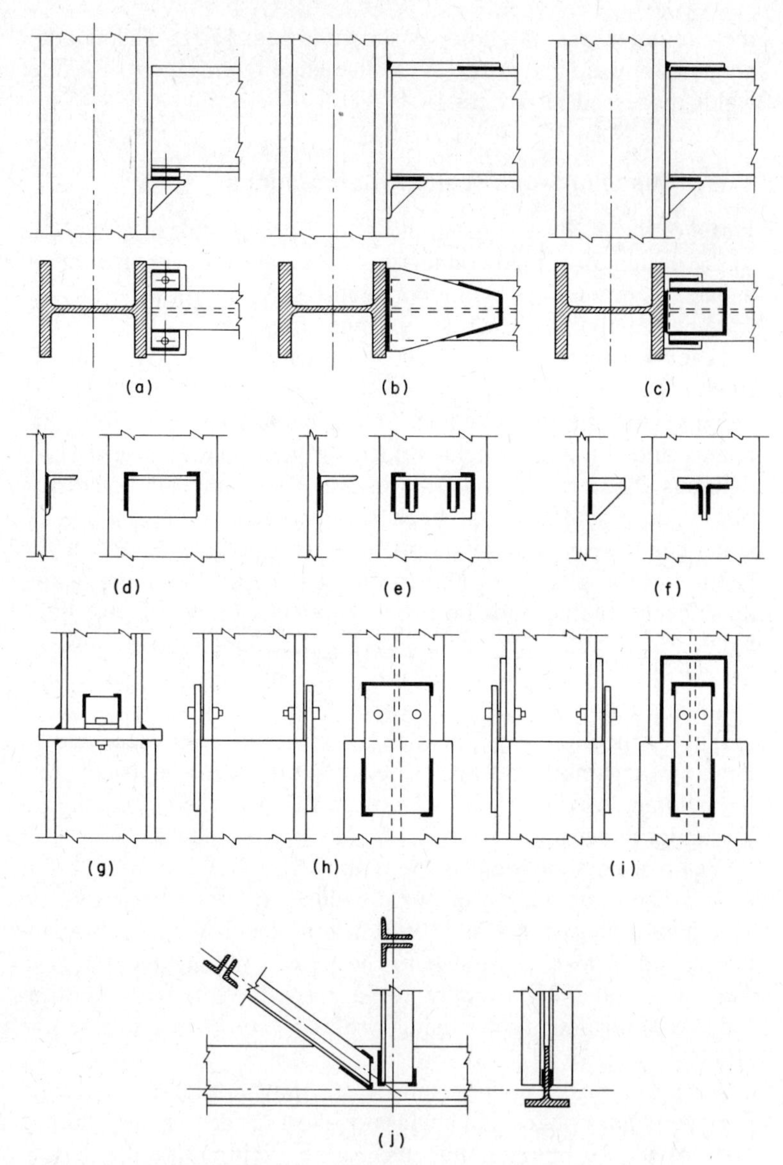

FIGURE 13-9. Welded framing connections.

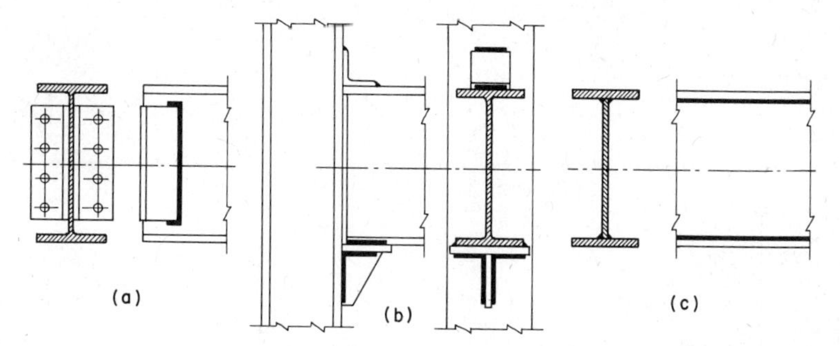

FIGURE 13-10. Some additional welded framing connections.

Figure 13-9*j* shows a type of welded connection used in light trusses in which the lower chord consists of a structural tee. Truss web members consisting of pairs of angles are welded to the stem of the tee chord. Other truss connections are shown in the details in Chapter 14.

Some additional connection details are given in Fig. 13-10. The detail in Fig. 13-10*a* is an arrangement for framing a beam to a girder in which welds are substituted for bolts or rivets. In this figure welds replace the fasteners that secure the connection angles to the web of the supported beam.

A welded connection for a stiffened seated beam connection to a column is shown in Fig. 13-10*b*. Figure 13-10*c* shows the simplicity of welding in connecting the upper and lower flanges of a plate girder to the web plate.

13-9. Symbols for Welds

Standard symbols are used in detail drawings of welded connections of structural elements. In addition to the type of weld, other information to be conveyed includes size, exact location, and finishes. Figure 13-11, reproduced from the AISC Manual, gives the standard symbols for welded joints. It will be noted that the symbol for a fillet weld is a triangle; this is drawn below the horizontal line if the weld is on the near side, above if it is on the

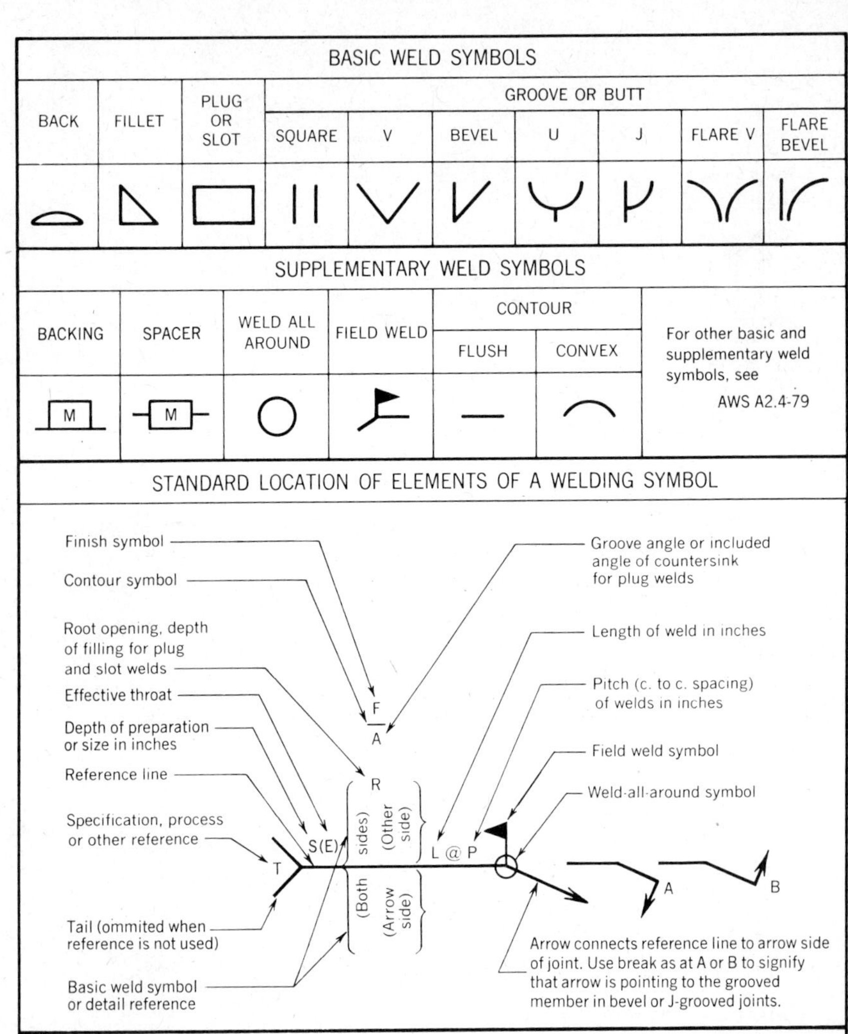

Note:

Size, weld symbol, length of weld and spacing must read in that order from left to right along the reference line. Neither orientation of reference line nor location of the arrow alter this rule.

The perpendicular leg of ◺, ⩗, ⊬, |⁄ weld symbols must be at left.

Arrow and Other Side welds are of the same size unless otherwise shown. Dimensions of fillet welds must be shown on both the Arrow Side and the Other Side Symbol.

The point of the field weld symbol must point toward the tail.

Symbols apply between abrupt changes in direction of welding unless governed by the "all around" symbol or otherwise dimensioned.

These symbols do not explicitly provide for the case that frequently occurs in structural work, where duplicate material (such as stiffeners) occurs on the far side of a web or gusset plate. The fabricating industry has adopted this convention: that when the billing of the detail material discloses the existence of a member on the far side as well as on the near side, the welding shown for the near side shall be duplicated on the far side.

FIGURE 13-11. Standard weld symbols used on construction drawings. Reprinted from the *Manual of Steel Construction*, 8th ed. (Ref. 1), with permission of the publishers. American Institute of Steel Construction.

far side; two triangles, one above and one below, are drawn for welds on both sides of the joint. The size of the weld is placed to the left of the vertical line of the triangle and the length to the right side of the hypotenuse. Figure 14-15 shows how this symbol is used to indicate the welding required in a lower-chord joint of a truss.

14

Roof Trusses

14-1. General

A truss is a framed structure, usually supported only at its ends, with a system of members so arranged and secured to one another that the stresses transmitted from one member to another are longitudinal only—that is, the members are in compression or tension. Basically, a truss is composed of a system of triangles because a triangle is the only polygon whose shape is incapable of being changed without changing the length of one or more of its sides.

The most common use of trusses is in the structural task of achieving a horizontal span. Figure 14-1 shows a typical single-span roof truss with a gable-form top and flat bottom, a type of truss extensively used for the dual function of providing for a pitched roof surface and a flat, horizontal ceiling surface. Some of the terminology used for the components of such a truss, as indicated in the illustration, are as follows:

Chord Members. These are the top and bottom boundary members of the truss, analogous to the top and bottom flanges of a steel beam. For trusses of modest size these members are often made of a single element that is continuous through several joints,

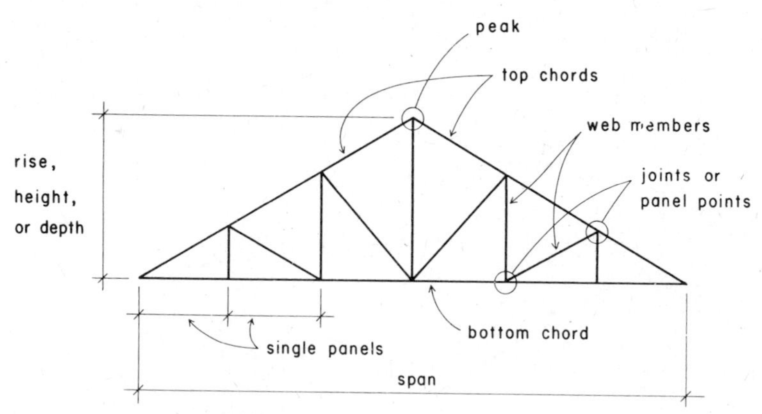

FIGURE 14-1. Elements of a truss.

with a total length limited only by the maximum ordinarily obtainable for the element selected.

Web Members. The interior members of the truss are called web members. Unless there are interior joints, these members are of a single piece between joints.

Panels. Most trusses have a pattern that consists of some repetitive, modular unit. This unit ordinarily is referred to as the panel of the truss; joints are sometimes referred to as panel points.

A critical dimension of a truss is its overall height, which is sometimes referred to as its rise or its depth. For the truss illustrated, this dimension relates to the establishment of the roof pitch and also determines the length of the web members. A critical concern with regard to the efficiency of the truss as a spanning structure is the ratio of the span of the truss to its height. Although beams and joists may be functional with span/height ratios as high as 20 to 30, trusses generally require much lower ratios.

Trusses may be used in a number of ways as part of the total structural system for a building. Figure 14-2 shows a series of single-span, planar trusses in the form shown in Figure 14-1 with the other elements of the building structure that develop the roof system and provide support for the trusses. In this example the

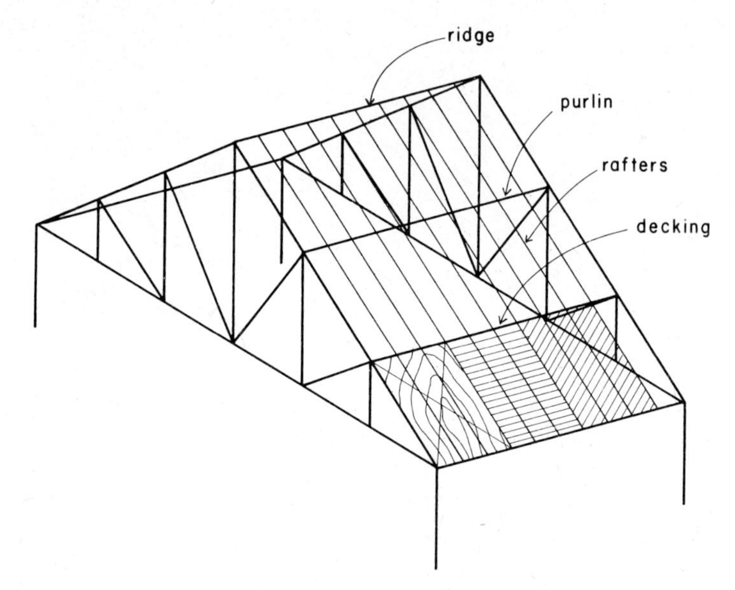

FIGURE 14-2. Elements of truss systems.

trusses are spaced a considerable distance apart. In this situation it is common to use purlins to span between the trusses, supported at the top chord joints of the trusses to avoid bending in the chords. The purlins, in turn, support a series of closely spaced rafters that are parallel to the trusses. The roof deck is then attached to the rafters so that the roof surface actually floats above the level of the top of the trusses.

Figure 14-3 shows a similar structural system of trusses with parallel chords. This system may be used for a floor or a flat roof.

When the trusses are slightly closer together it may be more practical to eliminate the purlins and to increase the size of the top chords to accommodate the additional bending due to the rafters. As an extension of this idea, if the trusses are really close, it may be possible to eliminate the rafters as well and to place the deck directly on the top chords of the trusses.

For various situations additional elements may be required for the complete structural system. If a ceiling is required, another framing system is used at the level of the bottom chords or sus-

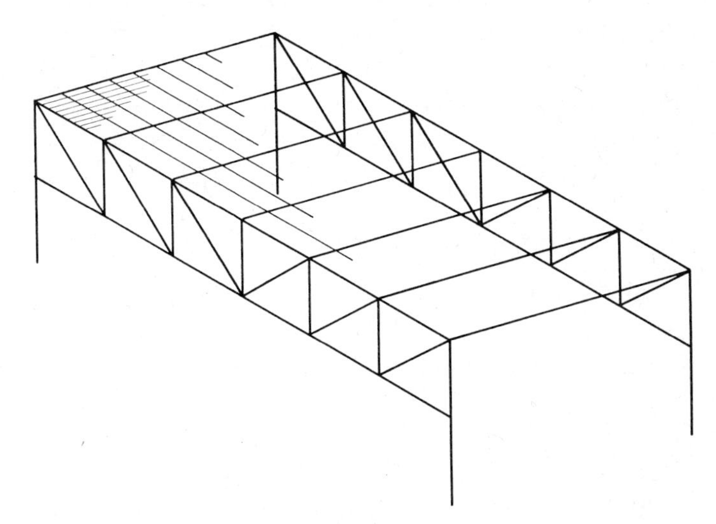

FIGURE 14-3. Structural system with flat chorded trusses.

pended some distance below it. If the roof and ceiling framing do not provide it adequately, it may be necessary to use some bracing system perpendicular to the trusses.

Truss patterns are derived from a number of considerations, starting with the basic profile of the truss. For various reasons a number of classic truss patterns have evolved and have become standard parts of our structural vocabulary. Some of these carry the names of the designers who first developed them. Several of these common truss forms are shown in Figure 14-4.

The two most common forms of steel trusses of small to medium size are those shown in Figure 14-5. In both cases the members may be connected by rivets, bolts, or welds. The most common practice is to use welding for connections that are assembled in the fabricating shop and high-strength bolts (torque tensioned) for field connections.

14-2. Loads on Trusses

The first step in the design of a roof truss consists of computing the loads the truss will be required to support. These are dead and

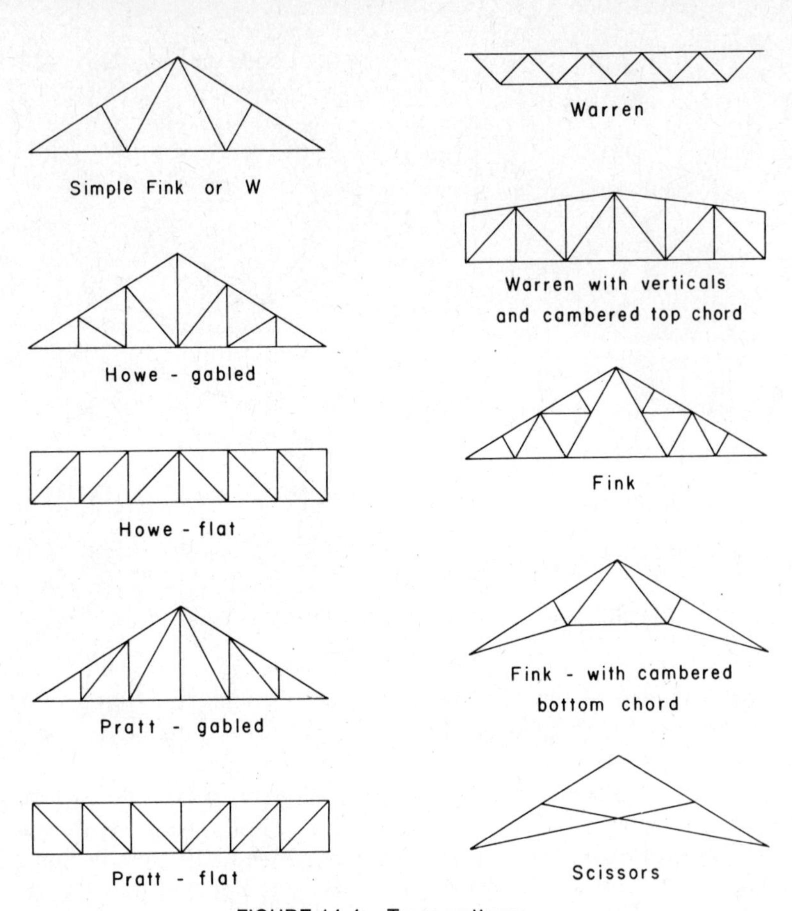

FIGURE 14-4. Truss patterns.

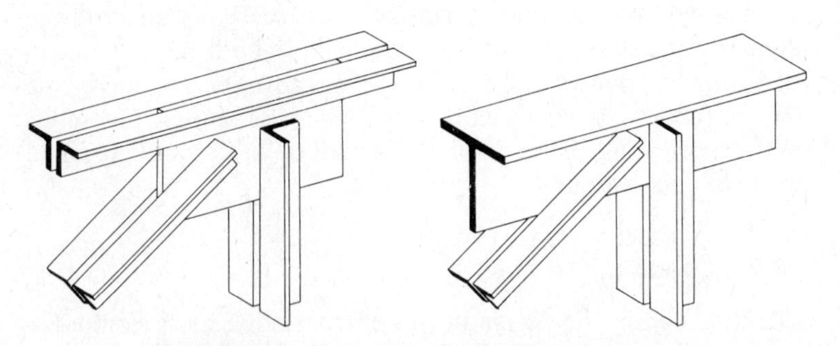

FIGURE 14-5. Typical details for light steel trusses.

live loads. The former includes the weight of all construction materials supported by the truss; the latter includes loads resulting from snow and wind, and, on flat roofs, occupancy loads and an allowance for the possible ponding of water due to impaired drainage.

The following items constitute the materials to be considered in computing the dead loads: roof covering and roof deck, purlins and sway bracing, ceiling and any suspended loads, and the weight of the truss itself. Obviously all cannot be determined exactly before the truss is designed, but all may be checked later to see whether a sufficient allowance has been made. The dead loads are downward vertical forces, hence the end reactions of the truss are also vertical with respect to these loads. Table 10-1 gives the weights of certain roofing materials and Table 14-1 provides estimated weights of steel trusses for various spans and pitches. With respect to the latter, one procedure is to establish an estimate in pounds per square foot of roof surface and consider this load as acting at the panel points of the upper chord. A more exact method would be to apportion a part of such loads to the panel points of the lower chord, but this is customary only in trusses with exceptionally long spans. After the truss has been designed its actual weight may be computed and compared with the estimated weight.

The weight allowance for snow load depends primarily on the geographical location of the structure and the roof slope. Freshly fallen snow may weigh as much as 10 lb per cu ft [0.13 kg/m^3] and

TABLE 14-1. Approximate Weight of Steel Trusses in Pounds per Square Foot of Roof Surface

Span	Slope of roof			
Feet	45°	30°	25°	Flat
Up to 40	5	6	7	8
40–50	6	7	7	8
50–60	7	8	9	10
60–70	7	8	9	10
70–80	8	9	10	11

accumulations of wet or packed snow may exceed this value. The amount of snow retained on a roof over a given period depends on the type of roofing as well as the slope; for example, snow slides off a metal or slate roof more readily than from a wood shingle surface; also, the amount of insulation in the roof construction will influence the period of retention.

Required design live loads for roofs are specified by local building codes. When snow is a potential problem the load is usually based on anticipated snow accumulation. Otherwise the specified load is intended essentially to provide some capacity for sustaining loads experienced during construction and maintenance of the roof. The basic required load can usually be modified when the roof slope is of some significant angle and on the basis of the total roof surface area supported by the structure. Table 10-2 gives the minimum roof live loads specified by the *Uniform Building Code*, 1982 ed. (Ref. 2), which are based on the situation in which snow load is not the critical concern.

Magnitudes of design wind pressures and various other requirements for wind design are specified by local building codes. The code in force for a specific building location should be used for any design work. For a general explanation of the analysis and design of the effects of wind and earthquake forces on buildings the reader is referred to *Simplified Building Design for Wind and Earthquake Forces* by Ambrose and Vergun (Ref. 5). For an illustration of the analysis of wind effects on a roof truss see *Simplified Design of Building Trusses* by Parker (Ref. 4).

14-3. Graphical Analysis for Internal Forces in Planar Trusses

Figure 14-6 shows a single span, planar truss that is subjected to vertical gravity loads. We use this example to illustrate the procedures for determining the internal forces in the truss; that is, the tension and compression forces in the individual members of the truss. The space diagram in the figure shows the truss form, the support conditions, and the loads. The letters on the space diagram identify individual forces at the truss joints. The sequence of placement of the letters is arbitrary, the only necessary consid-

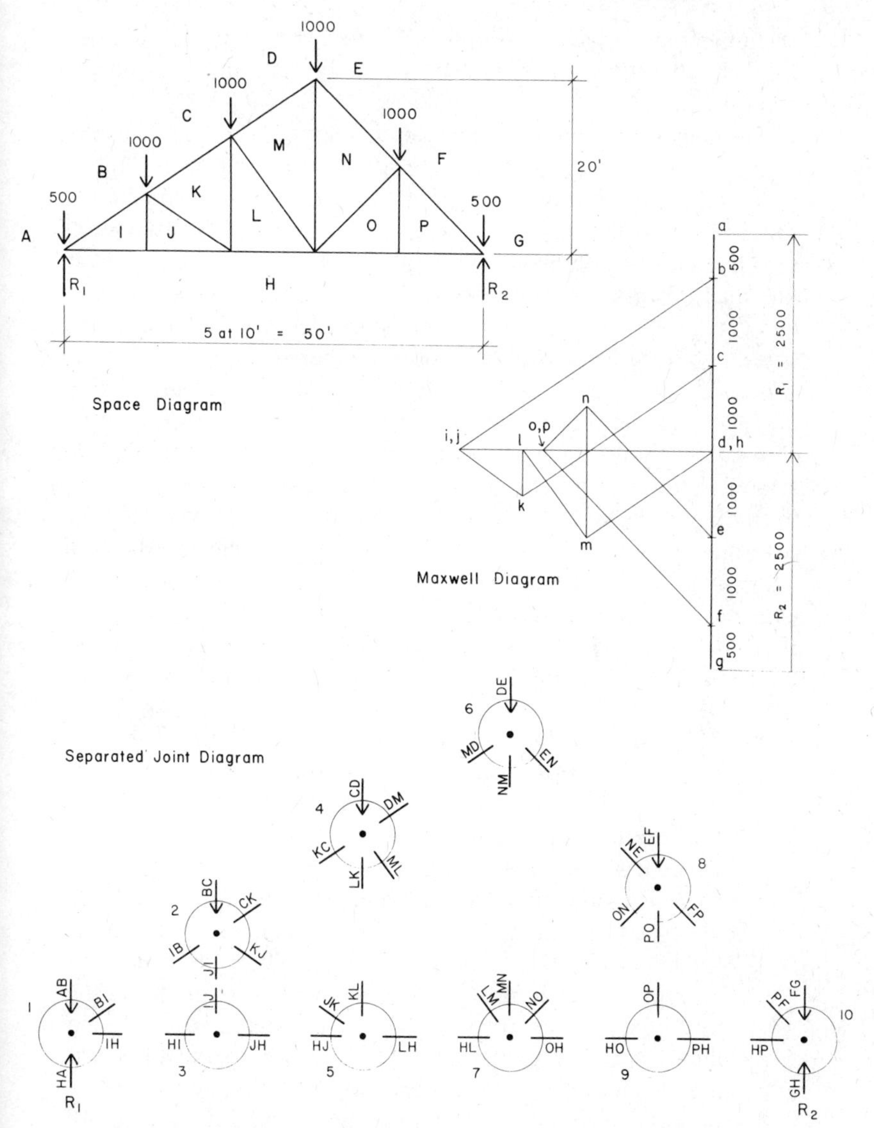

FIGURE 14-6. Examples of graphic diagrams for a planar truss.

eration being to place a letter in each space between the loads and the individual truss members so that each force at a joint can be identified by a two-letter symbol.

The separated joint diagram in the figure provides a useful means of visualizing the complete force system at each joint as well as the interrelation of the joints through the truss members. The individual forces at each joint are designated by two-letter symbols that are obtained by simple reading around the joint in the space diagram in a clockwise direction. Note that the two-letter symbols are reversed at the opposite ends of each of the truss members. Thus the top chord member at the left end of the truss is designated as *BI* when shown in the joint at the left support (joint 1) and as *IB* when shown in the first interior upper chord joint (joint 2). The purpose of this procedure is demonstrated in the following explanation of the graphical analysis. The third diagram in Figure 14-6 is a composite force polygon for the external and internal forces in the truss. It is called a Maxwell diagram after its originator, James Clerk Maxwell, an English engineer. The construction of this diagram constitutes a complete solution for the magnitudes and senses of the internal forces in the truss. The procedure for this construction is as follows.

1. *Construct the force polygon for the external forces*. Before this can be done the values for the reactions must be found. There are graphic techniques for finding the reactions, but it is usually much simpler and faster to find them with an algebraic solution. In this example, although the truss is not symmetrical, the loading is, and it may be observed that each of the reactions is equal to one-half the total load on the truss, or $\frac{5000}{2} = 2500$ lb. Because the external forces in this case are all in a single direction, the force polygon for the external forces is actually a straight line. Using the two-letter symbols for the forces and starting with letter *A* at the left end, we read the force sequence by moving in a clockwise direction around the outside of the truss. Thus the loads are read as *AB*, *BC*, *CD*, *DE*, *EF*, and *FG* and the two reactions are read as *GH* and *HA*. By beginning at *A* on the Maxwell diagram the force vector

sequence for the external forces is read from *A* to *B*, *B* to *C*, *C* to *D*, and so on, ending back at *A*, which shows that the force polygon closes and the external forces are in the necessary state of static equilibrium. Note that we have pulled the vectors for the reactions off to the side in the diagram to indicate them more clearly. Note also that we have used lowercase letters for the vector ends in the Maxwell diagram, whereas uppercase letters appear on the space diagram. The alphabetic correlation is retained (*A* to *a*) and any possible confusion between the two diagrams is prevented. The letters on the space diagram designate spaces, whereas the letters on the Maxwell diagram designate points of intersection of lines.

2. *Construct the force polygons for the individual joints*. The graphic procedure for this consists of locating the points on the Maxwell diagram that correspond to the remaining letters, *I* through *P*, on the space diagram. When all the lettered points on the diagram are located, the complete force polygon for each joint may be read on the diagram. To locate these points we use two relationships. The first is that the truss members can resist only those forces that are parallel to the members' positioned directions. Thus we know the directions of all the internal forces. The second relationship is a simple one from plane geometry: A point may be located as the intersection of two lines. Consider the forces at joint 1, as shown in the separated joint diagram in Figure 14-6. Note that there are four forces and that two of them are known (the load and the reaction) and two are unknown (the internal forces in the truss members). The force polygon for this joint, as shown on the Maxwell diagram, is read as *ABIHA*. *AB* represents the load, *BI*, the force in the upper chord member, *IH*, the force in the lower chord member, and *HA*, the reaction. Thus the location of point *I* on the Maxwell diagram is determined by noting that *I* must be in a horizontal direction from *H* (corresponding to the horizontal position of the lower chord) and in a direction from *B* parallel to the position of the upper chord.

The remaining points on the Maxwell diagram are found by the same process, using two known points on the diagram to project lines of known direction whose intersection will determine the location of another point. Once all the points are located the diagram is complete and can be used to find the magnitude and sense of each internal force. The process for construction of the Maxwell diagram typically consists of moving from joint to joint along the truss. Once one of the letters for an internal space is determined on the Maxwell diagram it may be used as a known point for finding the letter for an adjacent space on the space diagram. The only limitation of the process is that it is not possible to find more than one unknown point on the Maxwell diagram for any single joint. Consider joint 7 on the separated joint diagram in Figure 14-6. If we attempt to solve this joint first, knowing only the locations of letters *A* through *H* on the Maxwell diagram, we must locate four unknown points: *L*, *M*, *N*, and *O*. This is three more unknowns than we can determine in a single step, and we must first solve for three of the unknowns by using other joints.

Solving for a single unknown point on the Maxwell diagram corresponds to finding two unknown forces at a joint because each letter on the space diagram is used twice in the force identifications for the internal forces. Thus for joint 1 in the previous example the letter *I* is part of the identity for forces *BI* and *IH*, as shown on the separated joint diagram. The graphic determination of single points on the Maxwell diagram is therefore analogous to finding two unknown quantities in an algebraic solution. As discussed previously, two unknowns are the maximum that can be solved for in the equilibrium of a coplanar, concurrent force system, which is the condition of the individual joints in the truss.

When the Maxwell diagram is completed the internal forces can be read from the diagram as follows:

1. The magnitude is determined by measuring the length of the line in the diagram with the scale that was used to plot the vectors for the external forces.
2. The sense of individual forces is determined by reading the

forces in clockwise sequence around a single joint in the space diagram and tracing the same letter sequences on the Maxwell diagram.

The degree of accuracy attainable from a graphical analysis depends on the size of the construction and the accuracy of the drafting. The results of the analysis of the truss shown in Fig. 14-6 are displayed on the truss form in Fig. 14-7; *C* indicates compression and *T* indicates tension. Zero force members are indicated by placing a zero directly on the member. The values shown were actually determined from an algebraic analysis, since four place accuracy is not attainable from a graphical construction.

14-4. Design Forces for Truss Members

The primary concern in analysis of trusses is the determination of the critical forces for which each member of the truss must be designed. The first step in this process is the decision about which combinations of loading must be considered. In some cases the potential combinations may be quite numerous. When both wind and seismic actions are potentially critical and more than one type of live loading occurs (e.g., roof loads plus hanging loads) the theoretically possible combinations of loadings can be overwhelming. However, designers are usually able to exercise judgment in reducing the sensible combinations to a reasonable number; for example, it is statistically improbable that a violent windstorm will occur simultaneously with a major earthquake shock.

Once the required design loading conditions are established the usual procedure is to perform separate analyses for each of the loadings. The values obtained can then be combined at will for each member to ascertain the particular combination that establishes the critical result for the member. This means that in some cases certain members will be designed for one combination and others, for different combinations.

In most cases design codes permit an increase in allowable stress for design of members when the critical loading includes forces due to wind or seismic loads.

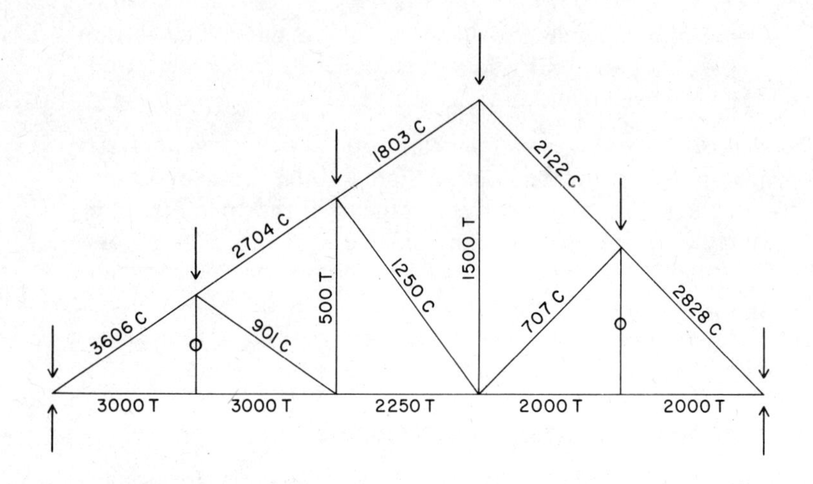

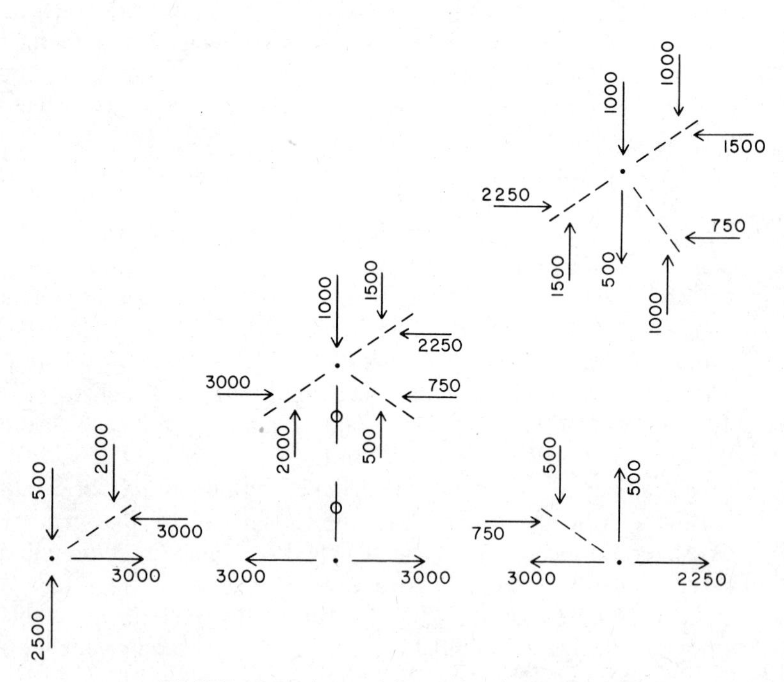

FIGURE 14-7. Internal forces for the truss.

14-5. Combined Stress in Truss Members

When analyzing trusses the usual procedure is to assume that the loads will be applied to the truss joints. This results in the members themselves being loaded only through the joints and thus having only direct tension or compression forces. In some cases, however, truss members may be directly loaded; for example, when the top chord of a truss supports a roof deck without benefit

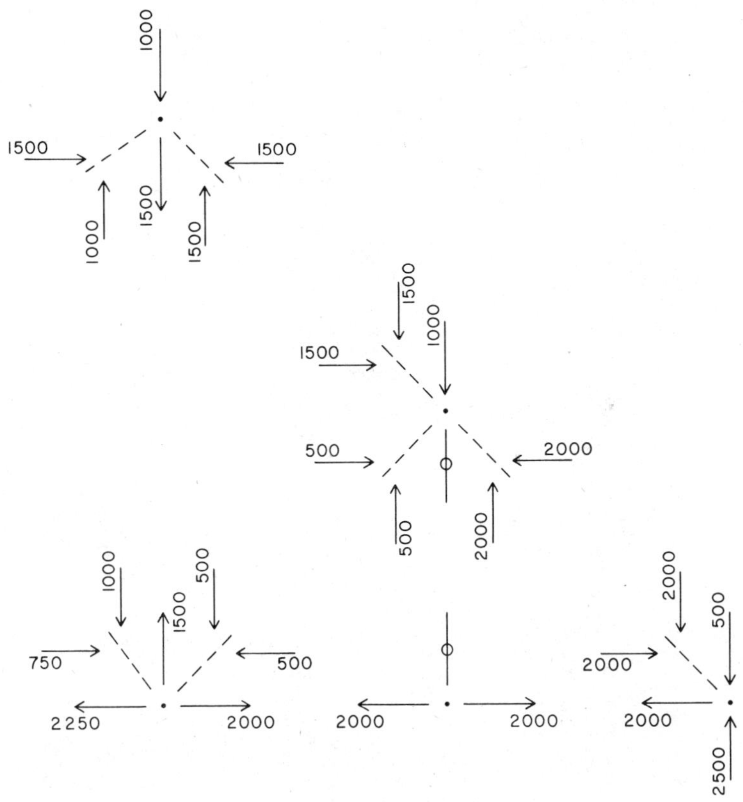

FIGURE 14-7. (*Continued*)

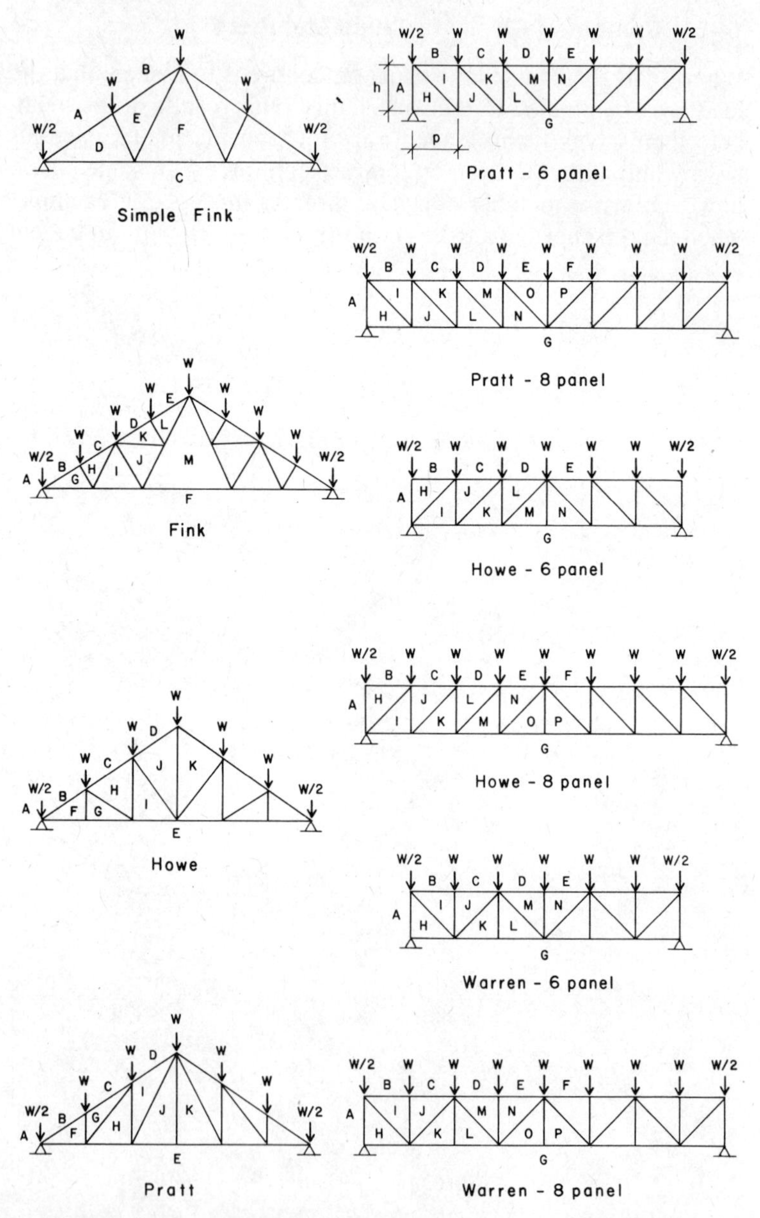

FIGURE 14-8. Simple trusses of parallel-chorded and gable form.

TABLE 14-2. Coefficients for Internal Forces in Simple Trusses

Force in members = (table coefficient) X (panel load, W)

T indicates tension, C indicates compression

Gable Form Trusses

Truss Member	Type of Force	Roof Slope 4/12	Roof Slope 6/12	Roof Slope 8/12
Truss 1 - Simple Fink				
AD	C	4.74	3.35	2.70
BE	C	3.95	2.80	2.26
DC	T	4.50	3.00	2.25
FC	T	3.00	2.00	1.50
DE	C	1.06	0.90	0.84
EF	T	1.06	0.90	0.84
Truss 2 - Fink				
BG	C	11.08	7.83	6.31
CH	C	10.76	7.38	5.76
DK	C	10.44	6.93	5.20
EL	C	10.12	6.48	4.65
FG	T	10.50	7.00	5.25
FI	T	9.00	6.00	4.50
FM	T	6.00	4.00	3.00
GH	C	0.95	0.89	0.83
HI	T	1.50	1.00	0.75
IJ	C	1.90	1.79	1.66
JK	T	1.50	1.00	0.75
KL	C	0.95	0.89	0.83
JM	T	3.00	2.00	1.50
LM	T	4.50	3.00	2.25
Truss 3 - Howe				
BF	C	7.90	5.59	4.51
CH	C	6.32	4.50	3.61
DJ	C	4.75	3.35	2.70
EF	T	7.50	5.00	3.75
EI	T	6.00	4.00	3.00
GH	C	1.58	1.12	0.90
HI	T	0.50	0.50	0.50
IJ	C	1.81	1.41	1.25
JK	T	2.00	2.00	2.00
Truss 4 - Pratt				
BF	C	7.90	5.59	4.51
CG	C	7.90	5.59	4.51
DI	C	6.32	4.50	3.61
EF	T	7.50	5.00	3.75
EH	T	6.00	4.00	3.00
EJ	T	4.50	3.00	2.25
FG	C	1.00	1.00	1.00
GH	T	1.81	1.41	1.25
HI	C	1.50	1.50	1.50
IJ	T	2.12	1.80	1.68

Flat - Chorded Trusses

Truss Member	Type of Force	6 Panel Truss $\frac{h}{p} = 1$	6 Panel Truss $\frac{h}{p} = \frac{3}{4}$	8 Panel Truss $\frac{h}{p} = 1$	8 Panel Truss $\frac{h}{p} = \frac{3}{4}$
Truss 5 - Pratt					
BI	C	2.50	3.33	3.50	4.67
CK	C	4.00	5.33	6.00	8.00
DM	C	4.50	6.00	7.50	10.00
EO	C	–	–	8.00	10.67
GH	O	0	0	0	0
GJ	T	2.50	3.33	3.50	4.67
GL	T	4.00	5.33	6.00	8.00
GN	T	–	–	7.50	10.00
AH	C	3.00	3.00	4.00	4.00
IJ	C	2.50	2.50	3.50	3.50
KL	C	1.50	1.50	2.50	2.50
MN	C	1.00	1.00	1.50	1.50
OP	C	–	–	1.00	1.00
HI	T	3.53	4.17	4.95	5.83
JK	T	2.12	2.50	3.54	4.17
LM	T	0.71	0.83	2.12	2.50
NO	T	–	–	0.71	0.83
Truss 6 - Howe					
BH	O	0	0	0	0
CJ	C	2.50	3.33	3.50	4.67
DL	C	4.00	5.33	6.00	8.00
EN	C	–	–	7.50	10.00
GI	T	2.50	3.33	3.50	4.67
GK	T	4.00	5.33	6.00	8.00
GM	T	4.50	6.00	7.50	10.00
GO	T	–	–	8.00	10.67
AH	C	0.50	0.50	0.50	0.50
IJ	T	1.50	1.50	2.50	2.50
KL	T	0.50	0.50	1.50	1.50
MN	T	0	0	0.50	0.50
OP	O	–	–	0	0
HI	C	3.53	4.17	4.95	5.83
JK	C	2.12	2.50	3.54	4.17
LM	C	0.71	0.83	2.12	2.50
NO	C	–	–	0.71	0.83
Truss 7 - Warren					
BI	C	2.50	3.33	3.50	4.67
DM	C	4.50	6.00	7.50	10.00
GH	O	0	0	0	0
GK	T	4.00	5.33	6.00	8.00
GO	T	–	–	8.00	10.67
AH	C	3.00	3.00	4.00	4.00
IJ	C	1.00	1.00	1.00	1.00
KL	O	0	0	0	0
MN	C	1.00	1.00	1.00	1.00
OP	O	–	–	0	0
HI	T	3.53	4.17	4.95	5.83
JK	C	2.12	2.50	3.54	4.17
LM	T	0.71	0.83	2.12	2.50
NO	C	–	–	0.71	0.83

of joists. Thus the chord member is directly loaded with a linear uniform load and functions as a beam between its end joints.

The usual procedure in these situations is to accumulate the loads at the truss joints and analyze the truss as a whole for the typical joint loading arrangement. The truss members that sustain the direct loading are then designed for the combined effects of the axial force caused by the truss action and the bending caused by the direct loading.

A typical situation for a roof truss is one in which the actual loading consists of the roof load distributed continuously along the top chords and a ceiling loading distributed continuously along the bottom chords. The top chords are thus designed for a combination of axial compression and bending and the bottom chords for a combination of axial tension plus bending. This will of course result in somewhat larger members being required for both chords and any estimate of the truss weight should account for this anticipated additional requirement.

14-6. Internal Forces Found by Coefficients

Figure 14-8 shows a number of simple trusses of both parallel-chorded and gable form. Table 14-2 lists coefficients that may be used to find the values for the internal forces in these trusses. For the gable-form trusses coefficients are given for three different slopes of the top chord: 4 in 12, 6 in 12, and 8 in 12. For the parallel-chorded trusses coefficients are given for two different ratios of the truss depth to the truss panel length: 1 to 1 and 3 to 4. Loading results from vertical gravity loads and is assumed to be applied symmetrically to the truss; internal panel point loads are equal to W.

The table values are based on a value of $W = 1.0$. To use the tables it is necessary only to find the true value of the panel point loading and multiply it by the table coefficient to find the force in a truss member. Note that because of the symmetry of the trusses and the loads, the internal forces in the members are the same on each half of the truss; therefore we have given the coefficients for only the left half of each truss.

14-7. Design of a Steel Roof Truss

The following example is used to illustrate several of the issues and the general process in the design of short-to-medium-span roof trusses. The form of the truss is shown in Fig. 14-9. The principal loading consists of the concentrated panel point loads on the top, delivered by steel purlins that span between the trusses. The purlins support a roof of cement-bonded wood fiber

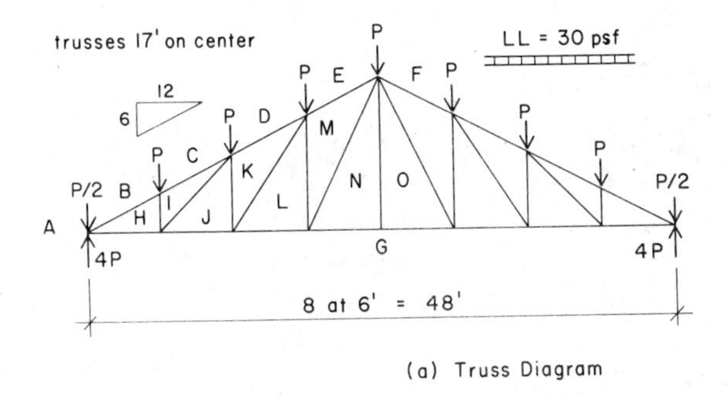

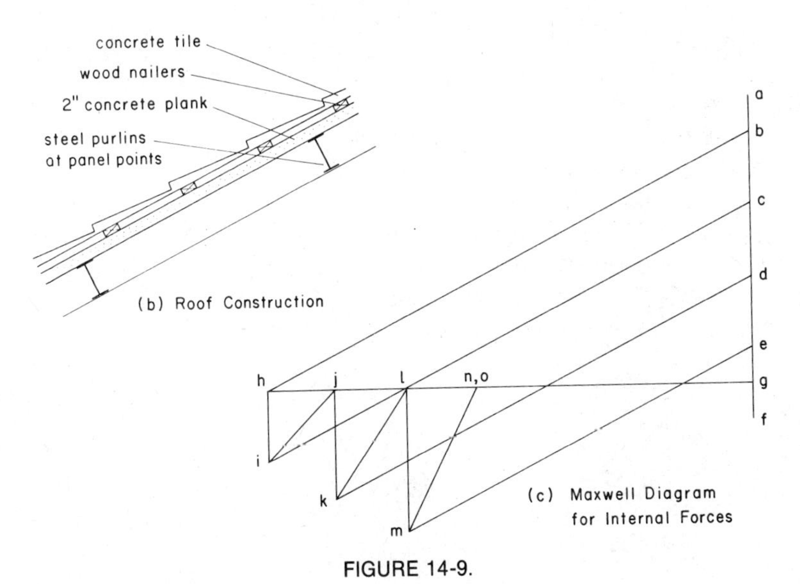

FIGURE 14-9.

deck units, wood nailing strips, and cement tile roofing. The dead weight of this roof construction is estimated for design:

Concrete tile	10 psf	0.48 kN/m²
Wood nailing strips	1	0.05
Deck units	8	0.38
Purlins (estimate)	3	0.14
Total dead load	22 psf	1.05 kN/m²

14-8. Design of the Purlin

Selection of the shape to be used for the purlin depends on a number of considerations, including the need for attachment of the deck to the purlins and the purlins to the trusses. Several possibilities for the choice of a purlin are shown in Fig. 14-10. If a W shape is placed directly on the top chord of the truss as shown in Fig. 14-10*a*, a problem that must be considered is the bending of the purlin on its weak axis because of its tilted position. Two modifications that may be used to eliminate this problem are shown at *b* and *c* in Fig. 14-10. Sag rods and channels were common in earlier times but other options are now considered more favorable.

A consideration that must be made is whether the deck provides lateral bracing for the purlins. The type of deck described for this construction probably does not offer good bracing and the best choice for the purlin may be the tube shown in Fig. 14-10*d*. Without further belaboring the decision we choose the tube and proceed with its design for the biaxial bending.

As shown in Fig. 14-11, the dead weight of the roof is distrib-

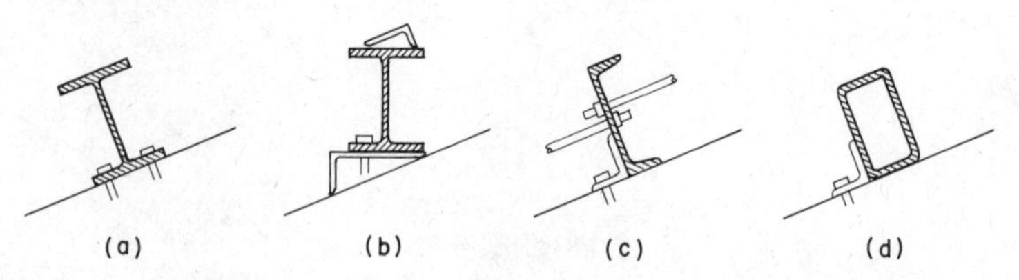

FIGURE 14-10. Alternatives for the steel purlins.

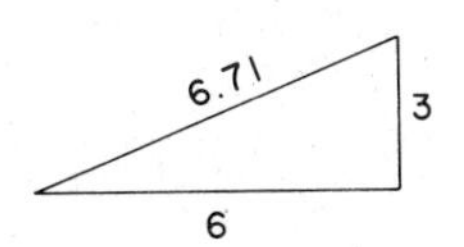

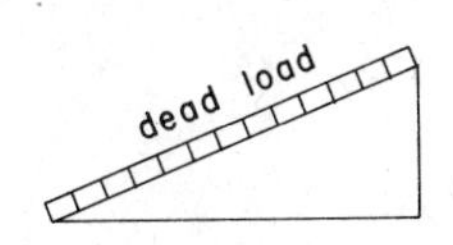

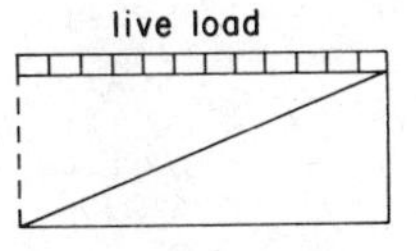

FIGURE 14-11. Roof load distribution on the truss.

uted over the actual roof surface. The load carried by a single interior purlin is thus

$$w_d = 22 \times 6.71 \times 17 = 2510 \text{ lb } [11.16 \text{ kN}]$$

The live load is distributed on the horizontal plane and the purlin load is

$$w_L = 30 \times 6 \times 17 = 3060 \text{ lb } [13.61 \text{ kN}]$$

The total load is 2510 + 3060 = 5570 lb [24.77 kN].

If the member is placed in a tilted position, as shown at *a, b,* and *d* in Fig. 14-10, this vertical load must be divided into two components that relate to the major and minor axes of the member. The bending moments developed by these two load components are determined and the combined effect is investigated by use of the combined action formula

$$\frac{f_{bx}}{F_{bx}} + \frac{f_{by}}{F_{by}} \leq 1.0$$

where f_{bx} = the actual bending stress about the x axis,
F_{bx} = the allowable bending stress for the major axis of the shape,
f_{by} = the actual bending stress about the y axis,
F_{by} = the allowable bending stress for the minor axis of the shape.

For W and S shapes a higher allowable stress is permitted for the minor axis; $F_{by} = 0.75F_y$. For the tube, however, the allowable stresses are the same.

For a first trial we consider the use of a rectangular steel tube, TS 6 × 4 × 0.3125, for which Table A-13 gives $S_x = 8.72$ in.^{3} and $S_y = 6.92$ in.^{3} Using the components of the load as shown in Fig.

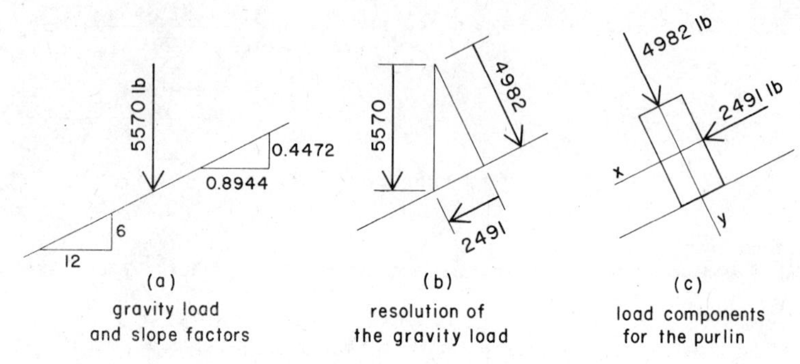

FIGURE 14-12. Resolution of forces on the purlin.

14-12, we find

$$M_x = \frac{WL}{8} = \frac{4982 \times 17}{8} = 10{,}587 \text{ lb-ft } [14.35 \text{ kN-m}]$$

$$M_y = \frac{WL}{8} = \frac{2491 \times 17}{8} = 5293 \text{ lb-ft } [7.17 \text{ kN-m}]$$

The corresponding maximum bending stresses are

$$f_{bx} = \frac{M_x}{S_x} = \frac{10{,}587 \times 12}{8.72} = 14{,}572 \text{ psi } [100.4 \text{ MPa}]$$

$$f_{by} = \frac{M_y}{S_y} = \frac{5293 \times 12}{6.92} = 9179 \text{ psi } [63.3 \text{ MPa}]$$

Using the formula for the combined action analysis, we obtain

$$\frac{f_{bx}}{F_{bx}} + \frac{f_{by}}{F_{by}} = \frac{14{,}572}{24{,}000} + \frac{9179}{24{,}000} = 0.607 + 0.382 = 0.989$$

Because this is less than 1.0, the section is adequate.

We note from Table A-13 that this tube weighs 19.08 lb/ft, which means that the average weight of the purlins is 19.08/6 = 3.18 psf. This is close enough to our estimate of 3 psf to warrant no revision in the design loading.

14-9. Determination of the Truss Reactions and Internal Forces

The loading on the truss is assumed to be that shown in Fig. 14-9. A concentrated load of P occurs at each of the interior panel points of the top chord and a load of one-half P occurs at the truss ends. The total load is thus $8P$ and the reactions for the symmetrical truss are each $8P/2 = 4P$. The load P consists of the total load on one purlin plus a portion of the weight of the truss. From Table 14-1 we determine that the estimated weight should be 7 psf for the supported area. Thus the panel load increment becomes $7 \times 6 \times 17 = 714$ lb. By adding this to the total purlin load found in Art. 14-8 we obtain the value of P as $714 + 5570 = 6284$ lb.

For the truss analysis this value of P is rounded off to 6.3 kips. The Maxwell diagram for one-half the truss is shown in Fig. 14-9. The results of the analysis of internal forces are recorded on the truss diagram in Fig. 14-13a. For a truss of this size the top

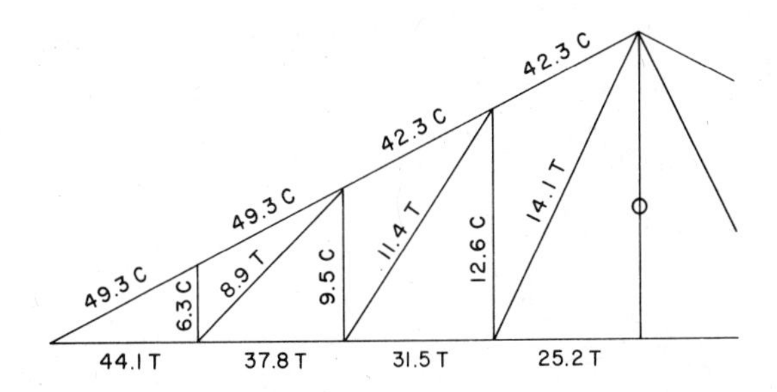

a) internal forces – in kips

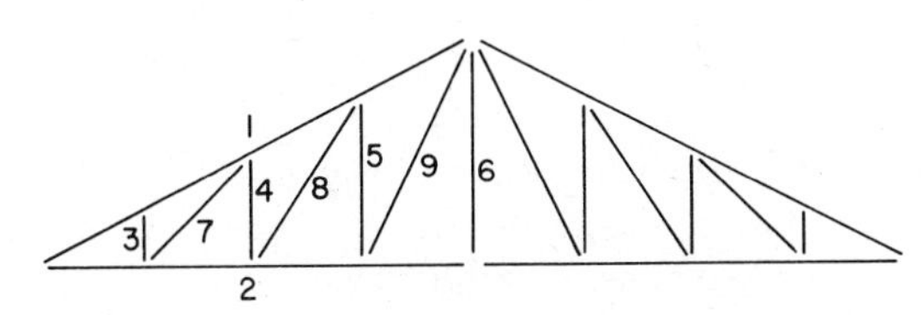

b) individual truss members

FIGURE 14-13. Internal forces and truss member arrangement.

chords on each side of the peak would probably be made of a single piece. The bottom chord is a bit long for a single piece and would likely be made with at least one splice. A scheme for the truss member configuration on the basis of these assumptions is shown in Fig. 14-13*b*.

Because of the relatively low slope of the roof and the magnitude of the dead load, it is not likely that this truss will be critically affected by the wind load unless an unusually high wind pressure must be sustained. We proceed with the assumption that design for gravity load is adequate.

14-10. Selection of the Truss Members

For a truss of this size the construction is most likely to be one of the following:

Top and bottom chords of structural tees with webs of double angles welded directly to the stems of the tees.

All double angles with joints using steel gusset plates; the angles are connected to the plates by welds or high strength bolts.

Another possibility is a truss of welded steel pipe or tube members, which are usually quite expensive and therefore not so favorable unless their neat appearance is considered important for trusses that are exposed to view.

For purpose of illustration we choose the members on the basis of an all-welded construction with double angles and gusset plates. Before proceeding to a consideration of the design of individual members some general limitations and requirements should be established:

1. *Minimum thickness of angle leg*. This is based on the minimum weld that is used. For this size truss it is advisable to select a minimum size of $\frac{1}{4}$ in. [6.35 mm] for the fillet welds. For this weld the minimum dimension recommended for the angle leg is $\frac{5}{16}$ in. [7.94 mm].
2. *Minimum radius of gyration*. Based on the recommendations of Sec. 1.8.4 of the AISC Specification, the limits for

slenderness ratios are $L/r = 200$ for compression and $L/240$ for tension members.

3. *Unbraced length of members.* For web members this is their actual length in either direction. For the chords the unbraced length in the plane of the truss is the panel length. However, the unbraced length perpendicular to the plane of the truss depends on the lateral bracing provided for the truss.

Considering the last point raised, we take the lateral unbraced length to be the panel length for the top chord because the purlins occur at the panel points. For the bottom chord we anticipate that the trusses will be braced at alternate panel points by some form of horizontal bracing. The critical unsupported length of the bottom chord, in a direction perpendicular to the plane of the truss, is therefore 12 ft.

The critical design factors for the truss members are summarized in Table 14-3. The members are referred to by the numbers shown on the proposed truss layout in Fig. 14-13. Members may be designed directly from the table data, subject to the limitations listed. For tension members the critical concern is the stress on the effective area discussed in Art. 12-4. For compression members the slenderness ratios are usually quite high, requiring a design with the reduced value of compression stress as discussed in Art. 11-5. The design of compression members is usually made easier with the use of column load tables, such as those in Chapter 11. Table 11-4 is an abbreviated example of the more extensive tables of double angles in the AISC Manual (Ref. 1).

There are three possibilities for combinations of double angles: equal leg angles, unequal leg angles with long legs back-to-back, and unequal leg angles with short legs back-to-back. The unequal leg angles with long legs back-to-back are often used because they tend to produce a combination that has its stiffness and radius of gyration on both axes closer in value. They also tend to be stronger in bending on the x-axis, offering more resistance to sag when occurring in other than a vertical position. For the bottom chord in this example problem, however, it may be more desirable to use unequal leg angles with short legs back-to-back be-

TABLE 14-3. Data for the Truss Members

Truss member (See Fig. 14-13)	Design force (kips)	Slenderness considerations: Critical length and axis (ft)	Slenderness considerations: Minimum r and axis[a] (in.)	Selection and properties: Angles long legs back-to-back	r_x (in.)	r_y (in.)	Weight (lb/ft)	Capacity[b] (kips)
1	49.3 C	6.71	0.40	$3\frac{1}{2} \times 2\frac{1}{2} \times \frac{5}{16}$	1.11	1.10	12.2	56
2	44.1 T	$6 - x$	$0.30 - x$	$3\frac{1}{2} \times 2\frac{1}{2} \times \frac{5}{16}$	1.11	1.10	12.2	48
		$12 - y$	$0.60 - y$					
3	6.3 C	3	0.18	$2 \times 2 \times \frac{5}{16}$	0.601	1.0	7.84	40
4	9.5 C	6	0.36	$2 \times 2 \times \frac{5}{16}$	0.601	1.0	7.84	24
5	12.6 C	9	0.54	$2 \times 2 \times \frac{5}{16}$	0.601	1.0	7.84	11
6	0	12	0.60	$2 \times 2 \times \frac{5}{16}$	0.601	1.0	7.84	27 T
7	8.9 T	8.48	0.42	$2 \times 2 \times \frac{5}{16}$	0.601	1.0	7.84	27
8	11.4 T	10.8	0.52	$2 \times 2 \times \frac{5}{16}$	0.601	1.0	7.84	27
9	14.1 T	13.4	0.65	$2\frac{1}{2} \times 2 \times \frac{5}{16}$	0.776	0.948	9.0	34

[a] Required for both axes if separate values are not given.
[b] Column load for compression members. Area times 22 ksi for tension members, using the area of the connected legs only.

cause they will have greater stiffness in the direction perpendicular to the plane of the truss.

Although other choices are possible, the angle combinations shown in Table 14-3 satisfy the various data and special requirements discussed in this example. The total weight of the truss members, determined by multiplying the member weights by the member lengths, is 2162 lb, to which must be added the weight of gusset plates, lacing, support connections, purlin seat connections, and cross bracing. The total weight is thus likely to be as much as 50% more than that of the members alone. If this is true, the weight per sq ft of supported area is

$$w = \frac{1.5 \times 2162}{48 \times 17} = 3.97 \text{ psf}$$

which is somewhat less than the value of 7 psf obtained for estimation of the truss weight from Table 14-1. This is due in part to the fact that the truss is welded; bolted connections are generally somewhat heavier.

14-11. Design of the Truss Joints

We now consider the design of the welded truss joints. Although there are 10 separate joints, we illustrate the process by considering only two: the left support and the bottom chord joint at 12 ft from the left support.

Figure 14-14 shows a possible layout for the bottom chord joint. Four truss members intersect at this joint: GJ, JK, KL, and LG. However, because the chord is made continuous through the joint, only three sets of angles are connected to the gusset plate.

A basic consideration in the layout of this joint is the need to avoid twisting, which is generally accomplished by having the action lines of the axial forces in the members intersect at a common point: the panel point of the truss. If the stresses in the members are uniformly distributed on the full cross sections of the members, the action lines will coincide with the centroids of the double-angle combinations. The connections to the gusset plates, however, do not fully develop the stresses in the outstand-

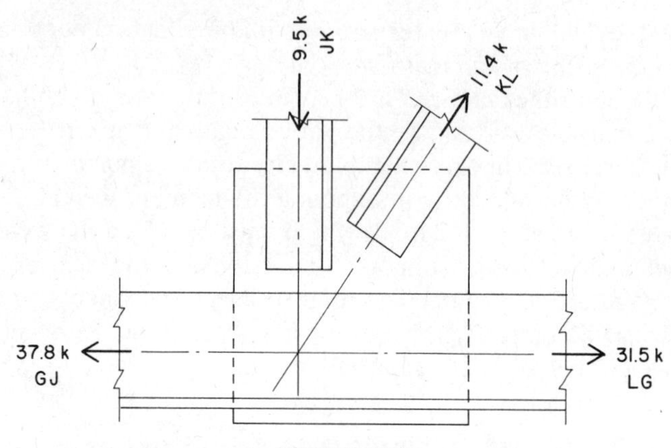

FIGURE 14-14.

ing legs of the angles. Therefore some designers prefer to consider that the forces at the joint occur symmetrically with respect to the connected legs, as indicated in the layout in Fig. 14-14.

If the forces are axial to the full cross sections of the members, the joint design will develop as illustrated in Fig. 13-6 and explained in Art. 13-5. If the forces at the joint are symmetrical with respect to the connected leg, the welds will be placed symmetrically, as shown in Fig. 14-15. The design of the former is already explained in Art. 13-5; therefore we describe the latter approach in this example.

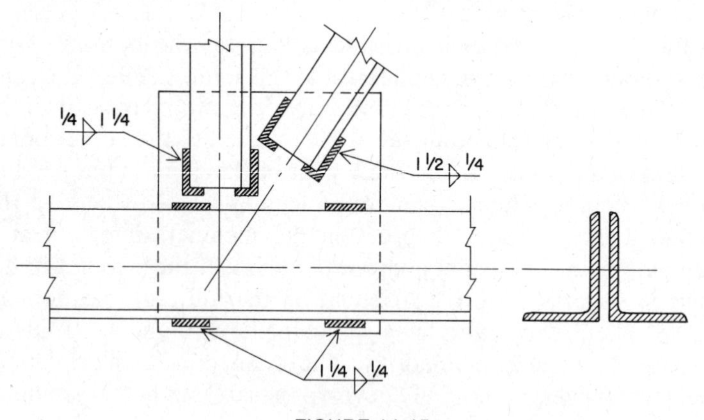

FIGURE 14-15.

Using the previously selected size of $\frac{1}{4}$ in. for the fillet welds and assuming welds made with E 60 XX electrodes, we find in Table 13-1 that the weld has a capacity of 3.2 k/in. [0.56 kN/mm]. The total length of weld required for the attachment of an individual truss member is determined by dividing the internal force in the member by the unit capacity of the weld. Because the bottom chord is continuous at the joint under consideration, the attachment of these angles need develop only the difference in force between the chords on each side of the joint. The weld lengths required for this joint are

$$\text{Bottom chord: } L = \frac{37.8 - 31.5}{3.2} = 1.97 \text{ in.}$$

$$\text{Vertical web: } L = \frac{9.5}{3.2} = 3.17 \text{ in.}$$

$$\text{Diagonal web: } L = \frac{11.4}{3.2} = 3.56 \text{ in.}$$

Before selecting the size and layout of individual welds, we must consider a number of other factors;

1. *Minimum weld length.* As discussed in Art. 13-4, the minimum length is four times the weld size, plus an extra $\frac{1}{4}$ in. for starting and stopping. We consider the use of a minimum length of 1.25 in. for an individual weld.
2. Section 1.15 of the AISC Specification requires that connections be designed for a minimum of 6 kips or 50% of the capacity of the connected member, whichever is greater.
3. Connections should be arranged on the gusset plates to maximize the general stability of the joint.

Considering point 2 and using the actual member capacities given in Table 14-3, we find the following alternative lengths for the web members:

$$\text{Vertical web: } L = \frac{0.50 \times 24}{3.2} = 3.75 \text{ in.}$$

$$\text{Diagonal web: } L = \frac{0.50 \times 27}{3.2} = 4.22 \text{ in.}$$

Because both are slightly larger than the lengths determined for the actual internal forces, they become the critical requirements for the joint.

Consideration of point 3 reveals that it is desirable to place the welds for the bottom chords at the edges of the gusset plate, (Fig. 14-15). It should be observed that this arrangement produces greater twisting resistance for the gusset plate, both in and out of the plane of the truss.

Note that the welds shown in Fig. 14-15 occur on both sides of the gusset plate and that there is actually twice as much total weld for each truss member.

14-12. Truss Joint with Tee Chord

Figure 14-16 shows a construction that is often used with light steel trusses. In this case the double-angle chord is replaced by a single structural tee and the web members are welded directly to the stem of the tee, thus eliminating the need for the gusset plate. A critical concern in the selection of the tee is the need for sufficient height of the stem to accommodate the welds for the web members. For this type of joint it may be desirable to cut the ends of diagonal members at an angle or to arrange the welds in a different manner to make the joint layout more compact. In Figure 14-16 the weld is placed completely across the end of the diagonal member, which permits the use of a short return weld on the two sides.

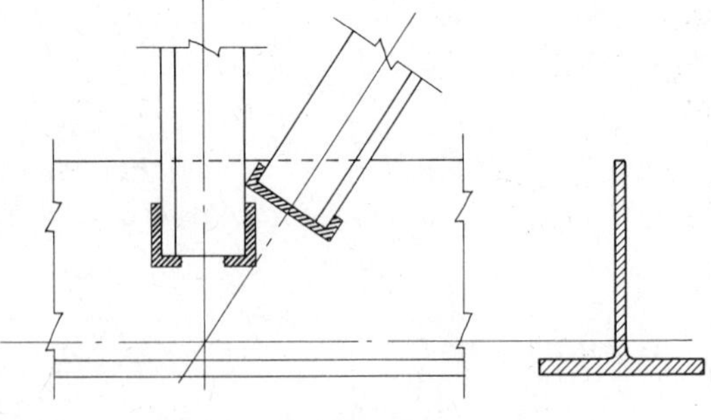

FIGURE 14-16.

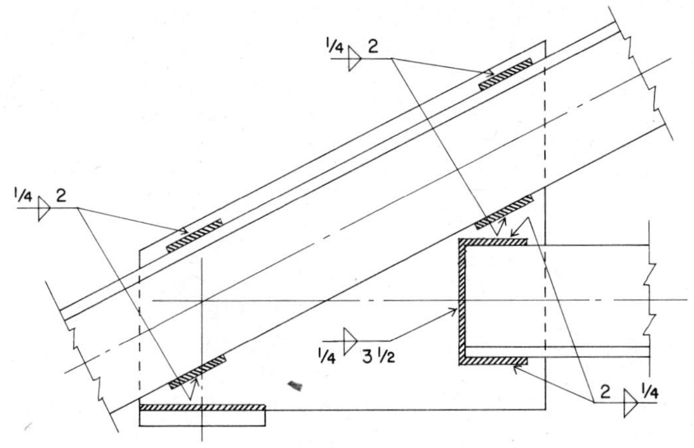

FIGURE 14-17.

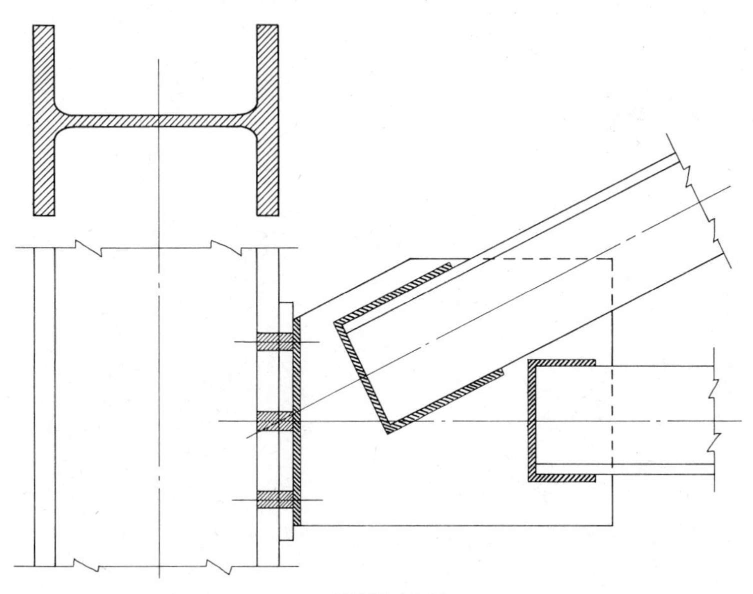
FIGURE 14-18.

When structural tees are used as truss members it is necessary to ensure that the width–thickness ratio of the flange and the depth–thickness ratio of the web comply with the minimum requirements of the AISC Specification. This does not apply to members that sustain only tension force but it does when compression or bending must be resisted.

14-13. Truss Support Joints

Figure 14-17 is a detail for the truss joint at the left support. In this layout the joint has been developed to facilitate support in the form of direct vertical bearing on top of a wall, on the top flange of a girder, or on the cap plate of a steel column. The joint has also been developed to allow for the extension of the top chord to form a roof overhang beyond the support. The internal forces are much greater in this joint than in the joint in Art. 14-12; thus the amount of weld required is considerably larger.

An alternative detail for the truss support is shown in Fig. 14-18. In this case the joint is developed to facilitate attachment to the face of a steel column with a bolted shear connection.

15

Plastic Behavior and Strength Design

15-1. Plastic versus Elastic Behavior

The discussions up to this point of the design of members in bending have been based on bending stresses well within the yield point stress. In general, allowable stresses are based on the *theory of elastic behavior*. However, it has been found by tests that steel members can carry loads much higher than anticipated, even when the yield point stress is reached at sections of maximum bending moment. This is particularly evident in continuous beams and in structures with rigid connections. An inherent property of structural steel is its ability to resist large deformations without failure. These large deformations occur chiefly in the *plastic range,* with no increase in the magnitude of the bending stress. Because of this phenomenon, the *plastic design theory,*

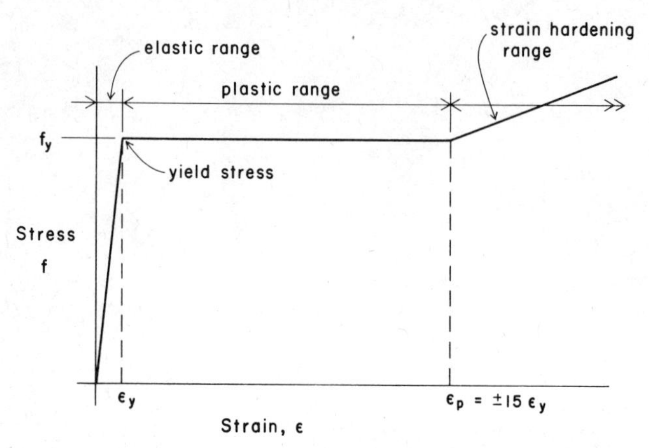

FIGURE 15-1. Form of the stress-strain response for ductile steel.

sometimes called the *ultimate strength design theory* (or more recently *strength design*), has been developed.

Figure 15-1 represents the typical form of a load-test response for a specimen of ductile steel. The graph shows that up to a stress f_y, the yield point, the deformations are directly proportional to the applied stresses and that beyond the yield point there is a deformation without an increase in stress. For A36 steel this additional deformation, called the *plastic range,* is approximately 15 times that produced elastically. Note that beyond this range *strain hardening* (loss of ductility) begins when further deformation can occur only with an increase in stress.

For plastic behavior to be significant the extent of the plastic range of deformation must be several times the elastic deformation. As the yield point is increased in magnitude, this ratio of deformations decreases, which is to say that higher strength steels tend to be less ductile. At present, the theory of plastic design is generally limited to steels with a yield point of not more than 65 ksi [450 MPa].

15-2. Plastic Moment and the Plastic Hinge

Article 7.1 explains the design of members in bending in accordance with the theory of elasticity. When the extreme fiber stress does not exceed the elastic limit, the bending stresses in the cross section of a beam are directly proportional to their distances from the neutral surface. In addition, the strains (deformations) in these fibers are also proportional to their distances from the neutral surface. Both stresses and strains are zero at the neutral surface and both increase to maximum magnitudes at the fibers farthest from the neutral surface.

The following example illustrates the analysis of a steel beam for bending, according to the theory of elastic behavior.

A simple steel beam has a span of 16 ft [4.88 m] with a concentrated load of 18 kips [80 kN] at the center of the span. The section used is an S 12 × 31.8, the beam is adequately braced throughout its length, and the beam weight is ignored in the computations. Let us compute the maximum extreme fiber stress. (See Fig. 15-2.)

To do this we use the flexure formula

$$f = \frac{M}{S} \qquad \text{(Art. 6-2)}$$

Then

$$M = \frac{PL}{4} = \frac{18 \times 16}{4} = 72 \text{ kip-ft}$$

$$\left[M = \frac{80 \times 4.88}{4} = 97.6 \text{ kN-m}\right]$$

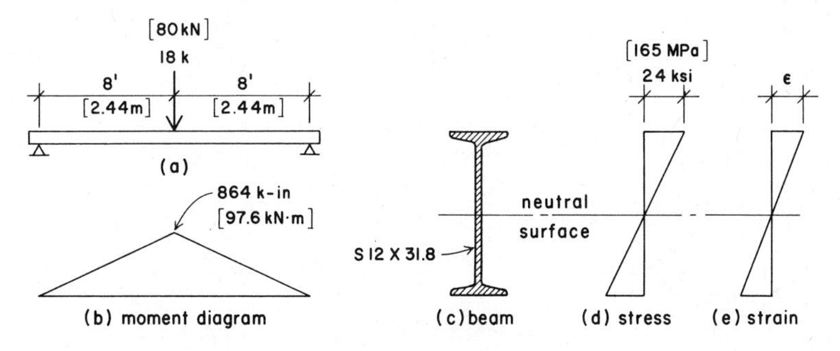

FIGURE 15-2. Elastic behavior of the beam.

which is the maximum bending moment. In Table A-1 we find S = 36.4 in.^{3} [597 × 10^3 mm^3]. Thus

$$f = \frac{M}{S} = \frac{72 \times 12}{36.4} = 23.7 \text{ ksi}$$

$$\left[f = \frac{97.6 \times 10^6}{597 \times 10^3} = 163.5 \text{ MPa}\right]$$

which is the stress on the fiber farthest from the neutral surface. (See Fig. 15-2*d*.)

Note that this stress occurs only at the beam section at the center of the span, where the bending moment has its maximum value. Figure 15-2*e* shows the deformations that accompany the stresses shown in Fig. 15-2*d*. Note that both stresses and deformations are directly proportional to their distances from the neutral surface in elastic analysis.

When a steel beam is loaded to produce an extreme fiber stress in excess of the yield point the property of the material's ductility affects the distribution of the stresses in the beam cross section. Elastic analysis does not suffice to explain this phenomenon because the beam will experience some plastic deformation.

Assume that the bending moment on a beam is of such magnitude that the extreme fiber stress is f_y, the yield stress. Then, if M_y is the elastic bending moment at the yield stress, $M = M_y$, and the distribution of the stresses in the cross section is as shown in Fig. 15-3*a*; the maximum bending stress f_y is at the extreme fiber.

Next consider that the loading and the resulting bending moment have been increased; M is now greater than M_y. The stress

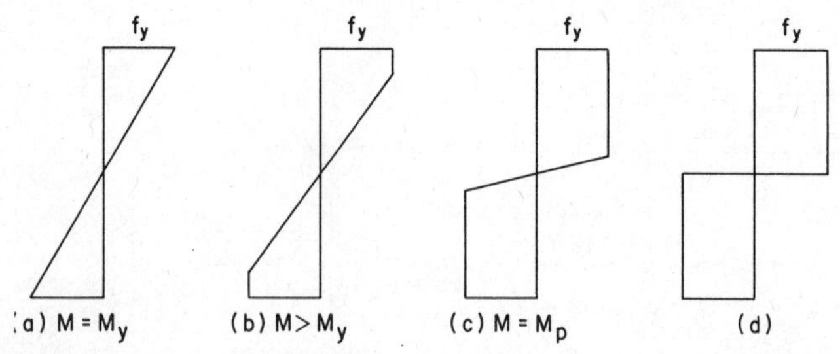

FIGURE 15-3. Progression of stress response—elastic to plastic.

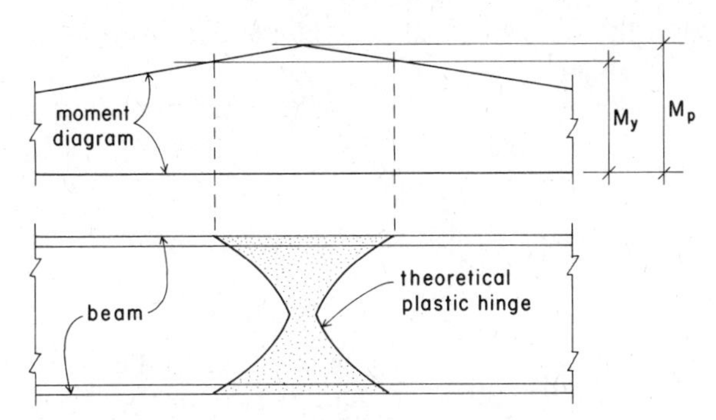

FIGURE 15-4. Development of the plastic hinge.

on the extreme fiber is still f_y, but *the material has yielded* and a greater area of the cross section is also stressed to f_y. The stress distribution is shown in Fig. 15-3*b*.

Now imagine that the load is further increased. The stress on the extreme fiber is still f_y and, theoretically, *all fibers in the cross section are stressed to* f_y. This idealized plastic stress distribution is shown in Fig. 15-3*d*. The bending moment that produces this condition is M_p, the plastic bending moment. In reality about 10% of the central portion of the cross section continues to resist in an elastic manner, as indicated in Fig. 15-3*c*. This small resistance is quite negligible, and we assume that the stresses on all fibers of the cross section are f_y, as shown in Fig. 15-3*d*. The section is now said to be fully plastic and any further increase in load will result in large deformations; the beam acts as if it were hinged at this section. We call this a plastic hinge at which free rotation is permitted only after M_p has been attained. (See Fig. 15-4.) At sections of a beam in which this condition prevails the bending resistance of the cross section has been exhausted.

15-3. Plastic Section Modulus

In elastic design the moment that produces the maximum allowable resisting moment may be found by the flexure formula

$$M = f \times S$$

where M = the maximum allowable bending moment in inch-pounds,
f = the maximum allowable bending stress in pounds per square inch,
S = the section modulus in inches to the third power.

If the extreme fiber is stressed to the yield stress

$$M_y = f_y \times S$$

where M_y = the elastic bending moment at yield stress,
f_y = the yield stress in pounds per square inch,
S = the section modulus in inches to the third power.

Now let us find a similar relation between the plastic moment and its plastic resisting moment. Refer to Fig. 15-5, which shows the cross section of a W or S section in which the bending stress f_y, the yield stress, is constant over the cross section. In the figure

A_u = the upper area of the cross section above the neutral axis in square inches,
y_u = the distance of the centroid of A_u from the neutral axis,
A_l = lower area of the cross section below the neutral axis in square inches,
y_l = the distance of the centroid of A_l from the neutral axis.

For equilibrium the algebraic sum of the horizontal forces must be zero. Then

$$\sum H = 0$$

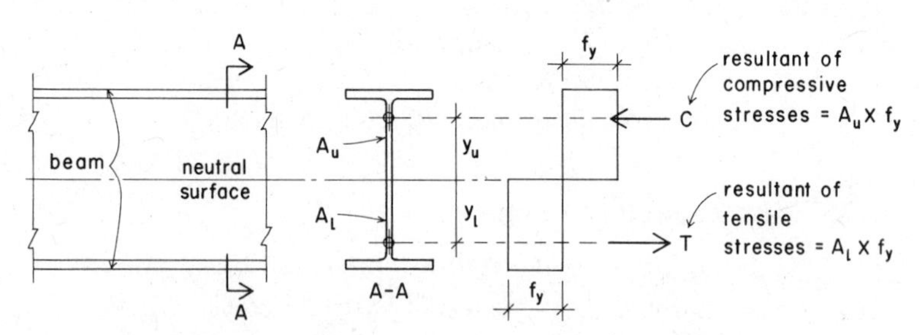

FIGURE 15-5. Development of the plastic resisting moment.

or

$$[A_u \times (+f_y)] + [A_l \times (-f_y)] = 0$$

and

$$A_u = A_l$$

This shows that the neutral axis divides the cross section into equal areas, which is apparent in symmetrical sections, but it applies to unsymmetrical sections as well. Also the bending moment equals the sum of the moments of the stresses in the section. Thus for M_p, the plastic moment,

$$M_p = (A_u \times f_y \times y_u) + (A_l \times f_y \times y_l)$$

or

$$M_p = f_y[(A_u \times y_u) + (A_l \times y_l)]$$

and

$$M_p = f_y \times Z$$

The quantity $(A_u y_u + A_l y_l)$ is called the *plastic section modulus* of the cross section and is designated by the letter Z; because it is an area multiplied by a distance, it is in units to the third power. If the area is in units of square inches and the distance is in linear inches, Z, the section modulus, is in units of inches to the third power.

The plastic section modulus is always larger than the elastic section modulus.

It is important to note that in plastic design the neutral axis for unsymmetrical cross sections does not pass through the centroid of the section. In plastic design the neutral axis divides the cross section into *equal* areas.

15-4. Computation of the Plastic Section Modulus

The notation used in Art. 15-3 is appropriate for both symmetrical and unsymmetrical sections. Consider now a symmetrical section such as a W or S shape, as shown in Fig. 15-5. $A_u = A_l$, $y_u = y_l$ and

$A_u + A_l = A$, the total area of the cross section. Then

$$M_p = (A_u \times f_y \times y_u) + (A_l \times f_y \times y_l)$$

and

$$M_p = f_y \times A \times y \quad \text{or} \quad M_p = f_y \times Z$$

where f_y = the yield stress,
A = the total area of the cross section,
y = the distance from the neutral axis to the centroid of the portion of the area on either side of the neutral axis,
Z = the plastic modulus of the section (in in.^3 or mm^3)

Now because $Z = A \times y$ we can readily compute the value of the plastic section modulus of a given cross section.

Consider a W 16 × 45. In Table A-1 we find that its total depth is 16.13 in. [410 mm] and its cross-sectional area is 13.3 in.^2 [8581 mm^2]. In Table A-7 we find that a WT 8 × 22.5 (which is one-half a W 16 × 45) has its centroid located 1.88 in. [48 mm] from the outside of the flange. Therefore the distance from the centroid of either half of the W shape to the neutral axis is one half the beam depth less the distance obtained for the tee. Thus the distance y is (see Fig. 15-6.)

$$y = (16.13/2) - 1.88 = 6.185 \text{ in.}$$

$$[y = (410/2) - 48 = 157 \text{ mm}]$$

Then the plastic modulus of the W 16 × 45 is

$$Z = A \times y = 13.3 \times 6.185 = 82.26 \text{ in.}^3$$

$$[Z = 8581 \times 157 = 1347 \times 10^3 \text{ mm}^3].$$

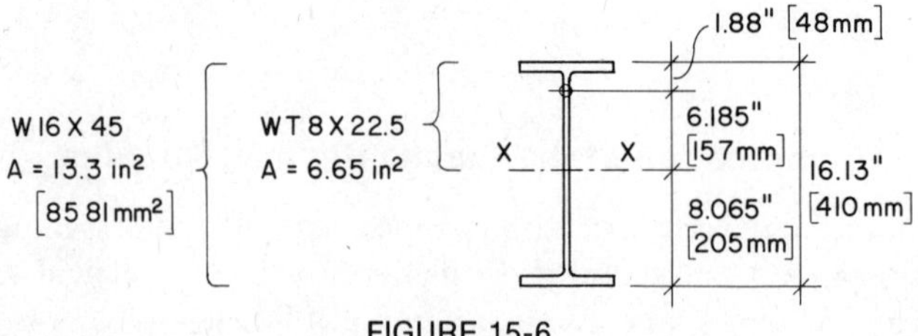

FIGURE 15-6.

Use of the full value of the plastic hinge moment requires the shape to have limited values for the width–thickness ratio of the flanges and the depth–thickness ratio of the web. These requirements are given in Section 2.7 of the AISC Specification.

15-5. Shape Factor

Consider a beam subjected to a bending moment that produces bending stresses for which the extreme fiber stress is f_y.

By elastic design

$$M_y = f_y \times S$$

By plastic design

$$M_p = f_y \times Z$$

Then

$$\frac{\text{plastic design}}{\text{elastic design}} = \frac{M_p}{M_y} = \frac{f_y \times Z}{f_y \times S} = \frac{Z}{S}$$

The relation between the two moments, M_p/M_y is called the *shape factor*. It is represented by the letter u. Thus

$$u = \frac{Z}{S}$$

Let us compute the value of u, the shape factor for a rectangle whose width is b and whose depth is d. (See Fig. 15-7.)

The elastic section modulus of this rectangle about the neutral axis parallel to the base is

$$S = \frac{bd^2}{6} \qquad \text{(Table 6-1)}$$

For the plastic section modulus of the rectangle

$$Z = \left(b \times \frac{d}{2} \times \frac{d}{4}\right) + \left(b \times \frac{d}{2} \times \frac{d}{4}\right)$$

or

$$Z = \frac{bd^2}{8} + \frac{bd^2}{8} = \frac{bd^2}{4}$$

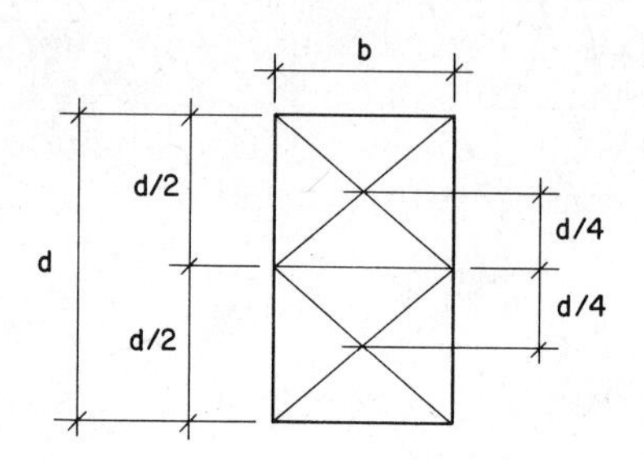

FIGURE 15-7.

Then

$$\frac{Z}{S} = \frac{bd^2}{4} \div \frac{bd^2}{6} = \frac{bd^2}{4} \times \frac{6}{bd^2} = \frac{3}{2} = 1.5$$

which is the shape factor for rectangles. Thus M_p, the plastic moment for a rectangular cross section, is 50% greater than M_y, the elastic moment at yield stress.

For the commonly used steel sections, such as W and S, the shape factor is approximately 1.12 for bending about the strong axis of the section. This means that these sections can support approximately 12% more than we expect them to carry on the basis of elastic design.

15-6. Restrained Beams

Figure 15-8 shows a uniformly distributed load of w lb per lin ft on a beam that is fixed (restrained from rotation) at both ends. The maximum bending moment for this condition is $wl^2/12$; it occurs at the supports and is a negative quantity. At the center of the span the moment is positive and its magnitude is $wl^2/24$.

Consider now that the load is increased from w to w_y, thus producing a stress of f_y in the extreme fibers of the cross section over the supports. At the supports

$$M_y = -\frac{w_y l^2}{12}$$

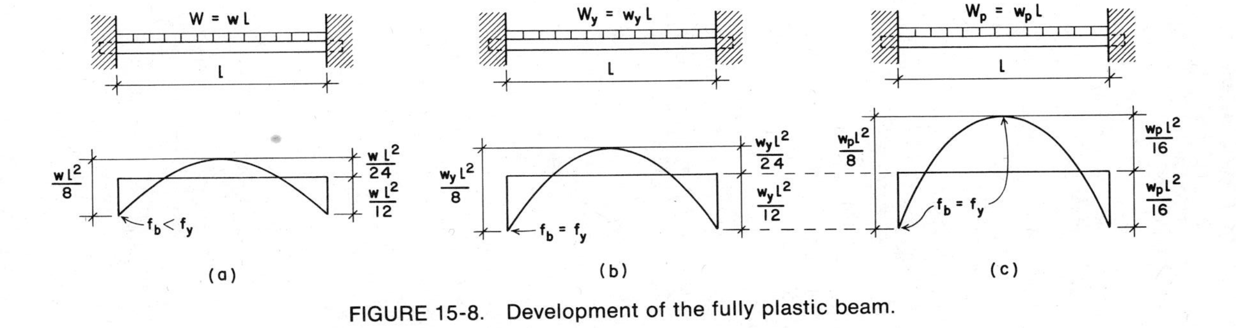

FIGURE 15-8. Development of the fully plastic beam.

and at the center of the span

$$M_y = + \frac{w_y l^2}{24}$$

as shown in Fig. 15-8*b*. This is the limit of the elastic behavior; the moments at the supports are still twice the magnitude of the moment at the midspan, at which none of the fibers of the cross section has reached the yield stress.

Now let us increase the load to w_p lb per lin ft. (See Fig. 15-8*c*.) The stresses in the fibers of the cross section *over the supports* will now begin to yield until *all* the fibers are stressed to f_y and plastic hinges are formed at the ends. The beam can no longer resist further rotation at its ends and the increase in load must be resisted by sections of the beam that are less stressed. The critical section lies at the center of the span. If the load w_p has been increased so that a plastic hinge is formed at the midspan, the beam will become a *mechanism*. A mechanism exists when sufficient plastic hinges have been formed and no further loading may be supported by the member. Note that the beam when loaded produces three plastic hinges, one at the midspan and two at the supports. The moments at the supports and at midspan are now equal and have magnitudes of $w_p l^2/16$, which means that the full strength of the beam is used at three sections instead of two. A load whose magnitude is greater than w_p lb per lin ft would result in a permanent deformation and consequently a practical failure.

15-7. Load Factor

Consider a beam of A36 steel laterally supported throughout its length. Its span is 24 ft [7.315 m] and it carries a concentrated load of 42 kips [186.8 kN] at the center. Let us determine the size of the beam in accordance with the theory of elastic behavior.

The maximum bending moment for this beam is

$$M = \frac{PL}{4} = \frac{42 \times 24}{4} = 252 \text{ kip-ft}$$

$$\left[M = \frac{186.8 \times 7.315}{4} = 341.6 \text{ kN-m}\right]$$

and the required section modulus is

$$S = \frac{M}{f_b} = \frac{252 \times 12}{24} = 126 \text{ in.}^3$$

$$\left[S = \frac{341.6 \times 10^6}{165} = 2070 \times 10^3 \text{ mm}^3\right]$$

Table A-1 shows that a W 21 × 62 has a section modulus of 127 in.^{3} and is acceptable.

Now let us compute the magnitude of the concentrated load at the center of the span that would produce a bending moment equal to the plastic resisting moment. In Table A-1 the plastic modulus for the W 21 × 62 is 144 in.^{3} Thus the plastic moment is

$$M_p = F_y \times Z_x = 36 \times 144 = 5184 \text{ kip-in., or } 432 \text{ kip-ft}$$

$$\left[M_p = 248 \times \frac{2360 \times 10^3}{10^6} = 585.3 \text{ kN-m}\right]$$

Then the load that corresponds to this moment is

$$M_p = 432 = \frac{PL}{4} = \frac{P \times 24}{4}$$

and

$$P = \frac{4 \times 432}{24} = 72 \text{ kips}$$

$$\left[M_p = 585.3 = \frac{P \times 7.315}{4}, P = 320 \text{ kN}\right]$$

This load would produce a plastic hinge at the center of the span and a slight increase in load would result in failure.

The term *load factor* is given to the ratio of the ultimate load to the design load. In this example it is 72/42 = 1.714.

Referring to Fig. 15-8c, we read in Art. 15-7 that a load greater than w_p lb/ft would result in a permanent deformation and a practical failure. In elastic design the allowable bending stress F_b is decreased to a fraction of F_y, the yield stress. For compact sections $F_b = 0.66\ F_y$. Therefore the implied factor of safety against yielding is 1/0.66, or 1.5. This factor is higher, of course, if the beam's limiting capacity is taken as the plastic moment.

In plastic design the concept of allowable stress is not used and computations are based strictly on the limit of the yield stress. Safety is produced by the use of the load factor, by which the beam is literally designed to fail, but at a load larger than that it actually must sustain. For simple and continuous beams the load factor is specified as 1.7; for rigid frames it is 1.85.

15-8. Design of a Simple Beam

The design of simple beams by the elastic or plastic theory will usually result in the same size beam, as illustrated in the following examples:

Example 1. A simple beam of A36 steel has a span of 20 ft [6.1 m] and supports a uniformly distributed load of 4.8 kips/ft [70 kN/m], including its own weight. Design this beam in accordance with the elastic theory, assuming that it is laterally supported throughout its length.

Solution: The maximum bending moment is $wL^2/8$. Then

$$M = \frac{4.8 \times (20)^2}{8} = 240 \text{ kip-ft}$$

$$\left[\frac{70 \times (6.1)^2}{8} = 325.6 \text{ kN-m}\right]$$

By using the allowable bending stress of 24 ksi [165 MPa] for a compact section the required section modulus is

$$S = \frac{M}{F_b} = \frac{240 \times 12}{24} = 120 \text{ in.}^3$$

$$\left[S = \frac{325.6 \times 10^6}{165} = 1973 \times 10^3 \text{ mm}^3\right]$$

In Table A-1 we find a W 21 × 62 with $S_x = 127$ in.^{3}

Example 2. Design the same beam in accordance with the plastic theory.

Solution: We adjust the load with the load factor, which is given in Art. 15-7 as 1.7. Thus

$$w_p = 4.8 \times 1.7 = 8.16 \text{ kips/ft}$$

$$[w_p = 70 \times 1.7 = 119 \text{ kN/m}]$$

and the maximum bending moment is

$$M_p = \frac{w_p L^2}{8} = \frac{8.16 \times (20)^2}{8} = 408 \text{ kip-ft}$$

$$\left[M_p = \frac{119 \times (6.1)^2}{8} = 553.5 \text{ kN-m}\right]$$

and

$$M_p = F_y \times Z, \quad Z = \frac{M_p}{F_y} = \frac{408 \times 12}{36} = 136 \text{ in.}^3$$

$$\left[Z = \frac{553.5 \times 10^6}{248} = 2232 \times 10^3 \text{ mm}^3\right]$$

which is the minimum plastic section modulus. From Table A-1 select a W 21 × 62 for which $Z_x = 144$ in.^{3} Note that the elastic and plastic theories have yielded the same result in this example, which is common for simple beams.

15-9. Design of a Beam with Fixed Ends

A beam that has both ends fixed or restrained is similar to an interior span of a fully continuous beam; the magnitudes of the maximum bending moments are the same. In structural steel it is economical to design these beams in accordance with plastic theory because there is a saving of material.

Example 1. A beam of A36 steel with both ends fixed has a clear span of 20 ft [6.1 m] and carries a uniformly distributed load of 4.8 kips/ft [70 kN/m], including its own weight. (This load and span are the same as those for the simple span beam in the example in Art. 15-8.) Design the beam in accordance with elastic theory, assuming that full lateral support is provided.

Solution: Referring to Fig. 15-8*a* we see that the maximum bending moment is at the supports; a negative moment of $wL^2/12$. Then

$$M = \frac{4.8 \times (20)^2}{12} = 160 \text{ kip-ft}$$

$$\left[M = \frac{70 \times (6.1)^2}{12} = 217 \text{ kN-m}\right]$$

Assuming a compact section so that $F_b = 24$ ksi [165 MPa],

$$S = \frac{M}{F_b} = \frac{160 \times 12}{24} = 80 \text{ in.}^3$$

$$\left[S = \frac{217 \times 10^6}{165} = 1315 \times 10^3 \text{ mm}^3\right]$$

which is the required section modulus. Referring to Table A-1, we find that a W 18 × 50 has $S_x = 88.9$ in.^{3} and is acceptable.

Example 2. Design the same beam in accordance with the plastic theory.

Solution: Using the load factor of 1.7 for a continuous beam

$$w_p = 4.8 \times 1.7 = 8.16 \text{ kips/ft}$$

$$[w_p = 70 \times 1.7 = 119 \text{ kN/m}]$$

Assuming plastic hinging at both the midspan and supports, as shown in Fig. 15-8*c*, we determine that the critical maximum moment is

$$M_p = \frac{wL^2}{16} = \frac{8.16 \times (20)^2}{16} = 204 \text{ kip-ft}$$

Then

$$Z = \frac{M_p}{F_y} = \frac{204 \times 12}{36} = 68 \text{ in.}^3$$

$$\left[M_p = \frac{119 \times (6.1)^2}{16} = 276.7 \text{ kN-m},\right.$$

$$\left.Z = \frac{276.7 \times 10^6}{248} = 1116 \times 10^3 \text{ mm}^3\right]$$

which is the required plastic section modulus. In Table A-1 we see that a W 18 × 40 has a plastic section modulus of 78.4 in.^{3} and is acceptable.

Note that a savings of 10 lb/ft in the weight of steel was achieved in this case by the use of the plastic theory. This represents a 20% reduction in material used.

Problem 15.9.A. A beam of A36 steel has a span of 22 ft [6.71 m] and is fixed at both ends. The load is 5 kips/ft [73 kN/m], which includes the beam weight, and full lateral support is provided. Using the elastic theory, determine the lightest weight W shape that will support the load.

Problem 15.9.B. Using the data given for the beam of the preceding problem, determine the lightest shape in accordance with plastic theory.

15-10. Scope of Plastic Design

The purpose of this brief chapter is to explain how the theory of plastic design in steel has evolved. The illustrative problems relate only to beams that are fixed at both ends; they show how the application of plastic design theory may result in economy of material. The theory, however, has applications far beyond fixed beams, particularly in the design of rigid frames in which combinations of bending and direct stress are involved. Because the design is based on higher stresses, adequate attention must be given to the possibility of local buckling and care must be taken to prevent excessive deflections that might occur because of the smaller sections that usually result from use of the theory. The reader who wishes to pursue the subject of plastic design is referred to *Commentary on Plastic Design in Steel* by the American Society of Civil Engineers or *Plastic Design in Steel* by the American Institute of Steel Construction.

References

1. *Manual of Steel Construction,* 8th ed., American Institute of Steel Construction, 400 N. Michigan Ave., Chicago, IL 60611, 1980. (Called the AISC Manual.)
2. *Uniform Building Code,* 1982 ed., International Conference of Building Officials, 5360 South Workman Mill Road, Whittier, CA 90601.
3. Charles G. Salmon and John E. Johnson, *Steel Structures: Design and Behavior,* 2nd ed., Harper & Row, New York, 1980.
4. Harry Parker, *Simplified Design of Building Trusses for Architects and Builders,* 3rd ed. prepared by James Ambrose, Wiley, New York, 1982.
5. James Ambrose and Dimitry Vergun, *Simplified Building Design for Wind and Earthquake Forces,* Wiley, New York, 1980.
6. *Standard Specifications, Load Tables, and Weight Tables,* Steel Joist Institute, 1703 Parham Rd., Suite 204, Richmond, VA 23229.
7. *Cold Formed Steel Design Manual,* American Iron and Steel Institute, 1000 16th St., N. W., Washington, DC 20036, 1977.
8. *Steel Deck Institute Design Manual for Composite Decks, Form Decks, and Roof Decks,* Publication 24, Steel Deck Institute, P. O. Box 3812, St. Louis, MO 63122.
9. *Steel Deck Institute Diaphragm Design Manual,* Steel Deck Institute, P. O. Box 3812, St. Louis, MO 63122, 1981.

Properties of Rolled Structural Shapes

The following tables contain data for standard rolled structural shapes. This material has been reproduced and adapted from the *Manual of Steel Construction,* 8th ed. (Ref. 1) with the permission of the publishers, the American Institute of Steel Construction. Reference may be made to Tables 4-1 and 4-2 to determine the availability of shapes in the various grades of steel.

To convert table units of	in.	$in.^2$	$in.^3$	$in.^4$
to SI units of	mm	mm^2	mm^3	mm^4
multiply by	25.4	645.2	$16.39(10)^3$	$0.4162(10)^6$

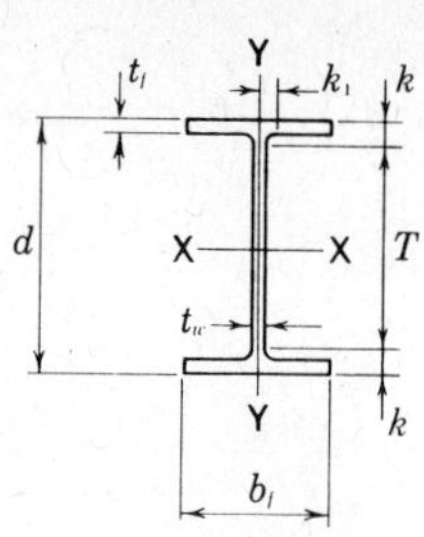

TABLE A-1. Properties of Wide Flange (W) Shapes

Designation	Area A	Depth d		Web Thickness t_w		Web $\frac{t_w}{2}$	Flange Width b_f		Flange Thickness t_f		Distance T	Distance k	Distance k_1
	In.^{2}	In.		In.		In.	In.		In.		In.	In.	In.
W 36x300	88.3	36.74	36¾	0.945	15/16	½	16.655	16⅝	1.680	1 11/16	31⅛	2 13/16	1½
x280	82.4	36.52	36½	0.885	⅞	7/16	16.595	16⅝	1.570	1 9/16	31⅛	2 11/16	1½
x260	76.5	36.26	36¼	0.840	13/16	7/16	16.550	16½	1.440	1 7/16	31⅛	2 9/16	1½
x245	72.1	36.08	36⅛	0.800	13/16	7/16	16.510	16½	1.350	1⅜	31⅛	2½	1 7/16
x230	67.6	35.90	35⅞	0.760	¾	⅜	16.470	16½	1.260	1¼	31⅛	2⅜	1 7/16
W 36x210	61.8	36.69	36¾	0.830	13/16	7/16	12.180	12⅛	1.360	1⅜	32⅛	2 5/16	1¼
x194	57.0	36.49	36½	0.765	¾	⅜	12.115	12⅛	1.260	1¼	32⅛	2 3/16	1 3/16
x182	53.6	36.33	36⅜	0.725	¾	⅜	12.075	12⅛	1.180	1 3/16	32⅛	2⅛	1 3/16
x170	50.0	36.17	36⅛	0.680	11/16	⅜	12.030	12	1.100	1⅛	32⅛	2	1 3/16
x160	47.0	36.01	36	0.650	⅝	5/16	12.000	12	1.020	1	32⅛	1 15/16	1⅛
x150	44.2	35.85	35⅞	0.625	⅝	5/16	11.975	12	0.940	15/16	32⅛	1⅞	1⅛
x135	39.7	35.55	35½	0.600	⅝	5/16	11.950	12	0.790	13/16	32⅛	1 11/16	1⅛
W 33x241	70.9	34.18	34⅛	0.830	13/16	7/16	15.860	15⅞	1.400	1⅜	29¾	2 3/16	1 3/16
x221	65.0	33.93	33⅞	0.775	¾	⅜	15.805	15¾	1.275	1¼	29¾	2 1/16	1 3/16
x201	59.1	33.68	33⅝	0.715	11/16	⅜	15.745	15¾	1.150	1⅛	29¾	1 15/16	1⅛
W 33x152	44.7	33.49	33½	0.635	⅝	5/16	11.565	11⅝	1.055	1 1/16	29¾	1⅞	1⅛
x141	41.6	33.30	33¼	0.605	⅝	5/16	11.535	11½	0.960	15/16	29¾	1¾	1 1/16
x130	38.3	33.09	33⅛	0.580	9/16	5/16	11.510	11½	0.855	⅞	29¾	1 11/16	1 1/16
x118	34.7	32.86	32⅞	0.550	9/16	5/16	11.480	11½	0.740	¾	29¾	1 9/16	1 1/16
W 30x211	62.0	30.94	31	0.775	¾	⅜	15.105	15⅛	1.315	1 5/16	26¾	2⅛	1⅛
x191	56.1	30.68	30⅝	0.710	11/16	⅜	15.040	15	1.185	1 3/16	26¾	1 15/16	1 1/16
x173	50.8	30.44	30½	0.655	⅝	5/16	14.985	15	1.065	1 1/16	26¾	1⅞	1 1/16
W 30x132	38.9	30.31	30¼	0.615	⅝	5/16	10.545	10½	1.000	1	26¾	1¾	1 1/16
x124	36.5	30.17	30⅛	0.585	9/16	5/16	10.515	10½	0.930	15/16	26¾	1 11/16	1
x116	34.2	30.01	30	0.565	9/16	5/16	10.495	10½	0.850	⅞	26¾	1⅝	1
x108	31.7	29.83	29⅞	0.545	9/16	5/16	10.475	10½	0.760	¾	26¾	1 9/16	1
x 99	29.1	29.65	29⅝	0.520	½	¼	10.450	10½	0.670	11/16	26¾	1 7/16	1
W 27x178	52.3	27.81	27¾	0.725	¾	⅜	14.085	14⅛	1.190	1 3/16	24	1⅞	1 1/16
x161	47.4	27.59	27⅝	0.660	11/16	⅜	14.020	14	1.080	1 1/16	24	1 13/16	1
x146	42.9	27.38	27⅜	0.605	⅝	5/16	13.965	14	0.975	1	24	1 11/16	1
W 27x114	33.5	27.29	27¼	0.570	9/16	5/16	10.070	10⅛	0.930	15/16	24	1⅝	15/16
x102	30.0	27.09	27⅛	0.515	½	¼	10.015	10	0.830	13/16	24	1 9/16	15/16
x 94	27.7	26.92	26⅞	0.490	½	¼	9.990	10	0.745	¾	24	1 7/16	15/16
x 84	24.8	26.71	26¾	0.460	7/16	¼	9.960	10	0.640	⅝	24	1⅜	15/16

TABLE A-1. (*Continued*)

Nominal Wt. per Ft.	Compact Section Criteria				r_T	$\frac{d}{A_f}$	Elastic Properties						Torsional constant	Plastic Modulus	
							Axis X-X			Axis Y-Y					
	$\frac{b_f}{2t_f}$	F_y'	$\frac{d}{t_w}$	F_y'''			I	S	r	I	S	r	J	Z_x	Z_y
Lb.		Ksi		Ksi	In.		In.^{4}	In.^{3}	In.	In.^{4}	In.^{3}	In.	In.^{4}	In.^{3}	In.^{3}
300	5.0	—	38.9	43.7	4.39	1.31	20300	1110	15.2	1300	156	3.83	64.2	1260	241
280	5.3	—	41.3	38.8	4.37	1.40	18900	1030	15.1	1200	144	3.81	52.6	1170	223
260	5.7	—	43.2	35.4	4.34	1.52	17300	953	15.0	1090	132	3.78	41.5	1080	204
245	6.1	—	45.1	32.5	4.32	1.62	16100	895	15.0	1010	123	3.75	34.6	1010	190
230	6.5	—	47.2	29.6	4.30	1.73	15000	837	14.9	940	114	3.73	28.6	943	176
210	4.5	—	44.2	33.8	3.09	2.21	13200	719	14.6	411	67.5	2.58	28.0	833	107
194	4.8	—	47.7	29.0	3.07	2.39	12100	664	14.6	375	61.9	2.56	22.2	767	97.7
182	5.1	—	50.1	26.3	3.05	2.55	11300	623	14.5	347	57.6	2.55	18.4	718	90.7
170	5.5	—	53.2	23.3	3.04	2.73	10500	580	14.5	320	53.2	2.53	15.1	668	83.8
160	5.9	—	55.4	21.5	3.02	2.94	9750	542	14.4	295	49.1	2.50	12.4	624	77.3
150	6.4	—	57.4	20.1	2.99	3.18	9040	504	14.3	270	45.1	2.47	10.1	581	70.9
135	7.6	—	59.3	18.8	2.93	3.77	7800	439	14.0	225	37.7	2.38	6.99	509	59.7
241	5.7	—	41.2	38.9	4.17	1.54	14200	829	14.1	932	118	3.63	35.8	939	182
221	6.2	—	43.8	34.5	4.15	1.68	12800	757	14.1	840	106	3.59	27.5	855	164
201	6.8	—	47.1	29.8	4.12	1.86	11500	684	14.0	749	95.2	3.56	20.5	772	147
152	5.5	—	52.7	23.7	2.94	2.74	8160	487	13.5	273	47.2	2.47	12.4	559	73.9
141	6.0	—	55.0	21.8	2.92	3.01	7450	448	13.4	246	42.7	2.43	9.70	514	66.9
130	6.7	—	57.1	20.3	2.88	3.36	6710	406	13.2	218	37.9	2.39	7.37	467	59.5
118	7.8	—	59.7	18.5	2.84	3.87	5900	359	13.0	187	32.6	2.32	5.30	415	51.3
211	5.7	—	39.9	41.4	3.99	1.56	10300	663	12.9	757	100	3.49	27.9	749	154
191	6.3	—	43.2	35.4	3.97	1.72	9170	598	12.8	673	89.5	3.46	20.6	673	138
173	7.0	—	46.5	30.6	3.94	1.91	8200	539	12.7	598	79.8	3.43	15.3	605	123
132	5.3	—	49.3	27.2	2.68	2.87	5770	380	12.2	196	37.2	2.25	9.72	437	58.4
124	5.7	—	51.6	24.8	2.66	3.09	5360	355	12.1	181	34.4	2.23	7.99	408	54.0
116	6.2	—	53.1	23.4	2.64	3.36	4930	329	12.0	164	31.3	2.19	6.43	378	49.2
108	6.9	—	54.7	22.0	2.61	3.75	4470	299	11.9	146	27.9	2.15	4.99	346	43.9
99	7.8	—	57.0	20.3	2.57	4.23	3990	269	11.7	128	24.5	2.10	3.77	312	38.6
178	5.9	—	38.4	44.9	3.72	1.66	6990	502	11.6	555	78.8	3.26	19.5	567	122
161	6.5	—	41.8	37.8	3.70	1.82	6280	455	11.5	497	70.9	3.24	14.7	512	109
146	7.2	—	45.3	32.2	3.68	2.01	5630	411	11.4	443	63.5	3.21	10.9	461	97.5
114	5.4	—	47.9	28.8	2.58	2.91	4090	299	11.0	159	31.5	2.18	7.33	343	49.3
102	6.0	—	52.6	23.9	2.56	3.26	3620	267	11.0	139	27.8	2.15	5.29	305	43.4
94	6.7	—	54.9	21.9	2.53	3.62	3270	243	10.9	124	24.8	2.12	4.03	278	38.8
84	7.8	—	58.1	19.6	2.49	4.19	2850	213	10.7	106	21.2	2.07	2.81	244	33.2

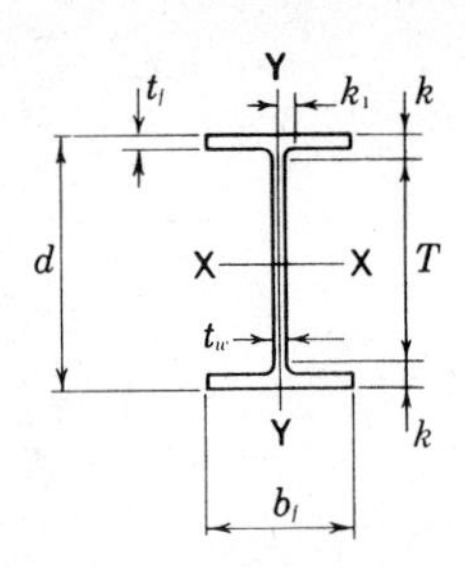

TABLE A-1. Properties of Wide Flange (W) Shapes

Designation	Area A	Depth d		Web			Flange				Distance		
				Thickness t_w		$\frac{t_w}{2}$	Width b_f		Thickness t_f		T	k	k_1
	In.^{2}	In.		In.		In.	In.		In.		In.	In.	In.
W 24x162	47.7	25.00	25	0.705	11/16	3/8	12.955	13	1.220	1 1/4	21	2	1 1/16
x146	43.0	24.74	24 3/4	0.650	5/8	5/16	12.900	12 7/8	1.090	1 1/16	21	1 7/8	1 1/16
x131	38.5	24.48	24 1/2	0.605	5/8	5/16	12.855	12 7/8	0.960	15/16	21	1 3/4	1 1/16
x117	34.4	24.26	24 1/4	0.550	9/16	5/16	12.800	12 3/4	0.850	7/8	21	1 5/8	1
x104	30.6	24.06	24	0.500	1/2	1/4	12.750	12 3/4	0.750	3/4	21	1 1/2	1
W 24x 94	27.7	24.31	24 1/4	0.515	1/2	1/4	9.065	9 1/8	0.875	7/8	21	1 5/8	1
x 84	24.7	24.10	24 1/8	0.470	1/2	1/4	9.020	9	0.770	3/4	21	1 9/16	15/16
x 76	22.4	23.92	23 7/8	0.440	7/16	1/4	8.990	9	0.680	11/16	21	1 7/16	15/16
x 68	20.1	23.73	23 3/4	0.415	7/16	1/4	8.965	9	0.585	9/16	21	1 3/8	15/16
W 24x 62	18.2	23.74	23 3/4	0.430	7/16	1/4	7.040	7	0.590	9/16	21	1 3/8	15/16
x 55	16.2	23.57	23 5/8	0.395	3/8	3/16	7.005	7	0.505	1/2	21	1 5/16	15/16
W 21x147	43.2	22.06	22	0.720	3/4	3/8	12.510	12 1/2	1.150	1 1/8	18 1/4	1 7/8	1 1/16
x132	38.8	21.83	21 7/8	0.650	5/8	5/16	12.440	12 1/2	1.035	1 1/16	18 1/4	1 13/16	1
x122	35.9	21.68	21 5/8	0.600	5/8	5/16	12.390	12 3/8	0.960	15/16	18 1/4	1 11/16	1
x111	32.7	21.51	21 1/2	0.550	9/16	5/16	12.340	12 3/8	0.875	7/8	18 1/4	1 5/8	15/16
x101	29.8	21.36	21 3/8	0.500	1/2	1/4	12.290	12 1/4	0.800	13/16	18 1/4	1 9/16	15/16
W 21x 93	27.3	21.62	21 5/8	0.580	9/16	5/16	8.420	8 3/8	0.930	15/16	18 1/4	1 11/16	1
x 83	24.3	21.43	21 3/8	0.515	1/2	1/4	8.355	8 3/8	0.835	13/16	18 1/4	1 9/16	15/16
x 73	21.5	21.24	21 1/4	0.455	7/16	1/4	8.295	8 1/4	0.740	3/4	18 1/4	1 1/2	15/16
x 68	20.0	21.13	21 1/8	0.430	7/16	1/4	8.270	8 1/4	0.685	11/16	18 1/4	1 7/16	7/8
x 62	18.3	20.99	21	0.400	3/8	3/16	8.240	8 1/4	0.615	5/8	18 1/4	1 3/8	7/8
W 21x 57	16.7	21.06	21	0.405	3/8	3/16	6.555	6 1/2	0.650	5/8	18 1/4	1 3/8	7/8
x 50	14.7	20.83	20 7/8	0.380	3/8	3/16	6.530	6 1/2	0.535	9/16	18 1/4	1 5/16	7/8
x 44	13.0	20.66	20 5/8	0.350	3/8	3/16	6.500	6 1/2	0.450	7/16	18 1/4	1 3/16	7/8
W 18x119	35.1	18.97	19	0.655	5/8	5/16	11.265	11 1/4	1.060	1 1/16	15 1/2	1 3/4	15/16
x106	31.1	18.73	18 3/4	0.590	9/16	5/16	11.200	11 1/4	0.940	15/16	15 1/2	1 5/8	15/16
x 97	28.5	18.59	18 5/8	0.535	9/16	5/16	11.145	11 1/8	0.870	7/8	15 1/2	1 9/16	7/8
x 86	25.3	18.39	18 3/8	0.480	1/2	1/4	11.090	11 1/8	0.770	3/4	15 1/2	1 7/16	7/8
x 76	22.3	18.21	18 1/4	0.425	7/16	1/4	11.035	11	0.680	11/16	15 1/2	1 3/8	13/16
W 18x 71	20.8	18.47	18 1/2	0.495	1/2	1/4	7.635	7 5/8	0.810	13/16	15 1/2	1 1/2	7/8
x 65	19.1	18.35	18 3/8	0.450	7/16	1/4	7.590	7 5/8	0.750	3/4	15 1/2	1 7/16	7/8
x 60	17.6	18.24	18 1/4	0.415	7/16	1/4	7.555	7 1/2	0.695	11/16	15 1/2	1 3/8	13/16
x 55	16.2	18.11	18 1/8	0.390	3/8	3/16	7.530	7 1/2	0.630	5/8	15 1/2	1 5/16	13/16
x 50	14.7	17.99	18	0.355	3/8	3/16	7.495	7 1/2	0.570	9/16	15 1/2	1 1/4	13/16

TABLE A-1. (*Continued*)

Nominal Wt. per Ft.	Compact Section Criteria				r_T	$\frac{d}{A_f}$	Elastic Properties						Torsional constant J	Plastic Modulus	
	$\frac{b_f}{2t_f}$	F_y'	$\frac{d}{t_w}$	F_y'''			Axis X-X			Axis Y-Y				Z_x	Z_y
							I	S	r	I	S	r			
Lb.		Ksi		Ksi	In.		In.^{4}	In.^{3}	In.	In.^{4}	In.^{3}	In.	In.^{4}	In.^{3}	In.^{3}
162	5.3	—	35.5	52.5	3.45	1.58	5170	414	10.4	443	68.4	3.05	18.5	468	105.
146	5.9	—	38.1	45.6	3.43	1.76	4580	371	10.3	391	60.5	3.01	13.4	418	93.2
131	6.7	—	40.5	40.3	3.40	1.98	4020	329	10.2	340	53.0	2.97	9.50	370	81.5
117	7.5	—	44.1	33.9	3.37	2.23	3540	291	10.1	297	46.5	2.94	6.72	327	71.4
104	8.5	58.5	48.1	28.5	3.35	2.52	3100	258	10.1	259	40.7	2.91	4.72	289	62.4
94	5.2	—	47.2	29.6	2.33	3.06	2700	222	9.87	109	24.0	1.98	5.26	254	37.5
84	5.9	—	51.3	25.1	2.31	3.47	2370	196	9.79	94.4	20.9	1.95	3.70	224	32.6
76	6.6	—	54.4	22.3	2.29	3.91	2100	176	9.69	82.5	18.4	1.92	2.68	200	28.6
68	7.7	—	57.2	20.2	2.26	4.52	1830	154	9.55	70.4	15.7	1.87	1.87	177	24.5
62	6.0	—	55.2	21.7	1.71	5.72	1550	131	9.23	34.5	9.80	1.38	1.71	153	15.7
55	6.9	—	59.7	18.5	1.68	6.66	1350	114	9.11	29.1	8.30	1.34	1.18	134	13.3
147	5.4	—	30.6	—	3.34	1.53	3630	329	9.17	376	60.1	2.95	15.4	373	92.6
132	6.0	—	33.6	58.6	3.31	1.70	3220	295	9.12	333	53.5	2.93	11.3	333	82.3
122	6.5	—	36.1	50.6	3.30	1.82	2960	273	9.09	305	49.2	2.92	8.98	307	75.6
111	7.1	—	39.1	43.2	3.28	1.99	2670	249	9.05	274	44.5	2.90	6.83	279	68.2
101	7.7	—	42.7	36.2	3.27	2.17	2420	227	9.02	248	40.3	2.89	5.21	253	61.7
93	4.5	—	37.3	47.5	2.17	2.76	2070	192	8.70	92.9	22.1	1.84	6.03	221	34.7
83	5.0	—	41.6	38.1	2.15	3.07	1830	171	8.67	81.4	19.5	1.83	4.34	196	30.5
73	5.6	—	46.7	30.3	2.13	3.46	1600	151	8.64	70.6	17.0	1.81	3.02	172	26.6
68	6.0	—	49.1	27.4	2.12	3.73	1480	140	8.60	64.7	15.7	1.80	2.45	160	24.4
62	6.7	—	52.5	24.0	2.10	4.14	1330	127	8.54	57.5	13.9	1.77	1.83	144	21.7
57	5.0	—	52.0	24.4	1.64	4.94	1170	111	8.36	30.6	9.35	1.35	1.77	129	14.8
50	6.1	—	54.8	22.0	1.60	5.96	984	94.5	8.18	24.9	7.64	1.30	1.14	110	12.2
44	7.2	—	59.0	19.0	1.57	7.06	843	81.6	8.06	20.7	6.36	1.26	0.77	95.4	10.2
119	5.3	—	29.0	—	3.02	1.59	2190	231	7.90	253	44.9	2.69	10.6	261	69.1
106	6.0	—	31.7	—	3.00	1.78	1910	204	7.84	220	39.4	2.66	7.48	230	60.5
97	6.4	—	34.7	54.7	2.99	1.92	1750	188	7.82	201	36.1	2.65	5.86	211	55.3
86	7.2	—	38.3	45.0	2.97	2.15	1530	166	7.77	175	31.6	2.63	4.10	186	48.4
76	8.1	64.2	42.8	36.0	2.95	2.43	1330	146	7.73	152	27.6	2.61	2.83	163	42.2
71	4.7	—	37.3	47.4	1.98	2.99	1170	127	7.50	60.3	15.8	1.70	3.48	145	24.7
65	5.1	—	40.8	39.7	1.97	3.22	1070	117	7.49	54.8	14.4	1.69	2.73	133	22.5
60	5.4	—	44.0	34.2	1.96	3.47	984	108	7.47	50.1	13.3	1.69	2.17	123	20.6
55	6.0	—	46.4	30.6	1.95	3.82	890	98.3	7.41	44.9	11.9	1.67	1.66	112	18.5
50	6.6	—	50.7	25.7	1.94	4.21	800	88.9	7.38	40.1	10.7	1.65	1.24	101	16.6

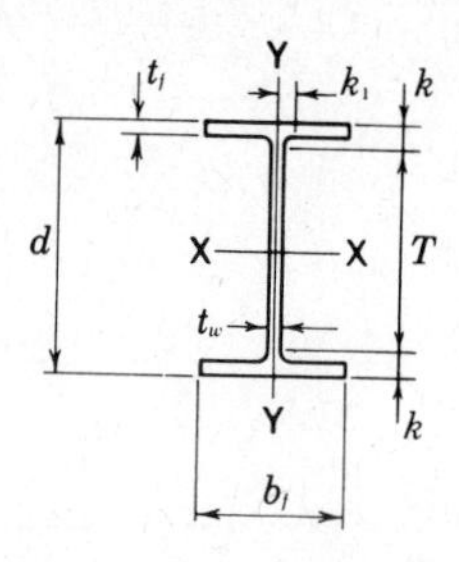

TABLE A-1. Properties of Wide Flange (W) Shapes

Designation	Area A	Depth d		Web Thickness t_w		Web $\frac{t_w}{2}$	Flange Width b_f		Flange Thickness t_f		Distance T	Distance k	Distance k_1
	In.^{2}	In.		In.		In.	In.		In.		In.	In.	In.
W 18x 46	13.5	18.06	18	0.360	3/8	3/16	6.060	6	0.605	5/8	15 1/2	1 1/4	13/16
x 40	11.8	17.90	17 7/8	0.315	5/16	3/16	6.015	6	0.525	1/2	15 1/2	1 3/16	13/16
x 35	10.3	17.70	17 3/4	0.300	5/16	3/16	6.000	6	0.425	7/16	15 1/2	1 1/8	3/4
W 16x100	29.4	16.97	17	0.585	9/16	5/16	10.425	10 3/8	0.985	1	13 5/8	1 11/16	15/16
x 89	26.2	16.75	16 3/4	0.525	1/2	1/4	10.365	10 3/8	0.875	7/8	13 5/8	1 9/16	7/8
x 77	22.6	16.52	16 1/2	0.455	7/16	1/4	10.295	10 1/4	0.760	3/4	13 5/8	1 7/16	7/8
x 67	19.7	16.33	16 3/8	0.395	3/8	3/16	10.235	10 1/4	0.665	11/16	13 5/8	1 3/8	13/16
W 16x 57	16.8	16.43	16 3/8	0.430	7/16	1/4	7.120	7 1/8	0.715	11/16	13 5/8	1 3/8	7/8
x 50	14.7	16.26	16 1/4	0.380	3/8	3/16	7.070	7 1/8	0.630	5/8	13 5/8	1 5/16	13/16
x 45	13.3	16.13	16 1/8	0.345	3/8	3/16	7.035	7	0.565	9/16	13 5/8	1 1/4	13/16
x 40	11.8	16.01	16	0.305	5/16	3/16	6.995	7	0.505	1/2	13 5/8	1 3/16	13/16
x 36	10.6	15.86	15 7/8	0.295	5/16	3/16	6.985	7	0.430	7/16	13 5/8	1 1/8	3/4
W 16x 31	9.12	15.88	15 7/8	0.275	1/4	1/8	5.525	5 1/2	0.440	7/16	13 5/8	1 1/8	3/4
x 26	7.68	15.69	15 3/4	0.250	1/4	1/8	5.500	5 1/2	0.345	3/8	13 5/8	1 1/16	3/4
W 14x730	215.0	22.42	22 3/8	3.070	3 1/16	1 9/16	17.890	17 7/8	4.910	4 15/16	11 1/4	5 9/16	2 3/16
x665	196.0	21.64	21 5/8	2.830	2 13/16	1 7/16	17.650	17 5/8	4.520	4 1/2	11 1/4	5 3/16	2 1/16
x605	178.0	20.92	20 7/8	2.595	2 5/8	1 5/16	17.415	17 3/8	4.160	4 3/16	11 1/4	4 13/16	1 15/16
x550	162.0	20.24	20 1/4	2.380	2 3/8	1 3/16	17.200	17 1/4	3.820	3 13/16	11 1/4	4 1/2	1 13/16
x500	147.0	19.60	19 5/8	2.190	2 3/16	1 1/8	17.010	17	3.500	3 1/2	11 1/4	4 3/16	1 3/4
x455	134.0	19.02	19	2.015	2	1	16.835	16 7/8	3.210	3 3/16	11 1/4	3 7/8	1 5/8
W 14x426	125.0	18.67	18 5/8	1.875	1 7/8	15/16	16.695	16 3/4	3.035	3 1/16	11 1/4	3 11/16	1 9/16
x398	117.0	18.29	18 1/4	1.770	1 3/4	7/8	16.590	16 5/8	2.845	2 7/8	11 1/4	3 1/2	1 1/2
x370	109.0	17.92	17 7/8	1.655	1 5/8	13/16	16.475	16 1/2	2.660	2 11/16	11 1/4	3 5/16	1 7/16
x342	101.0	17.54	17 1/2	1.540	1 9/16	13/16	16.360	16 3/8	2.470	2 1/2	11 1/4	3 1/8	1 3/8
x311	91.4	17.12	17 1/8	1.410	1 7/16	3/4	16.230	16 1/4	2.260	2 1/4	11 1/4	2 15/16	1 5/16
x283	83.3	16.74	16 3/4	1.290	1 5/16	11/16	16.110	16 1/8	2.070	2 1/16	11 1/4	2 3/4	1 1/4
x257	75.6	16.38	16 3/8	1.175	1 3/16	5/8	15.995	16	1.890	1 7/8	11 1/4	2 9/16	1 3/16
x233	68.5	16.04	16	1.070	1 1/16	9/16	15.890	15 7/8	1.720	1 3/4	11 1/4	2 3/8	1 3/16
x211	62.0	15.72	15 3/4	0.980	1	1/2	15.800	15 3/4	1.560	1 9/16	11 1/4	2 1/4	1 1/8
x193	56.8	15.48	15 1/2	0.890	7/8	7/16	15.710	15 3/4	1.440	1 7/16	11 1/4	2 1/8	1 1/16
x176	51.8	15.22	15 1/4	0.830	13/16	7/16	15.650	15 5/8	1.310	1 5/16	11 1/4	2	1 1/16
x159	46.7	14.98	15	0.745	3/4	3/8	15.565	15 5/8	1.190	1 3/16	11 1/4	1 7/8	1
x145	42.7	14.78	14 3/4	0.680	11/16	3/8	15.500	15 1/2	1.090	1 1/16	11 1/4	1 3/4	1

TABLE A-1. (*Continued*)

Nominal Wt. per Ft.	Compact Section Criteria				r_T	$\frac{d}{A_f}$	Elastic Properties						Torsional constant	Plastic Modulus	
							Axis X-X			Axis Y-Y					
	$\frac{b_f}{2t_f}$	F_y'	$\frac{d}{t_w}$	F_y'''			I	S	r	I	S	r	J	Z_x	Z_y
Lb.		Ksi		Ksi	In.		In.^{4}	In.^{3}	In.	In.^{4}	In.^{3}	In.	In.^{4}	In.^{3}	In.^{3}
46	5.0	—	50.2	26.2	1.54	4.93	712	78.8	7.25	22.5	7.43	1.29	1.22	90.7	11.7
40	5.7	—	56.8	20.5	1.52	5.67	612	68.4	7.21	19.1	6.35	1.27	0.81	78.4	9.95
35	7.1	—	59.0	19.0	1.49	6.94	510	57.6	7.04	15.3	5.12	1.22	0.51	66.5	8.06
100	5.3	—	29.0	—	2.81	1.65	1490	175	7.10	186	35.7	2.51	7.73	198	54.9
89	5.9	—	31.9	64.9	2.79	1.85	1300	155	7.05	163	31.4	2.49	5.45	175	48.1
77	6.8	—	36.3	50.1	2.77	2.11	1110	134	7.00	138	26.9	2.47	3.57	150	41.1
67	7.7	—	41.3	38.6	2.75	2.40	954	117	6.96	119	23.2	2.46	2.39	130	35.5
57	5.0	—	38.2	45.2	1.86	3.23	758	92.2	6.72	43.1	12.1	1.60	2.22	105	18.9
50	5.6	—	42.8	36.1	1.84	3.65	659	81.0	6.68	37.2	10.5	1.59	1.52	92.0	16.3
45	6.2	—	46.8	30.2	1.83	4.06	586	72.7	6.65	32.8	9.34	1.57	1.11	82.3	14.5
40	6.9	—	52.5	24.0	1.82	4.53	518	64.7	6.63	28.9	8.25	1.57	0.79	72.9	12.7
36	8.1	64.0	53.8	22.9	1.79	5.28	448	56.5	6.51	24.5	7.00	1.52	0.54	64.0	10.8
31	6.3	—	57.7	19.8	1.39	6.53	375	47.2	6.41	12.4	4.49	1.17	0.46	54.0	7.03
26	8.0	—	62.8	16.8	1.36	8.27	301	38.4	6.26	9.59	3.49	1.12	0.26	44.2	5.48
730	1.8	—	7.3	—	4.99	0.25	14300	1280	8.17	4720	527	4.69	1450	1660	816
665	2.0	—	7.6	—	4.92	0.27	12400	1150	7.98	4170	472	4.62	1120	1480	730
605	2.1	—	8.1	—	4.85	0.29	10800	1040	7.80	3680	423	4.55	870	1320	652
550	2.3	—	8.5	—	4.79	0.31	9430	931	7.63	3250	378	4.49	670	1180	583
500	2.4	—	8.9	—	4.73	0.33	8210	838	7.48	2880	339	4.43	514	1050	522
455	2.6	—	9.4	—	4.68	0.35	7190	756	7.33	2560	304	4.38	395	936	468
426	2.8	—	10.0	—	4.64	0.37	6600	707	7.26	2360	283	4.34	331	869	434
398	2.9	—	10.3	—	4.61	0.39	6000	656	7.16	2170	262	4.31	273	801	402
370	3.1	—	10.8	—	4.57	0.41	5440	607	7.07	1990	241	4.27	222	736	370
342	3.3	—	11.4	—	4.54	0.43	4900	559	6.98	1810	221	4.24	178	672	338
311	3.6	—	12.1	—	4.50	0.47	4330	506	6.88	1610	199	4.20	136	603	304
283	3.9	—	13.0	—	4.46	0.50	3840	459	6.79	1440	179	4.17	104	542	274
257	4.2	—	13.9	—	4.43	0.54	3400	415	6.71	1290	161	4.13	79.1	487	246
233	4.6	—	15.0	—	4.40	0.59	3010	375	6.63	1150	145	4.10	59.5	436	221
211	5.1	—	16.0	—	4.37	0.64	2660	338	6.55	1030	130	4.07	44.6	390	198
193	5.5	—	17.4	—	4.35	0.68	2400	310	6.50	931	119	4.05	34.8	355	180
176	6.0	—	18.3	—	4.32	0.74	2140	281	6.43	838	107	4.02	26.5	320	163
159	6.5	—	20.1	—	4.30	0.81	1900	254	6.38	748	96.2	4.00	19.8	287	146
145	7.1	—	21.7	—	4.28	0.88	1710	232	6.33	677	87.3	3.98	15.2	260	133

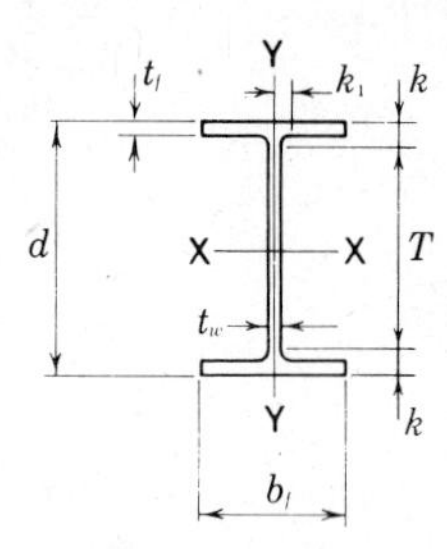

TABLE A-1. Properties of Wide Flange (W) Shapes

Designation	Area A	Depth d		Web: Thickness t_w		Web: $\frac{t_w}{2}$	Flange: Width b_f		Flange: Thickness t_f		Distance: T	Distance: k	Distance: k_1
	In.^{2}	In.		In.		In.	In.		In.		In.	In.	In.
W 14x132	38.8	14.66	14 5/8	0.645	5/8	5/16	14.725	14 3/4	1.030	1	11 1/4	1 11/16	15/16
x120	35.3	14.48	14 1/2	0.590	9/16	5/16	14.670	14 5/8	0.940	15/16	11 1/4	1 5/8	15/16
x109	32.0	14.32	14 3/8	0.525	1/2	1/4	14.605	14 5/8	0.860	7/8	11 1/4	1 9/16	7/8
x 99	29.1	14.16	14 1/8	0.485	1/2	1/4	14.565	14 5/8	0.780	3/4	11 1/4	1 7/16	7/8
x 90	26.5	14.02	14	0.440	7/16	1/4	14.520	14 1/2	0.710	11/16	11 1/4	1 3/8	7/8
W 14x 82	24.1	14.31	14 1/4	0.510	1/2	1/4	10.130	10 1/8	0.855	7/8	11	1 5/8	1
x 74	21.8	14.17	14 1/8	0.450	7/16	1/4	10.070	10 1/8	0.785	13/16	11	1 9/16	15/16
x 68	20.0	14.04	14	0.415	7/16	1/4	10.035	10	0.720	3/4	11	1 1/2	15/16
x 61	17.9	13.89	13 7/8	0.375	3/8	3/16	9.995	10	0.645	5/8	11	1 7/16	15/16
W 14x 53	15.6	13.92	13 7/8	0.370	3/8	3/16	8.060	8	0.660	11/16	11	1 7/16	15/16
x 48	14.1	13.79	13 3/4	0.340	5/16	3/16	8.030	8	0.595	5/8	11	1 3/8	7/8
x 43	12.6	13.66	13 5/8	0.305	5/16	3/16	7.995	8	0.530	1/2	11	1 5/16	7/8
W 14x 38	11.2	14.10	14 1/8	0.310	5/16	3/16	6.770	6 3/4	0.515	1/2	12	1 1/16	5/8
x 34	10.0	13.98	14	0.285	5/16	3/16	6.745	6 3/4	0.455	7/16	12	1	5/8
x 30	8.85	13.84	13 7/8	0.270	1/4	1/8	6.730	6 3/4	0.385	3/8	12	15/16	5/8
W 14x 26	7.69	13.91	13 7/8	0.255	1/4	1/8	5.025	5	0.420	7/16	12	15/16	9/16
x 22	6.49	13.74	13 3/4	0.230	1/4	1/8	5.000	5	0.335	5/16	12	7/8	9/16
W 12x336	98.8	16.82	16 7/8	1.775	1 3/4	7/8	13.385	13 3/8	2.955	2 15/16	9 1/2	3 11/16	1 1/2
x305	89.6	16.32	16 3/8	1.625	1 5/8	13/16	13.235	13 1/4	2.705	2 11/16	9 1/2	3 7/16	1 7/16
x279	81.9	15.85	15 7/8	1.530	1 1/2	3/4	13.140	13 1/8	2.470	2 1/2	9 1/2	3 3/16	1 3/8
x252	74.1	15.41	15 3/8	1.395	1 3/8	11/16	13.005	13	2.250	2 1/4	9 1/2	2 15/16	1 5/16
x230	67.7	15.05	15	1.285	1 5/16	11/16	12.895	12 7/8	2.070	2 1/16	9 1/2	2 3/4	1 1/4
x210	61.8	14.71	14 3/4	1.180	1 3/16	5/8	12.790	12 3/4	1.900	1 7/8	9 1/2	2 5/8	1 1/4
x190	55.8	14.38	14 3/8	1.060	1 1/16	9/16	12.670	12 5/8	1.735	1 3/4	9 1/2	2 7/16	1 3/16
x170	50.0	14.03	14	0.960	15/16	1/2	12.570	12 5/8	1.560	1 9/16	9 1/2	2 1/4	1 1/8
x152	44.7	13.71	13 3/4	0.870	7/8	7/16	12.480	12 1/2	1.400	1 3/8	9 1/2	2 1/8	1 1/16
x136	39.9	13.41	13 3/8	0.790	13/16	7/16	12.400	12 3/8	1.250	1 1/4	9 1/2	1 15/16	1
x120	35.3	13.12	13 1/8	0.710	11/16	3/8	12.320	12 3/8	1.105	1 1/8	9 1/2	1 13/16	1
x106	31.2	12.89	12 7/8	0.610	5/8	5/16	12.220	12 1/4	0.990	1	9 1/2	1 11/16	15/16
x 96	28.2	12.71	12 3/4	0.550	9/16	5/16	12.160	12 1/8	0.900	7/8	9 1/2	1 5/8	7/8
x 87	25.6	12.53	12 1/2	0.515	1/2	1/4	12.125	12 1/8	0.810	13/16	9 1/2	1 1/2	7/8
x 79	23.2	12.38	12 3/8	0.470	1/2	1/4	12.080	12 1/8	0.735	3/4	9 1/2	1 7/16	7/8
x 72	21.1	12.25	12 1/4	0.430	7/16	1/4	12.040	12	0.670	11/16	9 1/2	1 3/8	7/8
x 65	19.1	12.12	12 1/8	0.390	3/8	3/16	12.000	12	0.605	5/8	9 1/2	1 5/16	13/16

TABLE A-1. (*Continued*)

Nominal Wt. per Ft.	Compact Section Criteria				r_T	$\frac{d}{A_f}$	Elastic Properties						Torsional constant	Plastic Modulus	
							Axis X-X			Axis Y-Y					
	$\frac{b_f}{2t_f}$	F_y'	$\frac{d}{t_w}$	F_y'''			I	S	r	I	S	r	J	Z_x	Z_y
Lb.		Ksi		Ksi	In.		In.^{4}	In.^{3}	In.	In.^{4}	In.^{3}	In.	In.^{4}	In.^{3}	In.^{3}
132	7.1	—	22.7	—	4.05	0.97	1530	209	6.28	548	74.5	3.76	12.3	234	113
120	7.8	—	24.5	—	4.04	1.05	1380	190	6.24	495	67.5	3.74	9.37	212	102
109	8.5	58.6	27.3	—	4.02	1.14	1240	173	6.22	447	61.2	3.73	7.12	192	92.7
99	9.3	48.5	29.2	—	4.00	1.25	1110	157	6.17	402	55.2	3.71	5.37	173	83.6
90	10.2	40.4	31.9	—	3.99	1.36	999	143	6.14	362	49.9	3.70	4.06	157	75.6
82	5.9	—	28.1	—	2.74	1.65	882	123	6.05	148	29.3	2.48	5.08	139	44.8
74	6.4	—	31.5	—	2.72	1.79	796	112	6.04	134	26.6	2.48	3.88	126	40.6
68	7.0	—	33.8	57.7	2.71	1.94	723	103	6.01	121	24.2	2.46	3.02	115	36.9
61	7.7	—	37.0	48.1	2.70	2.15	640	92.2	5.98	107	21.5	2.45	2.20	102	32.8
53	6.1	—	37.6	46.7	2.15	2.62	541	77.8	5.89	57.7	14.3	1.92	1.94	87.1	22.0
48	6.7	—	40.6	40.2	2.13	2.89	485	70.3	5.85	51.4	12.8	1.91	1.46	78.4	19.6
43	7.5	—	44.8	32.9	2.12	3.22	428	62.7	5.82	45.2	11.3	1.89	1.05	69.6	17.3
38	6.6	—	45.5	31.9	1.77	4.04	385	54.6	5.87	26.7	7.88	1.55	0.80	61.5	12.1
34	7.4	—	49.1	27.4	1.76	4.56	340	48.6	5.83	23.3	6.91	1.53	0.57	54.6	10.6
30	8.7	55.3	51.3	25.1	1.74	5.34	291	42.0	5.73	19.6	5.82	1.49	0.38	47.3	8.99
26	6.0	—	54.5	22.2	1.28	6.59	245	35.3	5.65	8.91	3.54	1.08	0.36	40.2	5.54
22	7.5	—	59.7	18.5	1.25	8.20	199	29.0	5.54	7.00	2.80	1.04	0.21	33.2	4.39
336	2.3	—	9.5	—	3.71	0.43	4060	483	6.41	1190	177	3.47	243	603	274
305	2.4	—	10.0	—	3.67	0.46	3550	435	6.29	1050	159	3.42	185	537	244
279	2.7	—	10.4	—	3.64	0.49	3110	393	6.16	937	143	3.38	143	481	220
252	2.9	—	11.0	—	3.59	0.53	2720	353	6.06	828	127	3.34	108	428	196
230	3.1	—	11.7	—	3.56	0.56	2420	321	5.97	742	115	3.31	83.8	386	177
210	3.4	—	12.5	—	3.53	0.61	2140	292	5.89	664	104	3.28	64.7	348	159
190	3.7	—	13.6	—	3.50	0.65	1890	263	5.82	589	93.0	3.25	48.8	311	143
170	4.0	—	14.6	—	3.47	0.72	1650	235	5.74	517	82.3	3.22	35.6	275	126
152	4.5	—	15.8	—	3.44	0.79	1430	209	5.66	454	72.8	3.19	25.8	243	111
136	5.0	—	17.0	—	3.41	0.87	1240	186	5.58	398	64.2	3.16	18.5	214	98.0
120	5.6	—	18.5	—	3.38	0.96	1070	163	5.51	345	56.0	3.13	12.9	186	85.4
106	6.2	—	21.1	—	3.36	1.07	933	145	5.47	301	49.3	3.11	9.13	164	75.1
96	6.8	—	23.1	—	3.34	1.16	833	131	5.44	270	44.4	3.09	6.86	147	67.5
87	7.5	—	24.3	—	3.32	1.28	740	118	5.38	241	39.7	3.07	5.10	132	60.4
79	8.2	62.6	26.3	—	3.31	1.39	662	107	5.34	216	35.8	3.05	3.84	119	54.3
72	9.0	52.3	28.5	—	3.29	1.52	597	97.4	5.31	195	32.4	3.04	2.93	108	49.2
65	9.9	43.0	31.1	—	3.28	1.67	533	87.9	5.28	174	29.1	3.02	2.18	96.8	44.1

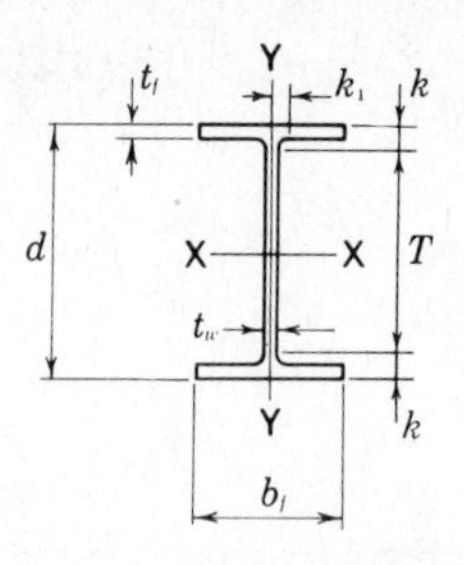

TABLE A-1. Properties of Wide Flange (W) Shapes

Designation	Area A	Depth d		Web Thickness t_w		Web $\frac{t_w}{2}$	Flange Width b_f		Flange Thickness t_f		Distance T	Distance k	Distance k_1
	In.^{2}	In.		In.		In.	In.		In.		In.	In.	In.
W 12x 58	17.0	12.19	12 1/4	0.360	3/8	3/16	10.010	10	0.640	5/8	9 1/2	1 3/8	13/16
x 53	15.6	12.06	12	0.345	3/8	3/16	9.995	10	0.575	9/16	9 1/2	1 1/4	13/16
W 12x 50	14.7	12.19	12 1/4	0.370	3/8	3/16	8.080	8 1/8	0.640	5/8	9 1/2	1 3/8	13/16
x 45	13.2	12.06	12	0.335	5/16	3/16	8.045	8	0.575	9/16	9 1/2	1 1/4	13/16
x 40	11.8	11.94	12	0.295	5/16	3/16	8.005	8	0.515	1/2	9 1/2	1 1/4	3/4
W 12x 35	10.3	12.50	12 1/2	0.300	5/16	3/16	6.560	6 1/2	0.520	1/2	10 1/2	1	9/16
x 30	8.79	12.34	12 3/8	0.260	1/4	1/8	6.520	6 1/2	0.440	7/16	10 1/2	15/16	1/2
x 26	7.65	12.22	12 1/4	0.230	1/4	1/8	6.490	6 1/2	0.380	3/8	10 1/2	7/8	1/2
W 12x 22	6.48	12.31	12 1/4	0.260	1/4	1/8	4.030	4	0.425	7/16	10 1/2	7/8	1/2
x 19	5.57	12.16	12 1/8	0.235	1/4	1/8	4.005	4	0.350	3/8	10 1/2	13/16	1/2
x 16	4.71	11.99	12	0.220	1/4	1/8	3.990	4	0.265	1/4	10 1/2	3/4	1/2
x 14	4.16	11.91	11 7/8	0.200	3/16	1/8	3.970	4	0.225	1/4	10 1/2	11/16	1/2
W 10x112	32.9	11.36	11 3/8	0.755	3/4	3/8	10.415	10 3/8	1.250	1 1/4	7 5/8	1 7/8	15/16
x100	29.4	11.10	11 1/8	0.680	11/16	3/8	10.340	10 3/8	1.120	1 1/8	7 5/8	1 3/4	7/8
x 88	25.9	10.84	10 7/8	0.605	5/8	5/16	10.265	10 1/4	0.990	1	7 5/8	1 5/8	13/16
x 77	22.6	10.60	10 5/8	0.530	1/2	1/4	10.190	10 1/4	0.870	7/8	7 5/8	1 1/2	13/16
x 68	20.0	10.40	10 3/8	0.470	1/2	1/4	10.130	10 1/8	0.770	3/4	7 5/8	1 3/8	3/4
x 60	17.6	10.22	10 1/4	0.420	7/16	1/4	10.080	10 1/8	0.680	11/16	7 5/8	1 5/16	3/4
x 54	15.8	10.09	10 1/8	0.370	3/8	3/16	10.030	10	0.615	5/8	7 5/8	1 1/4	11/16
x 49	14.4	9.98	10	0.340	5/16	3/16	10.000	10	0.560	9/16	7 5/8	1 3/16	11/16
W 10x 45	13.3	10.10	10 1/8	0.350	3/8	3/16	8.020	8	0.620	5/8	7 5/8	1 1/4	11/16
x 39	11.5	9.92	9 7/8	0.315	5/16	3/16	7.985	8	0.530	1/2	7 5/8	1 1/8	11/16
x 33	9.71	9.73	9 3/4	0.290	5/16	3/16	7.960	8	0.435	7/16	7 5/8	1 1/16	11/16
W 10x 30	8.84	10.47	10 1/2	0.300	5/16	3/16	5.810	5 3/4	0.510	1/2	8 5/8	15/16	1/2
x 26	7.61	10.33	10 3/8	0.260	1/4	1/8	5.770	5 3/4	0.440	7/16	8 5/8	7/8	1/2
x 22	6.49	10.17	10 1/8	0.240	1/4	1/8	5.750	5 3/4	0.360	3/8	8 5/8	3/4	1/2
W 10x 19	5.62	10.24	10 1/4	0.250	1/4	1/8	4.020	4	0.395	3/8	8 5/8	13/16	1/2
x 17	4.99	10.11	10 1/8	0.240	1/4	1/8	4.010	4	0.330	5/16	8 5/8	3/4	1/2
x 15	4.41	9.99	10	0.230	1/4	1/8	4.000	4	0.270	1/4	8 5/8	11/16	7/16
x 12	3.54	9.87	9 7/8	0.190	3/16	1/8	3.960	4	0.210	3/16	8 5/8	5/8	7/16

TABLE A-1. (*Continued*)

Nominal Wt. per Ft.	Compact Section Criteria				r_T	$\frac{d}{A_f}$	Elastic Properties						Torsional constant J	Plastic Modulus	
							Axis X-X			Axis Y-Y					
	$\frac{b_f}{2t_f}$	F_y'	$\frac{d}{t_w}$	F_y'''			I	S	r	I	S	r		Z_x	Z_y
Lb.		Ksi		Ksi	In.		In.^{4}	In.^{3}	In.	In.^{4}	In.^{3}	In.	In.^{4}	In.^{3}	In.^{3}
58	7.8	—	33.9	57.6	2.72	1.90	475	78.0	5.28	107	21.4	2.51	2.10	86.4	32.5
53	8.7	55.9	35.0	54.1	2.71	2.10	425	70.6	5.23	95.8	19.2	2.48	1.58	77.9	29.1
50	6.3	—	32.9	60.9	2.17	2.36	394	64.7	5.18	56.3	13.9	1.96	1.78	72.4	21.4
45	7.0	—	36.0	51.0	2.15	2.61	350	58.1	5.15	50.0	12.4	1.94	1.31	64.7	19.0
40	7.8	—	40.5	40.3	2.14	2.90	310	51.9	5.13	44.1	11.0	1.93	0.95	57.5	16.8
35	6.3	—	41.7	38.0	1.74	3.66	285	45.6	5.25	24.5	7.47	1.54	0.74	51.2	11.5
30	7.4	—	47.5	29.3	1.73	4.30	238	38.6	5.21	20.3	6.24	1.52	0.46	43.1	9.56
26	8.5	57.9	53.1	23.4	1.72	4.95	204	33.4	5.17	17.3	5.34	1.51	0.30	37.2	8.17
22	4.7	—	47.3	29.5	1.02	7.19	156	25.4	4.91	4.66	2.31	0.847	0.29	29.3	3.66
19	5.7	—	51.7	24.7	1.00	8.67	130	21.3	4.82	3.76	1.88	0.822	0.18	24.7	2.98
16	7.5	—	54.5	22.2	0.96	11.3	103	17.1	4.67	2.82	1.41	0.773	0.10	20.1	2.26
14	8.8	54.3	59.6	18.6	0.95	13.3	88.6	14.9	4.62	2.36	1.19	0.753	0.07	17.4	1.90
112	4.2	—	15.0	—	2.88	0.87	716	126	4.66	236	45.3	2.68	15.1	147	69.2
100	4.6	—	16.3	—	2.85	0.96	623	112	4.60	207	40.0	2.65	10.9	130	61.0
88	5.2	—	17.9	—	2.83	1.07	534	98.5	4.54	179	34.8	2.63	7.53	113	53.1
77	5.9	—	20.0	—	2.80	1.20	455	85.9	4.49	154	30.1	2.60	5.11	97.6	45.9
68	6.6	—	22.1	—	2.79	1.33	394	75.7	4.44	134	26.4	2.59	3.56	85.3	40.1
60	7.4	—	24.3	—	2.77	1.49	341	66.7	4.39	116	23.0	2.57	2.48	74.6	35.0
54	8.2	63.5	27.3	—	2.75	1.64	303	60.0	4.37	103	20.6	2.56	1.82	66.6	31.3
49	8.9	53.0	29.4	—	2.74	1.78	272	54.6	4.35	93.4	18.7	2.54	1.39	60.4	28.3
45	6.5	—	28.9	—	2.18	2.03	248	49.1	4.32	53.4	13.3	2.01	1.51	54.9	20.3
39	7.5	—	31.5	—	2.16	2.34	209	42.1	4.27	45.0	11.3	1.98	0.98	46.8	17.2
33	9.1	50.5	33.6	58.7	2.14	2.81	170	35.0	4.19	36.6	9.20	1.94	0.58	38.8	14.0
30	5.7	—	34.9	54.2	1.55	3.53	170	32.4	4.38	16.7	5.75	1.37	0.62	36.6	8.84
26	6.6	—	39.7	41.8	1.54	4.07	144	27.9	4.35	14.1	4.89	1.36	0.40	31.3	7.50
22	8.0	—	42.4	36.8	1.51	4.91	118	23.2	4.27	11.4	3.97	1.33	0.24	26.0	6.10
19	5.1	—	41.0	39.4	1.03	6.45	96.3	18.8	4.14	4.29	2.14	0.874	0.23	21.6	3.35
17	6.1	—	42.1	37.2	1.01	7.64	81.9	16.2	4.05	3.56	1.78	0.844	0.16	18.7	2.80
15	7.4	—	43.4	35.0	0.99	9.25	68.9	13.8	3.95	2.89	1.45	0.810	0.10	16.0	2.30
12	9.4	47.5	51.9	24.5	0.96	11.9	53.8	10.9	3.90	2.18	1.10	0.785	0.06	12.6	1.74

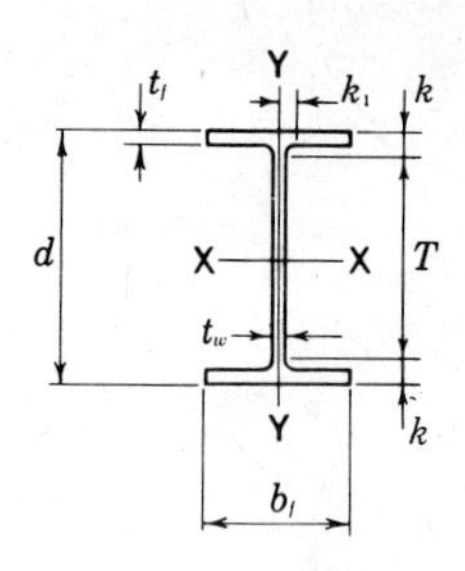

TABLE A-1. Properties of Wide Flange (W) Shapes

Designation	Area A	Depth d		Web Thickness t_w		Web $\frac{t_w}{2}$	Flange Width b_f		Flange Thickness t_f		Distance T	Distance k	Distance k_1
	In.^{2}	In.		In.		In.	In.		In.		In.	In.	In.
W 8x67	19.7	9.00	9	0.570	9/16	5/16	8.280	8 1/4	0.935	15/16	6 1/8	1 7/16	11/16
x58	17.1	8.75	8 3/4	0.510	1/2	1/4	8.220	8 1/4	0.810	13/16	6 1/8	1 5/16	11/16
x48	14.1	8.50	8 1/2	0.400	3/8	3/16	8.110	8 1/8	0.685	11/16	6 1/8	1 3/16	5/8
x40	11.7	8.25	8 1/4	0.360	3/8	3/16	8.070	8 1/8	0.560	9/16	6 1/8	1 1/16	5/8
x35	10.3	8.12	8 1/8	0.310	5/16	3/16	8.020	8	0.495	1/2	6 1/8	1	9/16
x31	9.13	8.00	8	0.285	5/16	3/16	7.995	8	0.435	7/16	6 1/8	15/16	9/16
W 8x28	8.25	8.06	8	0.285	5/16	3/16	6.535	6 1/2	0.465	7/16	6 1/8	15/16	9/16
x24	7.08	7.93	7 7/8	0.245	1/4	1/8	6.495	6 1/2	0.400	3/8	6 1/8	7/8	9/16
W 8x21	6.16	8.28	8 1/4	0.250	1/4	1/8	5.270	5 1/4	0.400	3/8	6 5/8	13/16	1/2
x18	5.26	8.14	8 1/8	0.230	1/4	1/8	5.250	5 1/4	0.330	5/16	6 5/8	3/4	7/16
W 8x15	4.44	8.11	8 1/8	0.245	1/4	1/8	4.015	4	0.315	5/16	6 5/8	3/4	1/2
x13	3.84	7.99	8	0.230	1/4	1/8	4.000	4	0.255	1/4	6 5/8	11/16	7/16
x10	2.96	7.89	7 7/8	0.170	3/16	1/8	3.940	4	0.205	3/16	6 5/8	5/8	7/16
W 6x25	7.34	6.38	6 3/8	0.320	5/16	3/16	6.080	6 1/8	0.455	7/16	4 3/4	13/16	7/16
x20	5.87	6.20	6 1/4	0.260	1/4	1/8	6.020	6	0.365	3/8	4 3/4	3/4	7/16
x15	4.43	5.99	6	0.230	1/4	1/8	5.990	6	0.260	1/4	4 3/4	5/8	3/8
W 6x16	4.74	6.28	6 1/4	0.260	1/4	1/8	4.030	4	0.405	3/8	4 3/4	3/4	7/16
x12	3.55	6.03	6	0.230	1/4	1/8	4.000	4	0.280	1/4	4 3/4	5/8	3/8
x 9	2.68	5.90	5 7/8	0.170	3/16	1/8	3.940	4	0.215	3/16	4 3/4	9/16	3/8
W 5x19	5.54	5.15	5 1/8	0.270	1/4	1/8	5.030	5	0.430	7/16	3 1/2	13/16	7/16
x16	4.68	5.01	5	0.240	1/4	1/8	5.000	5	0.360	3/8	3 1/2	3/4	7/16
W 4x13	3.83	4.16	4 1/8	0.280	1/4	1/8	4.060	4	0.345	3/8	2 3/4	11/16	7/16

TABLE A-1. (*Continued*)

Nominal Wt. per Ft.	Compact Section Criteria				r_T	$\frac{d}{A_f}$	Elastic Properties						Torsional constant J	Plastic Modulus	
							Axis X-X			Axis Y-Y					
	$\frac{b_f}{2t_f}$	F_y'	$\frac{d}{t_w}$	F_y'''			I	S	r	I	S	r		Z_x	Z_y
Lb.		Ksi		Ksi	In.		In.^{4}	In.^{3}	In.	In.^{4}	In.^{3}	In.	In.^{4}	In.^{3}	In.^{3}
67	4.4	—	15.8	—	2.28	1.16	272	60.4	3.72	88.6	21.4	2.12	5.06	70.2	32.7
58	5.1	—	17.2	—	2.26	1.31	228	52.0	3.65	75.1	18.3	2.10	3.34	59.8	27.9
48	5.9	—	21.3	—	2.23	1.53	184	43.3	3.61	60.9	15.0	2.08	1.96	49.0	22.9
40	7.2	—	22.9	—	2.21	1.83	146	35.5	3.53	49.1	12.2	2.04	1.12	39.8	18.5
35	8.1	64.4	26.2	—	2.20	2.05	127	31.2	3.51	42.6	10.6	2.03	0.77	34.7	16.1
31	9.2	50.0	28.1	—	2.18	2.30	110	27.5	3.47	37.1	9.27	2.02	0.54	30.4	14.1
28	7.0	—	28.3	—	1.77	2.65	98.0	24.3	3.45	21.7	6.63	1.62	0.54	27.2	10.1
24	8.1	64.1	32.4	63.0	1.76	3.05	82.8	20.9	3.42	18.3	5.63	1.61	0.35	23.2	8.57
21	6.6	—	33.1	60.2	1.41	3.93	75.3	18.2	3.49	9.77	3.71	1.26	0.28	20.4	5.69
18	8.0	—	35.4	52.7	1.39	4.70	61.9	15.2	3.43	7.97	3.04	1.23	0.17	17.0	4.66
15	6.4	—	33.1	60.3	1.03	6.41	48.0	11.8	3.29	3.41	1.70	0.876	0.14	13.6	2.67
13	7.8	—	34.7	54.7	1.01	7.83	39.6	9.91	3.21	2.73	1.37	0.843	0.09	11.4	2.15
10	9.6	45.8	46.4	30.7	0.99	9.77	30.8	7.81	3.22	2.09	1.06	0.841	0.04	8.87	1.66
25	6.7	—	19.9	—	1.66	2.31	53.4	16.7	2.70	17.1	5.61	1.52	0.46	18.9	8.56
20	8.2	62.1	23.8	—	1.64	2.82	41.4	13.4	2.66	13.3	4.41	1.50	0.24	14.9	6.72
15	11.5	31.8	26.0	—	1.61	3.85	29.1	9.72	2.56	9.32	3.11	1.46	0.10	10.8	4.75
16	5.0	—	24.2	—	1.08	3.85	32.1	10.2	2.60	4.43	2.20	0.966	0.22	11.7	3.39
12	7.1	—	26.2	—	1.05	5.38	22.1	7.31	2.49	2.99	1.50	0.918	0.09	8.30	2.32
9	9.2	50.3	34.7	54.8	1.03	6.96	16.4	5.56	2.47	2.19	1.11	0.905	0.04	6.23	1.72
19	5.8	—	19.1	—	1.38	2.38	26.2	10.2	2.17	9.13	3.63	1.28	0.31	11.6	5.53
16	6.9	—	20.9	—	1.37	2.78	21.3	8.51	2.13	7.51	3.00	1.27	0.19	9.59	4.57
13	5.9	—	14.9	—	1.10	2.97	11.3	5.46	1.72	3.86	1.90	1.00	0.15	6.28	2.92

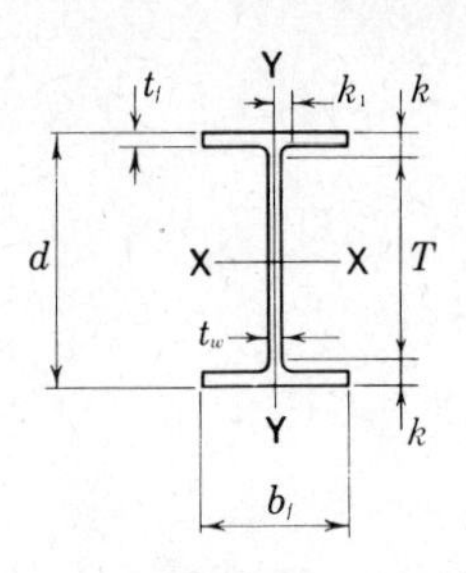

TABLE A-2. Properties of Miscellaneous (M) Shapes

Designation	Area A	Depth d		Web Thickness t_w		Web $\frac{t_w}{2}$	Flange Width b_f		Flange Thickness t_f		Distance T	Distance k	Grip	Max. Flge. Fastener
	In.^{2}	In.		In.		In.	In.		In.		In.	In.	In.	In.
M 14x18	5.10	14.00	14	0.215	3/16	1/8	4.000	4	0.270	1/4	12 3/4	5/8	1/4	3/4
M 12x11.8	3.47	12.00	12	0.177	3/16	1/8	3.065	3 1/8	0.225	1/4	10 7/8	9/16	1/4	—
M 10x9	2.65	10.00	10	0.157	3/16	1/8	2.690	2 3/4	0.206	3/16	8 7/8	9/16	3/16	—
M 8x6.5	1.92	8.00	8	0.135	1/8	1/16	2.281	2 1/4	0.189	3/16	7	1/2	3/16	—
M 6x20	5.89	6.00	6	0.250	1/4	1/8	5.938	6	0.379	3/8	4 1/4	7/8	3/8	7/8
M 6x4.4	1.29	6.00	6	0.114	1/8	1/16	1.844	1 7/8	0.171	3/16	5 1/8	7/16	3/16	—
M 5x18.9	5.55	5.00	5	0.316	5/16	3/16	5.003	5	0.416	7/16	3 1/4	7/8	7/16	7/8
M 4x13	3.81	4.00	4	0.254	1/4	1/8	3.940	4	0.371	3/8	2 3/8	13/16	3/8	3/4

TABLE A-2. (*Continued*)

Nominal Wt. per Ft	Compact Section Criteria				r_T	$\frac{d}{A_f}$	Elastic Properties						Torsional constant J	Plastic Modulus	
	$\frac{b_f}{2t_f}$	F_y'	$\frac{d}{t_w}$	F_y'''			Axis X-X I	Axis X-X S	Axis X-X r	Axis Y-Y I	Axis Y-Y S	Axis Y-Y r		Z_x	Z_y
Lb.		Ksi		Ksi	In.		In.^{4}	In.^{3}	In.	In.^{4}	In.^{3}	In.	In.^{4}	In.^{3}	In.^{3}
18	7.4	—	65.1	15.6	0.91	13.0	148	21.1	5.38	2.64	1.32	0.719	0.11	24.9	2.20
11.8	6.8	—	67.8	14.4	0.68	17.4	71.9	12.0	4.55	0.980	0.639	0.532	0.05	14.3	1.09
9	6.5	—	63.7	16.3	0.61	18.0	38.8	7.76	3.83	0.609	0.453	0.480	0.03	9.19	0.765
6.5	6.0	—	59.3	18.8	0.53	18.6	18.5	4.62	3.10	0.343	0.301	0.423	0.02	5.42	0.502
20	7.8	—	24.0	—	1.52	2.66	39.0	13.0	2.57	11.6	3.90	1.40	0.30	14.5	6.25
4.4	5.4	—	52.6	23.8	0.44	19.0	7.20	2.40	2.36	0.165	0.179	0.358	0.01	2.80	0.296
18.9	6.0	—	15.8	—	1.29	2.40	24.1	9.63	2.08	7.86	3.14	1.19	0.34	11.0	5.02
13	5.3	—	15.7	—	1.01	2.74	10.5	5.24	1.66	3.36	1.71	0.939	0.19	6.05	2.74

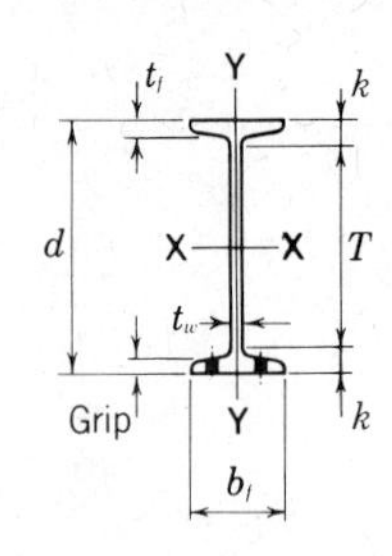

TABLE A-3. Properties of American Standard (S) Shapes

Designation	Area A	Depth d		Web Thickness t_w		Web $\frac{t_w}{2}$	Flange Width b_f		Flange Thickness t_f		Distance T	Distance k	Grip	Max. Flge. Fastener
	In.^{2}	In.		In.		In.	In.		In.		In.	In.	In.	In.
S 24x121	35.6	24.50	24 1/2	0.800	13/16	7/16	8.050	8	1.090	1 1/16	20 1/2	2	1 1/8	1
x106	31.2	24.50	24 1/2	0.620	5/8	5/16	7.870	7 7/8	1.090	1 1/16	20 1/2	2	1 1/8	1
S 24x100	29.3	24.00	24	0.745	3/4	3/8	7.245	7 1/4	0.870	7/8	20 1/2	1 3/4	7/8	1
x90	26.5	24.00	24	0.625	5/8	5/16	7.125	7 1/8	0.870	7/8	20 1/2	1 3/4	7/8	1
x80	23.5	24.00	24	0.500	1/2	1/4	7.000	7	0.870	7/8	20 1/2	1 3/4	7/8	1
S 20x96	28.2	20.30	20 1/4	0.800	13/16	7/16	7.200	7 1/4	0.920	15/16	16 3/4	1 3/4	15/16	1
x86	25.3	20.30	20 1/4	0.660	11/16	3/8	7.060	7	0.920	15/16	16 3/4	1 3/4	15/16	1
S 20x75	22.0	20.00	20	0.635	5/8	5/16	6.385	6 3/8	0.795	13/16	16 3/4	1 5/8	13/16	7/8
x66	19.4	20.00	20	0.505	1/2	1/4	6.255	6 1/4	0.795	13/16	16 3/4	1 5/8	13/16	7/8
S 18x70	20.6	18.00	18	0.711	11/16	3/8	6.251	6 1/4	0.691	11/16	15	1 1/2	11/16	7/8
x54.7	16.1	18.00	18	0.461	7/16	1/4	6.001	6	0.691	11/16	15	1 1/2	11/16	7/8
S 15x50	14.7	15.00	15	0.550	9/16	5/16	5.640	5 5/8	0.622	5/8	12 1/4	1 3/8	9/16	3/4
x42.9	12.6	15.00	15	0.411	7/16	1/4	5.501	5 1/2	0.622	5/8	12 1/4	1 3/8	9/16	3/4
S 12x50	14.7	12.00	12	0.687	11/16	3/8	5.477	5 1/2	0.659	11/16	9 1/8	1 7/16	11/16	3/4
x40.8	12.0	12.00	12	0.462	7/16	1/4	5.252	5 1/4	0.659	11/16	9 1/8	1 7/16	5/8	3/4
S 12x35	10.3	12.00	12	0.428	7/16	1/4	5.078	5 1/8	0.544	9/16	9 5/8	1 3/16	1/2	3/4
x31.8	9.35	12.00	12	0.350	3/8	3/16	5.000	5	0.544	9/16	9 5/8	1 3/16	1/2	3/4
S 10x35	10.3	10.00	10	0.594	5/8	5/16	4.944	5	0.491	1/2	7 3/4	1 1/8	1/2	3/4
x25.4	7.46	10.00	10	0.311	5/16	3/16	4.661	4 5/8	0.491	1/2	7 3/4	1 1/8	1/2	3/4
S 8x23	6.77	8.00	8	0.441	7/16	1/4	4.171	4 1/8	0.426	7/16	6	1	7/16	3/4
x18.4	5.41	8.00	8	0.271	1/4	1/8	4.001	4	0.426	7/16	6	1	7/16	3/4
S 7x20	5.88	7.00	7	0.450	7/16	1/4	3.860	3 7/8	0.392	3/8	5 1/8	15/16	3/8	5/8
x15.3	4.50	7.00	7	0.252	1/4	1/8	3.662	3 5/8	0.392	3/8	5 1/8	15/16	3/8	5/8
S 6x17.25	5.07	6.00	6	0.465	7/16	1/4	3.565	3 5/8	0.359	3/8	4 1/4	7/8	3/8	5/8
x12.5	3.67	6.00	6	0.232	1/4	1/8	3.332	3 3/8	0.359	3/8	4 1/4	7/8	3/8	—
S 5x14.75	4.34	5.00	5	0.494	1/2	1/4	3.284	3 1/4	0.326	5/16	3 3/8	13/16	5/16	—
x10	2.94	5.00	5	0.214	3/16	1/8	3.004	3	0.326	5/16	3 3/8	13/16	5/16	—
S 4x9.5	2.79	4.00	4	0.326	5/16	3/16	2.796	2 3/4	0.293	5/16	2 1/2	3/4	5/16	—
x7.7	2.26	4.00	4	0.193	3/16	1/8	2.663	2 5/8	0.293	5/16	2 1/2	3/4	5/16	—
S 3x7.5	2.21	3.00	3	0.349	3/8	3/16	2.509	2 1/2	0.260	1/4	1 5/8	11/16	1/4	—
x5.7	1.67	3.00	3	0.170	3/16	1/8	2.330	2 3/8	0.260	1/4	1 5/8	11/16	1/4	—

TABLE A-3. (*Continued*)

Nominal Wt. per Ft	Compact Section Criteria				r_T	$\frac{d}{A_f}$	Elastic Properties						Torsional constant J	Plastic Modulus	
	$\frac{b_f}{2t_f}$	F_y'	$\frac{d}{t_w}$	F_y'''			Axis X-X			Axis Y-Y				Z_x	Z_y
							I	S	r	I	S	r			
Lb.		Ksi		Ksi	In.		In.^{4}	In.^{3}	In.	In.^{4}	In.^{3}	In.	In.^{4}	In.^{3}	In.^{3}
121	3.7	—	30.6	—	1.86	2.79	3160	258	9.43	83.3	20.7	1.53	12.8	306	36.2
106	3.6	—	39.5	42.3	1.86	2.86	2940	240	9.71	77.1	19.6	1.57	10.1	279	33.2
100	4.2	—	32.2	63.6	1.59	3.81	2390	199	9.02	47.7	13.2	1.27	7.58	240	23.9
90	4.1	—	38.4	44.8	1.60	3.87	2250	187	9.21	44.9	12.6	1.30	6.04	222	22.3
80	4.0	—	48.0	28.7	1.61	3.94	2100	175	9.47	42.2	12.1	1.34	4.88	204	20.7
96	3.9	—	25.4	—	1.63	3.06	1670	165	7.71	50.2	13.9	1.33	8.39	198	24.9
86	3.8	—	30.8	—	1.63	3.13	1580	155	7.89	46.8	13.3	1.36	6.64	183	23.0
75	4.0	—	31.5	—	1.43	3.94	1280	128	7.62	29.8	9.32	1.16	4.59	153	16.7
66	3.9	—	39.6	42.1	1.44	4.02	1190	119	7.83	27.7	8.85	1.19	3.58	140	15.3
70	4.5	—	25.3	—	1.36	4.17	926	103	6.71	24.1	7.72	1.08	4.15	125	14.4
54.7	4.3	—	39.0	43.3	1.37	4.34	804	89.4	7.07	20.8	6.94	1.14	2.37	105	12.1
50	4.5	—	27.3	—	1.26	4.28	486	64.8	5.75	15.7	5.57	1.03	2.12	77.1	9.97
42.9	4.4	—	36.5	49.6	1.26	4.38	447	59.6	5.95	14.4	5.23	1.07	1.54	69.3	9.02
50	4.2	—	17.5	—	1.25	3.32	305	50.8	4.55	15.7	5.74	1.03	2.82	61.2	10.3
40.8	4.0	—	26.0	—	1.24	3.46	272	45.4	4.77	13.6	5.16	1.06	1.76	53.1	8.85
35	4.7	—	28.0	—	1.16	4.34	229	38.2	4.72	9.87	3.89	0.980	1.08	44.8	6.79
31.8	4.6	—	34.3	56.2	1.16	4.41	218	36.4	4.83	9.36	3.74	1.00	0.90	42.0	6.40
35	5.0	—	16.8	—	1.10	4.12	147	29.4	3.78	8.36	3.38	0.901	1.29	35.4	6.22
25.4	4.7	—	32.2	63.9	1.09	4.37	124	24.7	4.07	6.79	2.91	0.954	0.60	28.4	4.96
23	4.9	—	18.1	—	0.95	4.51	64.9	16.2	3.10	4.31	2.07	0.798	0.55	19.3	3.68
18.4	4.7	—	29.5	—	0.94	4.70	57.6	14.4	3.26	3.73	1.86	0.831	0.34	16.5	3.16
20	4.9	—	15.6	—	0.88	4.63	42.4	12.1	2.69	3.17	1.64	0.734	0.45	14.5	2.96
15.3	4.7	—	27.8	—	0.87	4.88	36.7	10.5	2.86	2.64	1.44	0.766	0.24	12.1	2.44
17.25	5.0	—	12.9	—	0.81	4.69	26.3	8.77	2.28	2.31	1.30	0.675	0.37	10.6	2.36
12.5	4.6	—	25.9	—	0.79	5.02	22.1	7.37	2.45	1.82	1.09	0.705	0.17	8.47	1.85
14.75	5.0	—	10.1	—	0.74	4.66	15.2	6.09	1.87	1.67	1.01	0.620	0.32	7.42	1.88
10	4.6	—	23.4	—	0.72	5.10	12.3	4.92	2.05	1.22	0.809	0.643	0.11	5.67	1.37
9.5	4.8	—	12.3	—	0.65	4.88	6.79	3.39	1.56	0.903	0.646	0.569	0.12	4.04	1.13
7.7	4.5	—	20.7	—	0.64	5.13	6.08	3.04	1.64	0.764	0.574	0.581	0.07	3.51	0.964
7.5	4.8	—	8.6	—	0.59	4.60	2.93	1.95	1.15	0.586	0.468	0.516	0.09	2.36	0.826
5.7	4.5	—	17.6	—	0.57	4.95	2.52	1.68	1.23	0.455	0.390	0.522	0.04	1.95	0.653

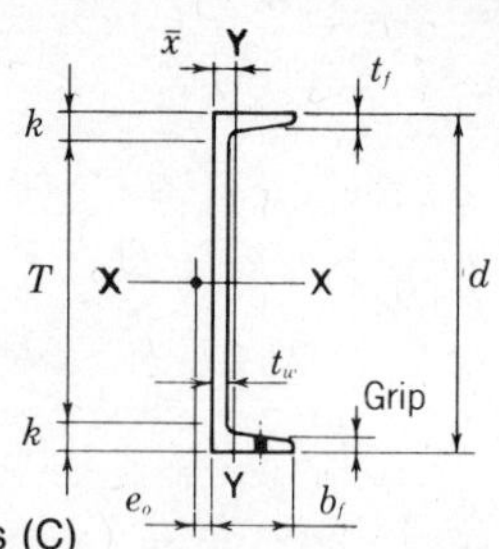

TABLE A-4. Properties of American Standard Channels (C)

Designation	Area A	Depth d	Web			Flange				Distance		Grip	Max. Flge. Fas- ten- er
			Thickness t_w		$\frac{t_w}{2}$	Width b_f		Average thickness t_f		T	k		
	In.²	In.	In.		In.	In.		In.		In.	In.	In.	In.
C 15×50	14.7	15.00	0.716	11/16	3/8	3.716	3 3/4	0.650	5/8	12 1/8	1 7/16	5/8	1
×40	11.8	15.00	0.520	1/2	1/4	3.520	3 1/2	0.650	5/8	12 1/8	1 7/16	5/8	1
×33.9	9.96	15.00	0.400	3/8	3/16	3.400	3 3/8	0.650	5/8	12 1/8	1 7/16	5/8	1
C 12×30	8.82	12.00	0.510	1/2	1/4	3.170	3 1/8	0.501	1/2	9 3/4	1 1/8	1/2	7/8
×25	7.35	12.00	0.387	3/8	3/16	3.047	3	0.501	1/2	9 3/4	1 1/8	1/2	7/8
×20.7	6.09	12.00	0.282	5/16	1/8	2.942	3	0.501	1/2	9 3/4	1 1/8	1/2	7/8
C 10×30	8.82	10.00	0.673	11/16	5/16	3.033	3	0.436	7/16	8	1	7/16	3/4
×25	7.35	10.00	0.526	1/2	1/4	2.886	2 7/8	0.436	7/16	8	1	7/16	3/4
×20	5.88	10.00	0.379	3/8	3/16	2.739	2 3/4	0.436	7/16	8	1	7/16	3/4
×15.3	4.49	10.00	0.240	1/4	1/8	2.600	2 5/8	0.436	7/16	8	1	7/16	3/4
C 9×20	5.88	9.00	0.448	7/16	1/4	2.648	2 5/8	0.413	7/16	7 1/8	15/16	7/16	3/4
×15	4.41	9.00	0.285	5/16	1/8	2.485	2 1/2	0.413	7/16	7 1/8	15/16	7/16	3/4
×13.4	3.94	9.00	0.233	1/4	1/8	2.433	2 3/8	0.413	7/16	7 1/8	15/16	7/16	3/4
C 8×18.75	5.51	8.00	0.487	1/2	1/4	2.527	2 1/2	0.390	3/8	6 1/8	15/16	3/8	3/4
×13.75	4.04	8.00	0.303	5/16	1/8	2.343	2 3/8	0.390	3/8	6 1/8	15/16	3/8	3/4
×11.5	3.38	8.00	0.220	1/4	1/8	2.260	2 1/4	0.390	3/8	6 1/8	15/16	3/8	3/4
C 7×14.75	4.33	7.00	0.419	7/16	3/16	2.299	2 1/4	0.366	3/8	5 1/4	7/8	3/8	5/8
×12.25	3.60	7.00	0.314	5/16	3/16	2.194	2 1/4	0.366	3/8	5 1/4	7/8	3/8	5/8
× 9.8	2.87	7.00	0.210	3/16	1/8	2.090	2 1/8	0.366	3/8	5 1/4	7/8	3/8	5/8
C 6×13	3.83	6.00	0.437	7/16	3/16	2.157	2 1/8	0.343	5/16	4 3/8	13/16	5/16	5/8
×10.5	3.09	6.00	0.314	5/16	3/16	2.034	2	0.343	5/16	4 3/8	13/16	3/8	5/8
× 8.2	2.40	6.00	0.200	3/16	1/8	1.920	1 7/8	0.343	5/16	4 3/8	13/16	5/16	5/8
C 5× 9	2.64	5.00	0.325	5/16	3/16	1.885	1 7/8	0.320	5/16	3 1/2	3/4	5/16	5/8
× 6.7	1.97	5.00	0.190	3/16	1/8	1.750	1 3/4	0.320	5/16	3 1/2	3/4	—	—
C 4× 7.25	2.13	4.00	0.321	5/16	3/16	1.721	1 3/4	0.296	5/16	2 5/8	11/16	5/16	5/8
× 5.4	1.59	4.00	0.184	3/16	1/16	1.584	1 5/8	0.296	5/16	2 5/8	11/16	—	—
C 3× 6	1.76	3.00	0.356	3/8	3/16	1.596	1 5/8	0.273	1/4	1 5/8	11/16	—	—
× 5	1.47	3.00	0.258	1/4	1/8	1.498	1 1/2	0.273	1/4	1 5/8	11/16	—	—
× 4.1	1.21	3.00	0.170	3/16	1/16	1.410	1 3/8	0.273	1/4	1 5/8	11/16	—	—

TABLE A-4. (*Continued*)

Nominal Weight per Ft.	$\bar{x}$	Shear Center Location e_o	$\frac{d}{A_f}$	Axis X-X			Axis Y-Y		
				I	S	r	I	S	r
	In.	In.		In.^{4}	In.^{3}	In.	In.^{4}	In.^{3}	In.
50	0.798	0.583	6.21	404	53.8	5.24	11.0	3.78	0.867
40	0.777	0.767	6.56	349	46.5	5.44	9.23	3.37	0.886
33.9	0.787	0.896	6.79	315	42.0	5.62	8.13	3.11	0.904
30	0.674	0.618	7.55	162	27.0	4.29	5.14	2.06	0.763
25	0.674	0.746	7.85	144	24.1	4.43	4.47	1.88	0.780
20.7	0.698	0.870	8.13	129	21.5	4.61	3.88	1.73	0.799
30	0.649	0.369	7.55	103	20.7	3.42	3.94	1.65	0.669
25	0.617	0.494	7.94	91.2	18.2	3.52	3.36	1.48	0.676
20	0.606	0.637	8.36	78.9	15.8	3.66	2.81	1.32	0.692
15.3	0.634	0.796	8.81	67.4	13.5	3.87	2.28	1.16	0.713
20	0.583	0.515	8.22	60.9	13.5	3.22	2.42	1.17	0.642
15	0.586	0.682	8.76	51.0	11.3	3.40	1.93	1.01	0.661
13.4	0.601	0.743	8.95	47.9	10.6	3.48	1.76	0.962	0.669
18.75	0.565	0.431	8.12	44.0	11.0	2.82	1.98	1.01	0.599
13.75	0.553	0.604	8.75	36.1	9.03	2.99	1.53	0.854	0.615
11.5	0.571	0.697	9.08	32.6	8.14	3.11	1.32	0.781	0.625
14.75	0.532	0.441	8.31	27.2	7.78	2.51	1.38	0.779	0.564
12.25	0.525	0.538	8.71	24.2	6.93	2.60	1.17	0.703	0.571
9.8	0.540	0.647	9.14	21.3	6.08	2.72	0.968	0.625	0.581
13	0.514	0.380	8.10	17.4	5.80	2.13	1.05	0.642	0.525
10.5	0.499	0.486	8.59	15.2	5.06	2.22	0.866	0.564	0.529
8.2	0.511	0.599	9.10	13.1	4.38	2.34	0.693	0.492	0.537
9	0.478	0.427	8.29	8.90	3.56	1.83	0.632	0.450	0.489
6.7	0.484	0.552	8.93	7.49	3.00	1.95	0.479	0.378	0.493
7.25	0.459	0.386	7.84	4.59	2.29	1.47	0.433	0.343	0.450
5.4	0.457	0.502	8.52	3.85	1.93	1.56	0.319	0.283	0.449
6	0.455	0.322	6.87	2.07	1.38	1.08	0.305	0.268	0.416
5	0.438	0.392	7.32	1.85	1.24	1.12	0.247	0.233	0.410
4.1	0.436	0.461	7.78	1.66	1.10	1.17	0.197	0.202	0.404

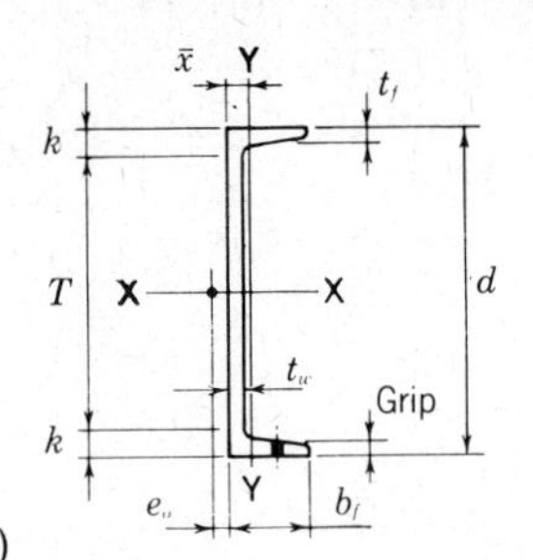

TABLE A-5. Properties of Miscellaneous Channels (MC)

Designation	Area A	Depth d	Web: Thickness t_w		Web: $\frac{t_w}{2}$	Flange: Width b_f		Flange: Average thickness t_f		Distance: T	Distance: k	Grip	Max. Flge. Fastener
	In.²	In.	In.		In.	In.		In.		In.	In.	In.	In.
MC 18x58	17.1	18.00	0.700	11/16	3/8	4.200	4 1/4	0.625	5/8	15 1/4	1 3/8	5/8	1
x51.9	15.3	18.00	0.600	5/8	5/16	4.100	4 1/8	0.625	5/8	15 1/4	1 3/8	5/8	1
x45.8	13.5	18.00	0.500	1/2	1/4	4.000	4	0.625	5/8	15 1/4	1 3/8	5/8	1
x42.7	12.6	18.00	0.450	7/16	1/4	3.950	4	0.625	5/8	15 1/4	1 3/8	5/8	1
MC 13x50	14.7	13.00	0.787	13/16	3/8	4.412	4 3/8	0.610	5/8	10 1/4	1 3/8	5/8	1
x40	11.8	13.00	0.560	9/16	1/4	4.185	4 1/8	0.610	5/8	10 1/4	1 3/8	9/16	1
x35	10.3	13.00	0.447	7/16	1/4	4.072	4 1/8	0.610	5/8	10 1/4	1 3/8	9/16	1
x31.8	9.35	13.00	0.375	3/8	3/16	4.000	4	0.610	5/8	10 1/4	1 3/8	9/16	1
MC 12x50	14.7	12.00	0.835	13/16	7/16	4.135	4 1/8	0.700	11/16	9 3/8	1 5/16	11/16	1
x45	13.2	12.00	0.712	11/16	3/8	4.012	4	0.700	11/16	9 3/8	1 5/16	11/16	1
x40	11.8	12.00	0.590	9/16	5/16	3.890	3 7/8	0.700	11/16	9 3/8	1 5/16	11/16	1
x35	10.3	12.00	0.467	7/16	1/4	3.767	3 3/4	0.700	11/16	9 3/8	1 5/16	11/16	1
MC 12x37	10.9	12.00	0.600	5/8	5/16	3.600	3 5/8	0.600	5/8	9 3/8	1 5/16	5/8	7/8
x32.9	9.67	12.00	0.500	1/2	1/4	3.500	3 1/2	0.600	5/8	9 3/8	1 5/16	9/16	7/8
x30.9	9.07	12.00	0.450	7/16	1/4	3.450	3 1/2	0.600	5/8	9 3/8	1 5/16	9/16	7/8
MC 12x10.6	3.10	12.00	0.190	3/16	1/8	1.500	1 1/2	0.309	5/16	10 5/8	11/16	—	—
MC 10x41.1	12.1	10.00	0.796	13/16	3/8	4.321	4 3/8	0.575	9/16	7 1/2	1 1/4	9/16	7/8
x33.6	9.87	10.00	0.575	9/16	5/16	4.100	4 1/8	0.575	9/16	7 1/2	1 1/4	9/16	7/8
x28.5	8.37	10.00	0.425	7/16	3/16	3.950	4	0.575	9/16	7 1/2	1 1/4	9/16	7/8
MC 10x28.3	8.32	10.00	0.477	1/2	1/4	3.502	3 1/2	0.575	9/16	7 1/2	1 1/4	9/16	7/8
x25.3	7.43	10.00	0.425	7/16	3/16	3.550	3 1/2	0.500	1/2	7 3/4	1 1/8	1/2	7/8
x24.9	7.32	10.00	0.377	3/8	3/16	3.402	3 3/8	0.575	9/16	7 1/2	1 1/4	9/16	7/8
x21.9	6.43	10.00	0.325	5/16	3/16	3.450	3 1/2	0.500	1/2	7 3/4	1 1/8	1/2	7/8
MC 10x 8.4	2.46	10.00	0.170	3/16	1/16	1.500	1 1/2	0.280	1/4	8 5/8	11/16	—	—
MC 10x 6.5	1.91	10.00	0.152	1/8	1/16	1.127	1 1/8	0.202	3/16	9 1/8	7/16	—	—

TABLE A-5. (*Continued*)

Nominal Weight per Ft.	$\bar{x}$	Shear Center Location e_o	$\frac{d}{A_f}$	Axis X-X			Axis Y-Y		
				I	S	r	I	S	r
	In.	In.		In.^{4}	In.^{3}	In.	In.^{4}	In.^{3}	In.
58	0.862	0.695	6.86	676	75.1	6.29	17.8	5.32	1.02
51.9	0.858	0.797	7.02	627	69.7	6.41	16.4	5.07	1.04
45.8	0.866	0.909	7.20	578	64.3	6.56	15.1	4.82	1.06
42.7	0.877	0.969	7.29	554	61.6	6.64	14.4	4.69	1.07
50	0.974	0.815	4.83	314	48.4	4.62	16.5	4.79	1.06
40	0.963	1.03	5.09	273	42.0	4.82	13.7	4.26	1.08
35	0.980	1.16	5.23	252	38.8	4.95	12.3	3.99	1.10
31.8	1.00	1.24	5.33	239	36.8	5.06	11.4	3.81	1.11
50	1.05	0.741	4.15	269	44.9	4.28	17.4	5.65	1.09
45	1.04	0.844	4.27	252	42.0	4.36	15.8	5.33	1.09
40	1.04	0.952	4.41	234	39.0	4.46	14.3	5.00	1.10
35	1.05	1.07	4.55	216	36.1	4.59	12.7	4.67	1.11
37	0.866	0.747	5.56	205	34.2	4.34	9.81	3.59	0.950
32.9	0.867	0.843	5.71	191	31.8	4.44	8.91	3.39	0.960
30.9	0.873	0.893	5.80	183	30.6	4.50	8.46	3.28	0.966
10.6	0.269	0.284	25.9	55.4	9.23	4.22	0.382	0.310	0.351
41.1	1.09	0.864	4.02	158	31.5	3.61	15.8	4.88	1.14
33.6	1.08	1.06	4.24	139	27.8	3.75	13.2	4.38	1.16
28.5	1.12	1.21	4.40	127	25.3	3.89	11.4	4.02	1.17
28.3	0.933	0.928	4.97	118	23.6	3.77	8.21	3.20	0.993
25.3	0.918	0.977	5.63	107	21.4	3.79	7.61	2.89	1.01
24.9	0.954	1.03	5.11	110	22.0	3.87	7.32	2.99	1.00
21.9	0.954	1.09	5.80	98.5	19.7	3.91	6.74	2.70	1.02
8.4	0.284	0.332	23.8	32.0	6.40	3.61	0.328	0.270	0.365
6.5	0.180	0.167	43.8	22.1	4.42	3.40	0.112	0.118	0.242

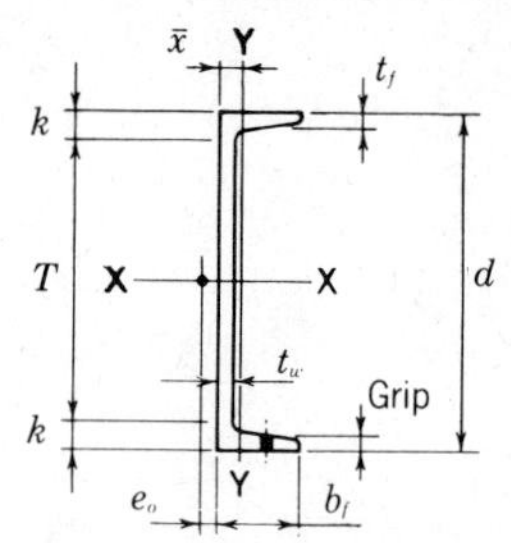

TABLE A-5. Properties of Miscellaneous Channels (MC)

Designation	Area A	Depth d	Web			Flange				Distance		Grip	Max. Flge. Fastener
			Thickness t_w		$\frac{t_w}{2}$	Width b_f		Average thickness t_f		T	k		
	In.^{2}	In.	In.		In.	In.		In.		In.	In.	In.	In.
MC 9x25.4	7.47	9.00	0.450	7/16	1/4	3.500	3 1/2	0.550	9/16	6 5/8	1 3/16	9/16	7/8
x23.9	7.02	9.00	0.400	3/8	3/16	3.450	3 1/2	0.550	9/16	6 5/8	1 3/16	9/16	7/8
MC 8x22.8	6.70	8.00	0.427	7/16	3/16	3.502	3 1/2	0.525	1/2	5 5/8	1 3/16	1/2	7/8
x21.4	6.28	8.00	0.375	3/8	3/16	3.450	3 1/2	0.525	1/2	5 5/8	1 3/16	1/2	7/8
MC 8x20	5.88	8.00	0.400	3/8	3/16	3.025	3	0.500	1/2	5 3/4	1 1/8	1/2	7/8
x18.7	5.50	8.00	0.353	3/8	3/16	2.978	3	0.500	1/2	5 3/4	1 1/8	1/2	7/8
MC 8x 8.5	2.50	8.00	0.179	3/16	1/16	1.874	1 7/8	0.311	5/16	6 1/2	3/4	5/16	5/8
MC 7x22.7	6.67	7.00	0.503	1/2	1/4	3.603	3 5/8	0.500	1/2	4 3/4	1 1/8	1/2	7/8
x19.1	5.61	7.00	0.352	3/8	3/16	3.452	3 1/2	0.500	1/2	4 3/4	1 1/8	1/2	7/8
MC 7x17.6	5.17	7.00	0.375	3/8	3/16	3.000	3	0.475	1/2	4 7/8	1 1/16	1/2	3/4
MC 6x18	5.29	6.00	0.379	3/8	3/16	3.504	3 1/2	0.475	1/2	3 7/8	1 1/16	1/2	7/8
x15.3	4.50	6.00	0.340	5/16	3/16	3.500	3 1/2	0.385	3/8	4 1/4	7/8	3/8	7/8
MC 6x16.3	4.79	6.00	0.375	3/8	3/16	3.000	3	0.475	1/2	3 7/8	1 1/16	1/2	3/4
x15.1	4.44	6.00	0.316	5/16	3/16	2.941	3	0.475	1/2	3 7/8	1 1/16	1/2	3/4
MC 6x12	3.53	6.00	0.310	5/16	1/8	2.497	2 1/2	0.375	3/8	4 3/8	13/16	3/8	5/8

TABLE A-5. (*Continued*)

Nominal Weight per Ft.	$\bar{x}$	Shear Center Location e_o	$\frac{d}{A_f}$	Axis X-X			Axis Y-Y		
				I	S	r	I	S	r
	In.	In.		In.^{4}	In.^{3}	In.	In.^{4}	In.^{3}	In.
25.4	0.970	0.986	4.68	88.0	19.6	3.43	7.65	3.02	1.01
23.9	0.981	1.04	4.74	85.0	18.9	3.48	7.22	2.93	1.01
22.8	1.01	1.04	4.35	63.8	16.0	3.09	7.07	2.84	1.03
21.4	1.02	1.09	4.42	61.6	15.4	3.13	6.64	2.74	1.03
20	0.840	0.843	5.29	54.5	13.6	3.05	4.47	2.05	0.872
18.7	0.849	0.889	5.37	52.5	13.1	3.09	4.20	1.97	0.874
8.5	0.428	0.542	13.7	23.3	5.83	3.05	0.628	0.434	0.501
22.7	1.04	1.01	3.89	47.5	13.6	2.67	7.29	2.85	1.05
19.1	1.08	1.15	4.06	43.2	12.3	2.77	6.11	2.57	1.04
17.6	0.873	0.890	4.91	37.6	10.8	2.70	4.01	1.89	0.881
18	1.12	1.17	3.60	29.7	9.91	2.37	5.93	2.48	1.06
15.3	1.05	1.16	4.45	25.4	8.47	2.38	4.97	2.03	1.05
16.3	0.927	0.930	4.21	26.0	8.68	2.33	3.82	1.84	0.892
15.1	0.940	0.982	4.29	25.0	8.32	2.37	3.51	1.75	0.889
12	0.704	0.725	6.41	18.7	6.24	2.30	1.87	1.04	0.728

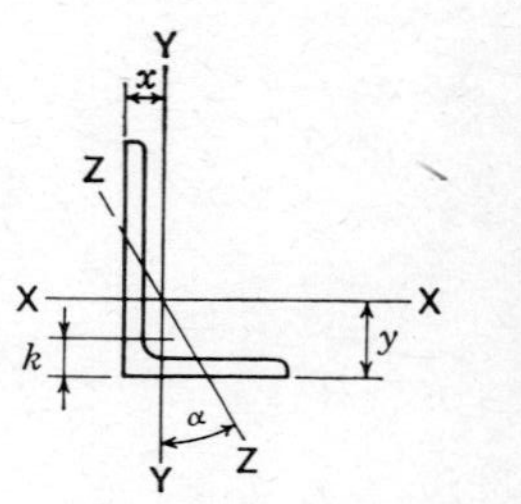

TABLE A-6. Properties of Angles

Size and Thickness	k	Weight per Foot	Area	AXIS X-X				AXIS Y-Y				AXIS Z-Z	
				I	S	r	y	I	S	r	x	r	Tan α
In.	In.	Lb.	In.^{2}	In.^{4}	In.^{3}	In.	In.	In.^{4}	In.^{3}	In.	In.	In.	
L 9 × 4 ×													
L 8 × 8 × $1\frac{1}{8}$	$1\frac{3}{4}$	56.9	16.7	98.0	17.5	2.42	2.41	98.0	17.5	2.42	2.41	1.56	1.000
1	$1\frac{5}{8}$	51.0	15.0	89.0	15.8	2.44	2.37	89.0	15.8	2.44	2.37	1.56	1.000
$\frac{7}{8}$	$1\frac{1}{2}$	45.0	13.2	79.6	14.0	2.45	2.32	79.6	14.0	2.45	2.32	1.57	1.000
$\frac{3}{4}$	$1\frac{3}{8}$	38.9	11.4	69.7	12.2	2.47	2.28	69.7	12.2	2.47	2.28	1.58	1.000
$\frac{5}{8}$	$1\frac{1}{4}$	32.7	9.61	59.4	10.3	2.49	2.23	59.4	10.3	2.49	2.23	1.58	1.000
$\frac{1}{2}$	$1\frac{1}{8}$	26.4	7.75	48.6	8.36	2.50	2.19	48.6	8.36	2.50	2.19	1.59	1.000
L 8 × 6 × 1	$1\frac{1}{2}$	44.2	13.0	80.8	15.1	2.49	2.65	38.8	8.92	1.73	1.65	1.28	0.543
$\frac{3}{4}$	$1\frac{1}{4}$	33.8	9.94	63.4	11.7	2.53	2.56	30.7	6.92	1.76	1.56	1.29	0.551
$\frac{1}{2}$	1	23.0	6.75	44.3	8.02	2.56	2.47	21.7	4.79	1.79	1.47	1.30	0.558
L8 × 4 × 1	$1\frac{1}{2}$	37.4	11.0	69.6	14.1	2.52	3.05	11.6	3.94	1.03	1.05	0.846	0.247
$\frac{3}{4}$	$1\frac{1}{4}$	28.7	8.44	54.9	10.9	2.55	2.95	9.36	3.07	1.05	0.953	0.852	0.258
$\frac{1}{2}$	1	19.6	5.75	38.5	7.49	2.59	2.86	6.74	2.15	1.08	0.859	0.865	0.267
L 7 × 4 × $\frac{3}{4}$	$1\frac{1}{4}$	26.2	7.69	37.8	8.42	2.22	2.51	9.05	3.03	1.09	1.01	0.860	0.324
$\frac{1}{2}$	1	17.9	5.25	26.7	5.81	2.25	2.42	6.53	2.12	1.11	0.917	0.872	0.335
$\frac{3}{8}$	$\frac{7}{8}$	13.6	3.98	20.6	4.44	2.27	2.37	5.10	1.63	1.13	0.870	0.880	0.340

L6 × 6 × 1	$1\frac{1}{2}$	37.4	11.0	35.5	8.57	1.80	1.86	35.5	8.57	1.80	1.86	1.17	1.000
$\frac{7}{8}$	$1\frac{3}{8}$	33.1	9.73	31.9	7.63	1.81	1.82	31.9	7.63	1.81	1.82	1.17	1.000
$\frac{3}{4}$	$1\frac{1}{4}$	28.7	8.44	28.2	6.66	1.83	1.78	28.2	6.66	1.83	1.78	1.17	1.000
$\frac{5}{8}$	$1\frac{1}{8}$	24.2	7.11	24.2	5.66	1.84	1.73	24.2	5.66	1.84	1.73	1.18	1.000
$\frac{1}{2}$	1	19.6	5.75	19.9	4.61	1.86	1.68	19.9	4.61	1.86	1.68	1.18	1.000
$\frac{3}{8}$	$\frac{7}{8}$	14.9	4.36	15.4	3.53	1.88	1.64	15.4	3.53	1.88	1.64	1.19	1.000
L 6 × 4 × $\frac{3}{4}$	$1\frac{1}{4}$	23.6	6.94	24.5	6.25	1.88	2.08	8.68	2.97	1.12	1.08	0.860	0.428
$\frac{5}{8}$	$1\frac{1}{8}$	20.0	5.86	21.1	5.31	1.90	2.03	7.52	2.54	1.13	1.03	0.864	0.435
$\frac{1}{2}$	1	16.2	4.75	17.4	4.33	1.91	1.99	6.27	2.08	1.15	0.987	0.870	0.440
$\frac{3}{8}$	$\frac{7}{8}$	12.3	3.61	13.5	3.32	1.93	1.94	4.90	1.60	1.17	0.941	0.877	0.446
L 6 × $3\frac{1}{2}$ × $\frac{3}{8}$	$\frac{7}{8}$	11.7	3.42	12.9	3.24	1.94	2.04	3.34	1.23	0.988	0.787	0.767	0.350
$\frac{5}{16}$	$\frac{13}{16}$	9.8	2.87	10.9	2.73	1.95	2.01	2.85	1.04	0.996	0.763	0.772	0.352
L 5 × 5 × $\frac{7}{8}$	$1\frac{3}{8}$	27.2	7.98	17.8	5.17	1.49	1.57	17.8	5.17	1.49	1.57	0.973	1.000
$\frac{3}{4}$	$1\frac{1}{4}$	23.6	6.94	15.7	4.53	1.51	1.52	15.7	4.53	1.51	1.52	0.975	1.000
$\frac{1}{2}$	1	16.2	4.75	11.3	3.16	1.54	1.43	11.3	3.16	1.54	1.43	0.983	1.000
$\frac{3}{8}$	$\frac{7}{8}$	12.3	3.61	8.74	2.42	1.56	1.39	8.74	2.42	1.56	1.39	0.990	1.000
$\frac{5}{16}$	$\frac{13}{16}$	10.3	3.03	7.42	2.04	1.57	1.37	7.42	2.04	1.57	1.37	0.994	1.000
L 5 × $3\frac{1}{2}$ × $\frac{3}{4}$	$1\frac{1}{4}$	19.8	5.81	13.9	4.28	1.55	1.75	5.55	2.22	0.977	0.996	0.748	0.464
$\frac{1}{2}$	1	13.6	4.00	9.99	2.99	1.58	1.66	4.05	1.56	1.01	0.906	0.755	0.479
$\frac{3}{8}$	$\frac{7}{8}$	10.4	3.05	7.78	2.29	1.60	1.61	3.18	1.21	1.02	0.861	0.762	0.486
$\frac{5}{16}$	$\frac{13}{16}$	8.7	2.56	6.60	1.94	1.61	1.59	2.72	1.02	1.03	0.838	0.766	0.489
L 5 × 3 × $\frac{1}{2}$	1	12.8	3.75	9.45	2.91	1.59	1.75	2.58	1.15	0.829	0.750	0.648	0.357
$\frac{3}{8}$	$\frac{7}{8}$	9.8	2.86	7.37	2.24	1.61	1.70	2.04	0.888	0.845	0.704	0.654	0.364
$\frac{5}{16}$	$\frac{13}{16}$	8.2	2.40	6.26	1.89	1.61	1.68	1.75	0.753	0.853	0.681	0.658	0.368
$\frac{1}{4}$	$\frac{3}{4}$	6.6	1.94	5.11	1.53	1.62	1.66	1.44	0.614	0.861	0.657	0.663	0.371
L 4 × 4 × $\frac{3}{4}$	$1\frac{1}{8}$	18.5	5.44	7.67	2.81	1.19	1.27	7.67	2.81	1.19	1.27	0.778	1.000
$\frac{5}{8}$	1	15.7	4.61	6.66	2.40	1.20	1.23	6.66	2.40	1.20	1.23	0.779	1.000
$\frac{1}{2}$	$\frac{7}{8}$	12.8	3.75	5.56	1.97	1.22	1.18	5.56	1.97	1.22	1.18	0.782	1.000
$\frac{3}{8}$	$\frac{3}{4}$	9.8	2.86	4.36	1.52	1.23	1.14	4.36	1.52	1.23	1.14	0.788	1.000
$\frac{5}{16}$	$\frac{11}{16}$	8.2	2.40	3.71	1.29	1.24	1.12	3.71	1.29	1.24	1.12	0.791	1.000
$\frac{1}{4}$	$\frac{5}{8}$	6.6	1.94	3.04	1.05	1.25	1.09	3.04	1.05	1.25	1.09	0.795	1.000

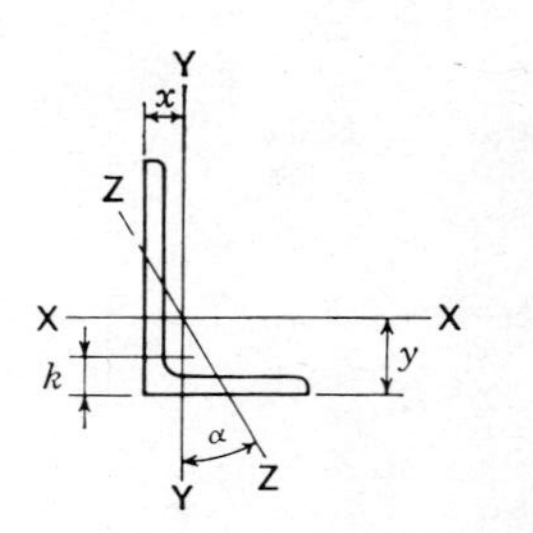

TABLE A-6. Properties of Angles (*Continued*)

Size and Thickness	k	Weight per Foot	Area	AXIS X-X				AXIS Y-Y				AXIS Z-Z	
				I	S	r	y	I	S	r	x	r	Tan α
In.	In.	Lb.	In.²	In.⁴	In.³	In.	In.	In.⁴	In.³	In.	In.	In.	
L 4 × 3½ × ½	15/16	11.9	3.50	5.32	1.94	1.23	1.25	3.79	1.52	1.04	1.00	0.722	0.750
3/8	13/16	9.1	2.67	4.18	1.49	1.25	1.21	2.95	1.17	1.06	0.955	0.727	0.755
5/16	3/4	7.7	2.25	3.56	1.26	1.26	1.18	2.55	0.994	1.07	0.932	0.730	0.757
1/4	11/16	6.2	1.81	2.91	1.03	1.27	1.16	2.09	0.808	1.07	0.909	0.734	0.759
L 4 × 3 × ½	15/16	11.1	3.25	5.05	1.89	1.25	1.33	2.42	1.12	0.864	0.827	0.639	0.543
3/8	13/16	8.5	2.48	3.96	1.46	1.26	1.28	1.92	0.866	0.879	0.782	0.644	0.551
5/16	3/4	7.2	2.09	3.38	1.23	1.27	1.26	1.65	0.734	0.887	0.759	0.647	0.554
1/4	11/16	5.8	1.69	2.77	1.00	1.28	1.24	1.36	0.599	0.896	0.736	0.651	0.558
L 3½ × 3½ × 3/8	3/4	8.5	2.48	2.87	1.15	1.07	1.01	2.87	1.15	1.07	1.01	0.687	1.000
5/16	11/16	7.2	2.09	2.45	0.976	1.08	0.990	2.45	0.976	1.08	0.990	0.690	1.000
1/4	5/8	5.8	1.69	2.01	0.794	1.09	0.968	2.01	0.794	1.09	0.968	0.694	1.000
L 3½ × 3 × 3/8	13/16	7.9	2.30	2.72	1.13	1.09	1.08	1.85	0.851	0.897	0.830	0.625	0.721
5/16	3/4	6.6	1.93	2.33	0.954	1.10	1.06	1.58	0.722	0.905	0.808	0.627	0.724
1/4	11/16	5.4	1.56	1.91	0.776	1.11	1.04	1.30	0.589	0.914	0.785	0.631	0.727

L 3½ × 2½ ×	3/8	13/16	7.2	2.11	2.56	1.09	1.10	1.16	1.09	0.592	0.719	0.660	0.537	0.496
	5/16	3/4	6.1	1.78	2.19	0.927	1.11	1.14	0.939	0.504	0.727	0.637	0.540	0.501
	1/4	11/16	4.9	1.44	1.80	0.755	1.12	1.11	0.777	0.412	0.735	0.614	0.544	0.506
L 3 × 3 ×	1/2	13/16	9.4	2.75	2.22	1.07	0.898	0.932	2.22	1.07	0.898	0.932	0.584	1.000
	3/8	11/16	7.2	2.11	1.76	0.833	0.913	0.888	1.76	0.833	0.913	0.888	0.587	1.000
	5/16	5/8	6.1	1.78	1.51	0.707	0.922	0.865	1.51	0.707	0.922	0.865	0.589	1.000
	1/4	9/16	4.9	1.44	1.24	0.577	0.930	0.842	1.24	0.577	0.930	0.842	0.592	1.000
	3/16	1/2	3.71	1.09	0.962	0.441	0.939	0.820	0.962	0.441	0.939	0.820	0.596	1.000
L 3 × 2½ ×	3/8	3/4	6.6	1.92	1.66	0.810	0.928	0.956	1.04	0.581	0.736	0.706	0.522	0.676
	1/4	5/8	4.5	1.31	1.17	0.561	0.945	0.911	0.743	0.404	0.753	0.661	0.528	0.684
	3/16	9/16	3.89	0.996	0.907	0.430	0.954	0.888	0.577	0.310	0.761	0.638	0.533	0.688
L 3 × 2 ×	3/8	11/16	5.9	1.73	1.53	0.781	0.940	1.04	0.543	0.371	0.559	0.539	0.430	0.428
	5/16	5/8	5.0	1.46	1.32	0.664	0.948	1.02	0.470	0.317	0.567	0.516	0.432	0.435
	1/4	9/16	4.1	1.19	1.09	0.542	0.957	0.993	0.392	0.260	0.574	0.493	0.435	0.440
	3/16	1/2	3.07	0.902	0.842	0.415	0.966	0.970	0.307	0.200	0.583	0.470	0.439	0.446
L 2^1× 2½ ×	3/8	11/16	5.9	1.73	0.984	0.566	0.753	0.762	0.984	0.566	0.753	0.762	0.487	1.000
	5/16	5/8	5.0	1.46	0.849	0.482	0.761	0.740	0.849	0.482	0.761	0.740	0.489	1.000
	1/4	9/16	4.1	1.19	0.703	0.394	0.769	0.717	0.703	0.394	0.769	0.717	0.491	1.000
	3/16	1/2	3.07	0.902	0.547	0.303	0.778	0.694	0.547	0.303	0.778	0.694	0.495	1.000
L 2½ × 2 ×	3/8	11/16	5.3	1.55	0.912	0.547	0.768	0.831	0.514	0.363	0.577	0.581	0.420	0.614
	5/16	5/8	4.5	1.31	0.788	0.466	0.776	0.809	0.46	0.310	0.584	0.559	0.422	0.620
	1/4	9/16	3.62	1.06	0.654	0.381	0.784	0.787	0.372	0.254	0.592	0.537	0.424	0.626
	3/16	1/2	2.75	0.809	0.509	0.293	0.793	0.764	0.291	0.196	0.600	0.514	0.427	0.631
L 2 × 2 ×	3/8	11/16	4.7	1.36	0.479	0.351	0.594	0.636	0.479	0.351	0.594	0.636	0.389	1.00
	5/16	5/8	3.92	1.15	0.416	0.300	0.601	0.614	0.416	0.300	0.601	0.614	0.390	1.000
	1/4	9/16	3.19	0.938	0.348	0.247	0.609	0.592	0.348	0.247	0.609	0.592	0.391	1.000
	3/16	1/2	2.44	0.715	0.272	0.190	0.617	0.569	0.272	0.190	0.617	0.569	0.394	1.000
	1/8	7/16	1.65	0.484	0.190	0.131	0.626	0.546	0.190	0.131	0.626	0.546	0.398	1.000

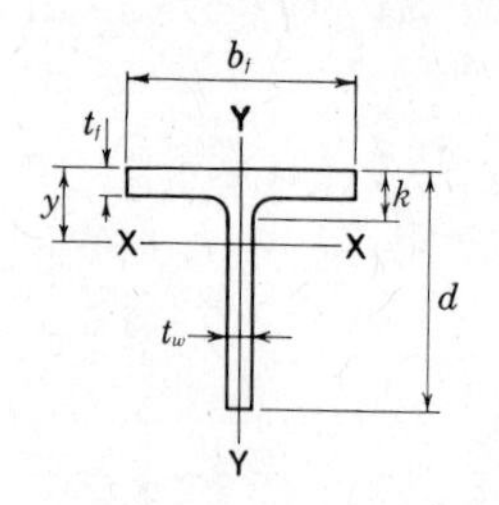

TABLE A-7. Properties of Structural Tees Cut from W Shapes (WT)

Designation	Area	Depth of Tee d		Stem Thickness t_w		Stem $\frac{t_w}{2}$	Area of Stem	Flange Width b_f		Flange Thickness t_f		Distance k
	In.^{2}	In.		In.		In.	In.^{2}	In.		In.		In.
WT 18 x150	44.1	18.370	$18\frac{3}{8}$	0.945	$\frac{15}{16}$	$\frac{1}{2}$	17.4	16.655	$16\frac{5}{8}$	1.680	$1\frac{11}{16}$	$2\frac{13}{16}$
x140	41.2	18.260	$18\frac{1}{4}$	0.885	$\frac{7}{8}$	$\frac{7}{16}$	16.2	16.595	$16\frac{5}{8}$	1.570	$1\frac{9}{16}$	$2\frac{11}{16}$
x130	38.2	18.130	$18\frac{1}{8}$	0.840	$\frac{13}{16}$	$\frac{7}{16}$	15.2	16.550	$16\frac{1}{2}$	1.440	$1\frac{7}{16}$	$2\frac{9}{16}$
x122.5	36.0	18.040	18	0.800	$\frac{13}{16}$	$\frac{7}{16}$	14.4	16.510	$16\frac{1}{2}$	1.350	$1\frac{3}{8}$	$2\frac{1}{2}$
x115	33.8	17.950	18	0.760	$\frac{3}{4}$	$\frac{3}{8}$	13.6	16.470	$16\frac{1}{2}$	1.260	$1\frac{1}{4}$	$2\frac{3}{8}$
WT 18 x105	30.9	18.345	$18\frac{3}{8}$	0.830	$\frac{13}{16}$	$\frac{7}{16}$	15.2	12.180	$12\frac{1}{8}$	1.360	$1\frac{3}{8}$	$2\frac{5}{16}$
x 97	28.5	18.245	$18\frac{1}{4}$	0.765	$\frac{3}{4}$	$\frac{3}{8}$	14.0	12.115	$12\frac{1}{8}$	1.260	$1\frac{1}{4}$	$2\frac{3}{16}$
x 91	26.8	18.165	$18\frac{1}{8}$	0.725	$\frac{3}{4}$	$\frac{3}{8}$	13.2	12.075	$12\frac{1}{8}$	1.180	$1\frac{3}{16}$	$2\frac{1}{8}$
x 85	25.0	18.085	$18\frac{1}{8}$	0.680	$\frac{11}{16}$	$\frac{3}{8}$	12.3	12.030	12	1.100	$1\frac{1}{8}$	2
x 80	23.5	18.005	18	0.650	$\frac{5}{8}$	$\frac{5}{16}$	11.7	12.000	12	1.020	1	$1\frac{15}{16}$
x 75	22.1	17.925	$17\frac{7}{8}$	0.625	$\frac{5}{8}$	$\frac{5}{16}$	11.2	11.975	12	0.940	$\frac{15}{16}$	$1\frac{7}{8}$
x 67.5	19.9	17.775	$17\frac{3}{4}$	0.600	$\frac{5}{8}$	$\frac{5}{16}$	10.7	11.950	12	0.790	$\frac{13}{16}$	$1\frac{11}{16}$
WT 16.5x120.5	35.4	17.090	$17\frac{1}{8}$	0.830	$\frac{13}{16}$	$\frac{7}{16}$	14.2	15.860	$15\frac{7}{8}$	1.400	$1\frac{3}{8}$	$2\frac{3}{16}$
x110.5	32.5	16.965	17	0.775	$\frac{3}{4}$	$\frac{3}{8}$	13.1	15.805	$15\frac{3}{4}$	1.275	$1\frac{1}{4}$	$2\frac{1}{16}$
x100.5	29.5	16.840	$16\frac{7}{8}$	0.715	$\frac{11}{16}$	$\frac{3}{8}$	12.0	15.745	$15\frac{3}{4}$	1.150	$1\frac{1}{8}$	$1\frac{15}{16}$
WT 16.5x 76	22.4	16.745	$16\frac{3}{4}$	0.635	$\frac{5}{8}$	$\frac{5}{16}$	10.6	11.565	$11\frac{5}{8}$	1.055	$1\frac{1}{16}$	$1\frac{7}{8}$
x 70.5	20.8	16.650	$16\frac{5}{8}$	0.605	$\frac{5}{8}$	$\frac{5}{16}$	10.1	11.535	$11\frac{1}{2}$	0.960	$\frac{15}{16}$	$1\frac{3}{4}$
x 65	19.2	16.545	$16\frac{1}{2}$	0.580	$\frac{9}{16}$	$\frac{5}{16}$	9.60	11.510	$11\frac{1}{2}$	0.855	$\frac{7}{8}$	$1\frac{11}{16}$
x 59	17.3	16.430	$16\frac{3}{8}$	0.550	$\frac{9}{16}$	$\frac{5}{16}$	9.04	11.480	$11\frac{1}{2}$	0.740	$\frac{3}{4}$	$1\frac{9}{16}$
WT 15 x105.5	31.0	15.470	$15\frac{1}{2}$	0.775	$\frac{3}{4}$	$\frac{3}{8}$	12.0	15.105	$15\frac{1}{8}$	1.315	$1\frac{5}{16}$	$2\frac{1}{8}$
x 95.5	28.1	15.340	$15\frac{3}{8}$	0.710	$\frac{11}{16}$	$\frac{3}{8}$	10.9	15.040	15	1.185	$1\frac{3}{16}$	$1\frac{15}{16}$
x 86.5	25.4	15.220	$15\frac{1}{4}$	0.655	$\frac{5}{8}$	$\frac{5}{16}$	9.97	14.985	15	1.065	$1\frac{1}{16}$	$1\frac{7}{8}$
WT 15 x 66	19.4	15.155	$15\frac{1}{8}$	0.615	$\frac{5}{8}$	$\frac{5}{16}$	9.32	10.545	$10\frac{1}{2}$	1.000	1	$1\frac{3}{4}$
x 62	18.2	15.085	$15\frac{1}{8}$	0.585	$\frac{9}{16}$	$\frac{5}{16}$	8.82	10.515	$10\frac{1}{2}$	0.930	$\frac{15}{16}$	$1\frac{11}{16}$
x 58	17.1	15.005	15	0.565	$\frac{9}{16}$	$\frac{5}{16}$	8.48	10.495	$10\frac{1}{2}$	0.850	$\frac{7}{8}$	$1\frac{5}{8}$
x 54	15.9	14.915	$14\frac{7}{8}$	0.545	$\frac{9}{16}$	$\frac{5}{16}$	8.13	10.475	$10\frac{1}{2}$	0.760	$\frac{3}{4}$	$1\frac{9}{16}$
x 49.5	14.5	14.825	$14\frac{7}{8}$	0.520	$\frac{1}{2}$	$\frac{1}{4}$	7.71	10.450	$10\frac{1}{2}$	0.670	$\frac{11}{16}$	$1\frac{7}{16}$

TABLE A-7. (*Continued*)

Nominal Weight per Ft.	$\frac{d}{t_w}$	AXIS X-X				AXIS Y-Y			$C_c' = \sqrt{\frac{2\pi^2 E}{Q_s Q_a F_y}}$, $Q_a = 1.0$			
		I	S	r	y	I	S	r	F_y = 36 ksi		F_y = 50 ksi	
Lb.		In.[4]	In.[3]	In.	In.	In.[4]	In.[3]	In.	Q_s	C_c'	Q_s	C_c'
150	19.4	1230	86.1	5.27	4.13	648	77.8	3.83	—	—	0.927	111
140	20.6	1140	80.0	5.25	4.07	599	72.2	3.81	—	—	0.867	115
130	21.6	1060	75.1	5.26	4.05	545	65.9	3.78	0.981	127	0.816	118
122.5	22.5	995	71.0	5.26	4.03	507	61.4	3.75	0.943	130	0.770	122
115.	23.6	934	67.0	5.25	4.01	470	57.1	3.73	0.896	133	0.715	127
105	22.1	985	73.1	5.65	4.87	206	33.8	2.58	0.960	129	0.791	120
97	23.8	901	67.0	5.62	4.80	187	30.9	2.56	0.887	134	0.705	127
91	25.1	845	63.1	5.62	4.77	174	28.8	2.55	0.831	138	0.635	134
85	26.6	786	58.9	5.61	4.73	160	26.6	2.53	0.767	144	0.565	142
80	27.7	740	55.8	5.61	4.74	147	24.6	2.50	0.720	149	0.521	148
75	28.7	698	53.1	5.62	4.78	135	22.5	2.47	0.677	153	0.486	154
67.5	29.6	636	49.7	5.66	4.96	113	18.9	2.38	0.634	158	0.457	158
120.5	20.6	871	65.8	4.96	3.85	466	58.8	3.63	—	—	0.867	115
110.5	21.9	799	60.8	4.96	3.81	420	53.2	3.59	0.968	128	0.801	120
100.5	23.6	725	55.5	4.95	3.78	375	47.6	3.56	0.896	133	0.715	127
76	26.4	592	47.4	5.14	4.26	136	23.6	2.47	0.775	143	0.574	141
70.5	27.5	552	44.7	5.15	4.29	123	21.3	2.43	0.728	148	0.529	147
65	28.5	513	42.1	5.18	4.36	109	18.9	2.39	0.685	152	0.492	152
59	29.9	469	39.2	5.20	4.47	93.6	16.3	2.32	0.621	160	0.447	160
105.5	20.0	610	50.5	4.43	3.40	378	50.1	3.49	—	—	0.897	113
95.5	21.6	549	45.7	4.42	3.35	336	44.7	3.46	0.981	127	0.816	118
86.5	23.2	497	41.7	4.42	3.31	299	39.9	3.43	0.913	132	0.735	125
66	24.6	421	37.4	4.66	3.90	98.0	18.6	2.25	0.853	137	0.664	131
62	25.8	396	35.3	4.66	3.90	90.4	17.2	2.23	0.801	141	0.601	138
58	26.6	373	33.7	4.67	3.94	82.1	15.7	2.19	0.767	144	0.565	142
54	27.4	349	32.0	4.69	4.01	73.0	13.9	2.15	0.733	147	0.533	147
49.5	28.5	322	30.0	4.71	4.09	63.9	12.2	2.10	0.685	152	0.492	152

Where no value of C_c' or Q_s is shown, the Tee complies with Specification Sect. 1.9.1.2.

TABLE A-7. Properties of Structural Tees Cut from W Shapes (WT)

Designation	Area	Depth of Tee d		Stem Thickness t_w		$\frac{t_w}{2}$	Area of Stem	Flange Width b_f		Flange Thickness t_f		Distance k
	In.^{2}	In.		In.		In.	In.^{2}	In.		In.		In.
WT 13.5x89	26.1	13.905	13 7/8	0.725	3/4	3/8	10.1	14.085	14 1/8	1.190	1 3/16	1 7/8
x80.5	23.7	13.795	13 3/4	0.660	11/16	3/8	9.10	14.020	14	1.080	1 1/16	1 13/16
x73	21.5	13.690	13 3/4	0.605	5/8	5/16	8.28	13.965	14	0.975	1	1 11/16
WT 13.5x57	16.8	13.645	13 5/8	0.570	9/16	5/16	7.78	10.070	10 1/8	0.930	15/16	1 5/8
x51	15.0	13.545	13 1/2	0.515	1/2	1/4	6.98	10.015	10	0.830	13/16	1 9/16
x47	13.8	13.460	13 1/2	0.490	1/2	1/4	6.60	9.990	10	0.745	3/4	1 7/16
x42	12.4	13.355	13 3/8	0.460	7/16	1/4	6.14	9.960	10	0.640	5/8	1 3/8
WT 12 x81	23.9	12.500	12 1/2	0.705	11/16	3/8	8.81	12.955	13	1.220	1 1/4	2
x73	21.5	12.370	12 3/8	0.650	5/8	5/16	8.04	12.900	12 7/8	1.090	1 1/16	1 7/8
x65.5	19.3	12.240	12 1/4	0.605	5/8	5/16	7.41	12.855	12 7/8	0.960	15/16	1 3/4
x58.5	17.2	12.130	12 1/8	0.550	9/16	5/16	6.67	12.800	12 3/4	0.850	7/8	1 5/8
x52	15.3	12.030	12	0.500	1/2	1/4	6.01	12.750	12 3/4	0.750	3/4	1 1/2
WT 12 x47	13.8	12.155	12 1/8	0.515	1/2	1/4	6.26	9.065	9 1/8	0.875	7/8	1 5/8
x42	12.4	12.050	12	0.470	1/2	1/4	5.66	9.020	9	0.770	3/4	1 9/16
x38	11.2	11.960	12	0.440	7/16	1/4	5.26	8.990	9	0.680	11/16	1 7/16
x34	10.0	11.865	11 7/8	0.415	7/16	1/4	4.92	8.965	9	0.585	9/16	1 3/8
WT 12 x31	9.11	11.870	11 7/8	0.430	7/16	1/4	5.10	7.040	7	0.590	9/16	1 3/8
x27.5	8.10	11.785	11 3/4	0.395	3/8	3/16	4.66	7.005	7	0.505	1/2	1 5/16
WT 10.5x73.5	21.6	11.030	11	0.720	3/4	3/8	7.94	12.510	12 1/2	1.150	1 1/8	1 7/8
x66	19.4	10.915	10 7/8	0.650	5/8	5/16	7.09	12.440	12 1/2	1.035	1 1/16	1 13/16
x61	17.9	10.840	10 7/8	0.600	5/8	5/16	6.50	12.390	12 3/8	0.960	15/16	1 11/16
x55.5	16.3	10.755	10 3/4	0.550	9/16	5/16	5.92	12.340	12 3/8	0.875	7/8	1 5/8
x50.5	14.9	10.680	10 5/8	0.500	1/2	1/4	5.34	12.290	12 1/4	0.800	13/16	1 9/16
WT 10.5x46.5	13.7	10.810	10 3/4	0.580	9/16	5/16	6.27	8.420	8 3/8	0.930	15/16	1 11/16
x41.5	12.2	10.715	10 3/4	0.515	1/2	1/4	5.52	8.355	8 3/8	0.835	13/16	1 9/16
x36.5	10.7	10.620	10 5/8	0.455	7/16	1/4	4.83	8.295	8 1/4	0.740	3/4	1 1/2
x34	10.0	10.565	10 5/8	0.430	7/16	1/4	4.54	8.270	8 1/4	0.685	11/16	1 7/16
x31	9.13	10.495	10 1/2	0.400	3/8	3/16	4.20	8.240	8 1/4	0.615	5/8	1 3/8
WT 10.5x28.5	8.37	10.530	10 1/2	0.405	3/8	3/16	4.26	6.555	6 1/2	0.650	5/8	1 3/8
x25	7.36	10.415	10 3/8	0.380	3/8	3/16	3.96	6.530	6 1/2	0.535	9/16	1 5/16
x22	6.49	10.330	10 3/8	0.350	3/8	3/16	3.62	6.500	6 1/2	0.450	7/16	1 3/16
WT 9x59.5	17.5	9.485	9 1/2	0.655	5/8	5/16	6.21	11.265	11 1/4	1.060	1 1/16	1 3/4
x53	15.6	9.365	9 3/8	0.590	9/16	5/16	5.53	11.200	11 1/4	0.940	15/16	1 5/8
x48.5	14.3	9.295	9 1/4	0.535	9/16	5/16	4.97	11.145	11 1/8	0.870	7/8	1 9/16
x43	12.7	9.195	9 1/4	0.480	1/2	1/4	4.41	11.090	11 1/8	0.770	3/4	1 7/16
x38	11.2	9.105	9 1/8	0.425	7/16	1/4	3.87	11.035	11	0.680	11/16	1 3/8

TABLE A-7. (*Continued*)

Nominal Weight per Ft.	$\frac{d}{t_w}$	AXIS X-X				AXIS Y-Y			$C_c' = \sqrt{\frac{2\pi^2 E}{Q_s Q_a F_y}}$, $Q_a = 1.0$			
		I	S	r	y	I	S	r	F_y = 36 ksi		F_y = 50 ksi	
Lb.		In.⁴	In.³	In.	In.	In.⁴	In.³	In.	Q_s	C_c'	Q_s	C_c'
89	19.2	414	38.2	3.98	3.05	278	39.4	3.26	—	—	0.937	111
80.5	20.9	372	34.4	3.96	2.99	248	35.4	3.24	—	—	0.851	116
73	22.6	336	31.2	3.95	2.95	222	31.7	3.21	0.938	130	0.765	122
57	23.9	289	28.3	4.15	3.42	79.4	15.8	2.18	0.883	134	0.700	128
51	26.3	258	25.3	4.14	3.37	69.6	13.9	2.15	0.780	143	0.578	141
47	27.5	239	23.8	4.16	3.41	62.0	12.4	2.12	0.728	148	0.529	147
42	29.0	216	21.9	4.18	3.48	52.8	10.6	2.07	0.664	155	0.476	155
81	17.7	293	29.9	3.50	2.70	221	34.2	3.05	—	—	—	—
73	19.0	264	27.2	3.50	2.66	195	30.3	3.01	—	—	0.947	110
65.5	20.2	238	24.8	3.52	2.65	170	26.5	2.97	—	—	0.887	114
58.5	22.1	212	22.3	3.51	2.62	149	23.2	2.94	0.960	129	0.791	120
52	24.1	189	20.0	3.51	2.59	130	20.3	2.91	0.874	135	0.690	129
47	23.6	186	20.3	3.67	2.99	54.5	12.0	1.98	0.896	133	0.715	127
42	25.6	166	18.3	3.67	2.97	47.2	10.5	1.95	0.810	140	0.610	137
38	27.2	151	16.9	3.68	3.00	41.3	9.18	1.92	0.741	146	0.541	146
34	28.6	137	15.6	3.70	3.06	35.2	7.85	1.87	0.681	153	0.489	153
31	27.6	131	15.6	3.79	3.46	17.2	4.90	1.38	0.724	148	0.525	148
27.5	29.8	117	14.1	3.80	3.50	14.5	4.15	1.34	0.626	159	0.450	159
73.5	15.3	204	23.7	3.08	2.39	188	30.0	2.95	—	—	—	—
66	16.8	181	21.1	3.06	2.33	166	26.7	2.93	—	—	—	—
61	18.1	166	19.3	3.04	2.28	152	24.6	2.92	—	—	0.993	107
55.5	19.6	150	17.5	3.03	2.23	137	22.2	2.90	—	—	0.917	112
50.5	21.4	135	15.8	3.01	2.18	124	20.2	2.89	0.990	127	0.826	118
46.5	18.6	144	17.9	3.25	2.74	46.4	11.0	1.84	—	—	0.968	109
41.5	20.8	127	15.7	3.22	2.66	40.7	9.75	1.83	—	—	0.856	116
36.5	23.3	110	13.8	3.21	2.60	35.3	8.51	1.81	0.908	132	0.730	125
34	24.6	103	12.9	3.20	2.59	32.4	7.83	1.80	0.853	137	0.664	131
31	26.2	93.8	11.9	3.21	2.58	28.7	6.97	1.77	0.784	142	0.583	140
28.5	26.0	90.4	11.8	3.29	2.85	15.3	4.67	1.35	0.793	142	0.592	139
25	27.4	80.3	10.7	3.30	2.93	12.5	3.82	1.30	0.733	147	0.533	147
22	29.5	71.1	9.68	3.31	2.98	10.3	3.18	1.26	0.638	158	0.460	158
59.5	14.5	119	15.9	2.60	2.03	126	22.5	2.69	—	—	—	—
53	15.9	104	14.1	2.59	1.97	110	19.7	2.66	—	—	—	—
48.5	17.4	93.8	12.7	2.56	1.91	100	18.0	2.65	—	—	—	—
43	19.2	82.4	11.2	2.55	1.86	87.6	15.8	2.63	—	—	0.937	111
38	21.4	71.8	9.83	2.54	1.80	76.2	13.8	2.61	0.990	127	0.826	118

Where no value of C_c' or Q_s is shown, the Tee complies with Specification Sect. 1.9.1.2.

TABLE A-7. Properties of Structural Tees Cut from W Shapes (WT)

Designation	Area	Depth of Tee d		Stem Thickness t_w		Stem $\frac{t_w}{2}$	Area of Stem	Flange Width b_f		Flange Thickness t_f		Distance k
	In.^{2}	In.		In.		In.	In.^{2}	In.		In.		In.
WT 9x35.5	10.4	9.235	9 1/4	0.495	1/2	1/4	4.57	7.635	7 5/8	0.810	13/16	1 1/2
x32.5	9.55	9.175	9 1/8	0.450	7/16	1/4	4.13	7.590	7 5/8	0.750	3/4	1 7/16
x30	8.82	9.120	9 1/8	0.415	7/16	1/4	3.78	7.555	7 1/2	0.695	11/16	1 3/8
x27.5	8.10	9.055	9	0.390	3/8	3/16	3.53	7.530	7 1/2	0.630	5/8	1 5/16
x25	7.33	8.995	9	0.355	3/8	3/16	3.19	7.495	7 1/2	0.570	9/16	1 1/4
WT 9x23	6.77	9.030	9	0.360	3/8	3/16	3.25	6.060	6	0.605	5/8	1 1/4
x20	5.88	8.950	9	0.315	5/16	3/16	2.82	6.015	6	0.525	1/2	1 3/16
x17.5	5.15	8.850	8 7/8	0.300	5/16	3/16	2.65	6.000	6	0.425	7/16	1 1/8
WT 8x50	14.7	8.485	8 1/2	0.585	9/16	5/16	4.96	10.425	10 3/8	0.985	1	1 11/16
x44.5	13.1	8.375	8 3/8	0.525	1/2	1/4	4.40	10.365	10 3/8	0.875	7/8	1 9/16
x38.5	11.3	8.260	8 1/4	0.455	7/16	1/4	3.76	10.295	10 1/4	0.760	3/4	1 7/16
x33.5	9.84	8.165	8 1/8	0.395	3/8	3/16	3.23	10.235	10 1/4	0.665	11/16	1 3/8
WT 8x28.5	8.38	8.215	8 1/4	0.430	7/16	1/4	3.53	7.120	7 1/8	0.715	11/16	1 3/8
x25	7.37	8.130	8 1/8	0.380	3/8	3/16	3.09	7.070	7 1/8	0.630	5/8	1 5/16
x22.5	6.63	8.065	8 1/8	0.345	3/8	3/16	2.78	7.035	7	0.565	9/16	1 1/4
x20	5.89	8.005	8	0.305	5/16	3/16	2.44	6.995	7	0.505	1/2	1 3/16
x18	5.28	7.930	7 7/8	0.295	5/16	3/16	2.34	6.985	7	0.430	7/16	1 1/8
WT 8x15.5	4.56	7.940	8	0.275	1/4	1/8	2.18	5.525	5 1/2	0.440	7/16	1 1/8
x13	3.84	7.845	7 7/8	0.250	1/4	1/8	1.96	5.500	5 1/2	0.345	3/8	1 1/16
WT 7x365	107	11.210	11 1/4	3.070	3 1/16	1 9/16	34.4	17.890	17 7/8	4.910	4 15/16	5 9/16
x332.5	97.8	10.820	10 7/8	2.830	2 13/16	1 7/16	30.6	17.650	17 5/8	4.520	4 1/2	5 3/16
x302.5	88.9	10.460	10 1/2	2.595	2 5/8	1 5/16	27.1	17.415	17 3/8	4.160	4 3/16	4 13/16
x275	80.9	10.120	10 1/8	2.380	2 3/8	1 3/16	24.1	17.200	17 1/4	3.820	3 13/16	4 1/2
x250	73.5	9.800	9 3/4	2.190	2 3/16	1 1/8	21.5	17.010	17	3.500	3 1/2	4 3/16
x227.5	66.9	9.510	9 1/2	2.015	2	1	19.2	16.835	16 7/8	3.210	3 3/16	3 7/8
x213	62.6	9.335	9 3/8	1.875	1 7/8	15/16	17.5	16.695	16 3/4	3.035	3 1/16	3 11/16
x199	58.5	9.145	9 1/8	1.770	1 3/4	7/8	16.2	16.590	16 5/8	2.845	2 7/8	3 1/2
x185	54.4	8.960	9	1.655	1 5/8	13/16	14.8	16.475	16 1/2	2.660	2 11/16	3 5/16
x171	50.3	8.770	8 3/4	1.540	1 9/16	13/16	13.5	16.360	16 3/8	2.470	2 1/2	3 1/8
x155.5	45.7	8.560	8 1/2	1.410	1 7/16	3/4	12.1	16.230	16 1/4	2.260	2 1/4	2 15/16
x141.5	41.6	8.370	8 3/8	1.290	1 5/16	11/16	10.8	16.110	16 1/8	2.070	2 1/16	2 3/4
x128.5	37.8	8.190	8 1/4	1.175	1 3/16	5/8	9.62	15.995	16	1.890	1 7/8	2 9/16
x116.5	34.2	8.020	8	1.070	1 1/16	9/16	8.58	15.890	15 7/8	1.720	1 3/4	2 3/8
x105.5	31.0	7.860	7 7/8	0.980	1	1/2	7.70	15.800	15 3/4	1.560	1 9/16	2 1/4
x 96.5	28.4	7.740	7 3/4	0.890	7/8	7/16	6.89	15.710	15 3/4	1.440	1 7/16	2 1/8
x 88	25.9	7.610	7 5/8	0.830	13/16	7/16	6.32	15.650	15 5/8	1.310	1 5/16	2
x 79.5	23.4	7.490	7 1/2	0.745	3/4	3/8	5.58	15.565	15 5/8	1.190	1 3/16	1 7/8
x 72.5	21.3	7.390	7 3/8	0.680	11/16	3/8	5.03	15.500	15 1/2	1.090	1 1/16	1 3/4

TABLE A-7. (*Continued*)

Nominal Weight per Ft.	$\frac{d}{t_w}$	AXIS X-X				AXIS Y-Y			$C_c' = \sqrt{\frac{2\pi^2 E}{Q_s Q_a F_y}}$, $Q_a = 1.0$			
		I	S	r	y	I	S	r	$F_y = 36$ ksi		$F_y = 50$ ksi	
Lb.		In.[4]	In.[3]	In.	In.	In.[4]	In.[3]	In.	Q_s	C_c'	Q_s	C_c'
35.5	18.7	78.2	11.2	2.74	2.26	30.1	7.89	1.70	—	—	0.963	109
32.5	20.4	70.7	10.1	2.72	2.20	27.4	7.22	1.69	—	—	0.877	114
30	22.0	64.7	9.29	2.71	2.16	25.0	6.63	1.69	0.964	128	0.796	120
27.5	23.2	59.5	8.63	2.71	2.16	22.5	5.97	1.67	0.913	132	0.735	125
25	25.3	53.5	7.79	2.70	2.12	20.0	5.35	1.65	0.823	139	0.625	135
23	25.1	52.1	7.77	2.77	2.33	11.3	3.72	1.29	0.831	138	0.635	134
20	28.4	44.8	6.73	2.76	2.29	9.55	3.17	1.27	0.690	152	0.496	152
17.5	29.5	40.1	6.21	2.79	2.39	7.67	2.56	1.22	0.638	158	0.460	158
50	14.5	76.8	11.4	2.28	1.76	93.1	17.9	2.51	—	—	—	—
44.5	16.0	67.2	10.1	2.27	1.70	81.3	15.7	2.49	—	—	—	—
38.5	18.2	56.9	8.59	2.24	1.63	69.2	13.4	2.47	—	—	0.988	108
33.5	20.7	48.6	7.36	2.22	1.56	59.5	11.6	2.46	—	—	0.861	115
28.5	19.1	48.7	7.77	2.41	1.94	21.6	6.06	1.60	—	—	0.942	110
25	21.4	42.3	6.78	2.40	1.89	18.6	5.26	1.59	0.990	127	0.826	118
22.5	23.4	37.8	6.10	2.39	1.86	16.4	4.67	1.57	0.904	133	0.725	126
20	26.2	33.1	5.35	2.37	1.81	14.4	4.12	1.57	0.784	142	0.583	140
18	26.9	30.6	5.05	2.41	1.88	12.2	3.50	1.52	0.754	145	0.553	144
15.5	28.9	27.4	4.64	2.45	2.02	6.20	2.24	1.17	0.668	154	0.479	155
13	31.4	23.5	4.09	2.47	2.09	4.80	1.74	1.12	0.563	168	0.406	168
365	3.7	739	95.4	2.62	3.47	2360	264	4.69	—	—	—	—
332.5	3.8	622	82.1	2.52	3.25	2080	236	4.62	—	—	—	—
302.5	4.0	524	70.6	2.43	3.05	1840	211	4.55	—	—	—	—
275	4.3	442	60.9	2.34	2.85	1630	189	4.49	—	—	—	—
250	4.5	375	52.7	2.26	2.67	1440	169	4.43	—	—	—	—
227.5	4.7	321	45.9	2.19	2.51	1280	152	4.38	—	—	—	—
213	5.0	287	41.4	2.14	2.40	1180	141	4.34	—	—	—	—
199	5.2	257	37.6	2.10	2.30	1090	131	4.31	—	—	—	—
185	5.4	229	33.9	2.05	2.19	994	121	4.27	—	—	—	—
171	5.7	203	30.4	2.01	2.09	903	110	4.24	—	—	—	—
155.5	6.1	176	26.7	1.96	1.97	807	99.4	4.20	—	—	—	—
141.5	6.5	153	23.5	1.92	1.86	722	89.7	4.17	—	—	—	—
128.5	7.0	133	20.7	1.88	1.75	645	80.7	4.13	—	—	—	—
116.5	7.5	116	18.2	1.84	1.65	576	72.5	4.10	—	—	—	—
105.5	8.0	102	16.2	1.81	1.57	513	65.0	4.07	—	—	—	—
96.5	8.7	89.8	14.4	1.78	1.49	466	59.3	4.05	—	—	—	—
88	9.2	80.5	13.0	1.76	1.43	419	53.5	4.02	—	—	—	—
79.5	10.1	70.2	11.4	1.73	1.35	374	48.1	4.00	—	—	—	—
72.5	10.9	62.5	10.2	1.71	1.29	338	43.7	3.98	—	—	—	—

Where no value of C_c' or Q_s is shown, the Tee complies with Specification Sect. 1.9.1.2.

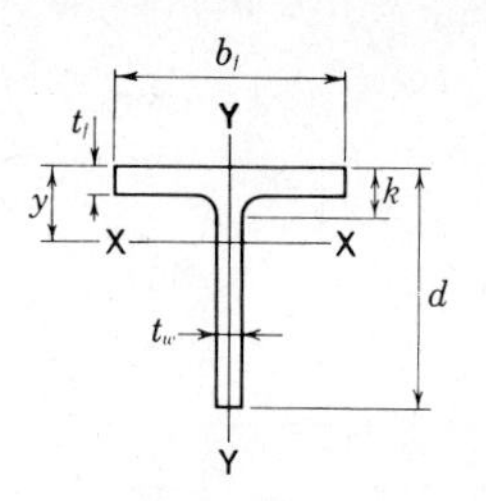

TABLE A-7. Properties of Structural Tees Cut from W Shapes (WT)

Designation	Area	Depth of Tee d		Stem			Area of Stem	Flange				Distance k
				Thickness t_w		$\frac{t_w}{2}$		Width b_f		Thickness t_f		
	In.^{2}	In.		In.		In.	In.^{2}	In.		In.		In.
WT 7x66	19.4	7.330	7 3/8	0.645	5/8	5/16	4.73	14.725	14 3/4	1.030	1	1 11/16
x60	17.7	7.240	7 1/4	0.590	9/16	5/16	4.27	14.670	14 5/8	0.940	15/16	1 5/8
x54.5	16.0	7.160	7 1/8	0.525	1/2	1/4	3.76	14.605	14 5/8	0.860	7/8	1 9/16
x49.5	14.6	7.080	7 1/8	0.485	1/2	1/4	3.43	14.565	14 5/8	0.780	3/4	1 7/16
x45	13.2	7.010	7	0.440	7/16	1/4	3.08	14.520	14 1/2	0.710	11/16	1 3/8
WT 7x41	12.0	7.155	7 1/8	0.510	1/2	1/4	3.65	10.130	10 1/8	0.855	7/8	1 5/8
x37	10.9	7.085	7 1/8	0.450	7/16	1/4	3.19	10.070	10 1/8	0.785	13/16	1 9/16
x34	9.99	7.020	7	0.415	7/16	1/4	2.91	10.035	10	0.720	3/4	1 1/2
x30.5	8.96	6.945	7	0.375	3/8	3/16	2.60	9.995	10	0.645	5/8	1 7/16
WT 7x26.5	7.81	6.960	7	0.370	3/8	3/16	2.58	8.060	8	0.660	11/16	1 7/16
x24	7.07	6.895	6 7/8	0.340	5/16	3/16	2.34	8.030	8	0.595	5/8	1 3/8
x21.5	6.31	6.830	6 7/8	0.305	5/16	3/16	2.08	7.995	8	0.530	1/2	1 5/16
WT 7x19	5.58	7.050	7	0.310	5/16	3/16	2.19	6.770	6 3/4	0.515	1/2	1 1/16
x17	5.00	6.990	7	0.285	5/16	3/16	1.99	6.745	6 3/4	0.455	7/16	1
x15	4.42	6.920	6 7/8	0.270	1/4	1/8	1.87	6.730	6 3/4	0.385	3/8	15/16
WT 7x13	3.85	6.955	7	0.255	1/4	1/8	1.77	5.025	5	0.420	7/16	15/16
x11	3.25	6.870	6 7/8	0.230	1/4	1/8	1.58	5.000	5	0.335	5/16	7/8
WT 6x168	49.4	8.410	8 3/8	1.775	1 3/4	7/8	14.9	13.385	13 3/8	2.955	2 15/16	3 11/16
x152.5	44.8	8.160	8 1/8	1.625	1 5/8	13/16	13.3	13.235	13 1/4	2.705	2 11/16	3 7/16
x139.5	41.0	7.925	7 7/8	1.530	1 1/2	3/4	12.1	13.140	13 1/8	2.470	2 1/2	3 3/16
x126	37.0	7.705	7 3/4	1.395	1 3/8	11/16	10.7	13.005	13	2.250	2 1/4	2 15/16
x115	33.9	7.525	7 1/2	1.285	1 5/16	11/16	9.67	12.895	12 7/8	2.070	2 1/16	2 3/4
x105	30.9	7.355	7 3/8	1.180	1 3/16	5/8	8.68	12.790	12 3/4	1.900	1 7/8	2 5/8
x 95	27.9	7.190	7 1/4	1.060	1 1/16	9/16	7.62	12.670	12 5/8	1.735	1 3/4	2 7/16
x 85	25.0	7.015	7	0.960	15/16	1/2	6.73	12.570	12 5/8	1.560	1 9/16	2 1/4
x 76	22.4	6.855	6 7/8	0.870	7/8	7/16	5.96	12.480	12 1/2	1.400	1 3/8	2 1/8
x 68	20.0	6.705	6 3/4	0.790	13/16	7/16	5.30	12.400	12 3/8	1.250	1 1/4	1 15/16
x 60	17.6	6.560	6 1/2	0.710	11/16	3/8	4.66	12.320	12 3/8	1.105	1 1/8	1 13/16
x 53	15.6	6.445	6 1/2	0.610	5/8	5/16	3.93	12.220	12 1/4	0.990	1	1 11/16
x 48	14.1	6.355	6 3/8	0.550	9/16	5/16	3.50	12.160	12 1/8	0.900	7/8	1 5/8
x 43.5	12.8	6.265	6 1/4	0.515	1/2	1/4	3.23	12.125	12 1/8	0.810	13/16	1 1/2
x 39.5	11.6	6.190	6 1/4	0.470	1/2	1/4	2.91	12.080	12 1/8	0.735	3/4	1 7/16
x 36	10.6	6.125	6 1/8	0.430	7/16	1/4	2.63	12.040	12	0.670	11/16	1 3/8
x 32.5	9.54	6.060	6	0.390	3/8	3/16	2.36	12.000	12	0.605	5/8	1 5/16

TABLE A-7. (*Continued*)

Nominal Weight per Ft.	$\frac{d}{t_w}$	AXIS X-X				AXIS Y-Y			$C_c' = \sqrt{\frac{2\pi^2 E}{Q_s Q_a F_y}}$, $Q_a = 1.0$			
		I	S	r	y	I	S	r	$F_y = 36$ ksi		$F_y = 50$ ksi	
Lb.		In.^{4}	In.^{3}	In.	In.	In.^{4}	In.^{3}	In.	Q_s	C_c'	Q_s	C_c'
66	11.4	57.8	9.57	1.73	1.29	274	37.2	3.76	—	—	—	—
60	12.3	51.7	8.61	1.71	1.24	247	33.7	3.74	—	—	—	—
54.5	13.6	45.3	7.56	1.68	1.17	223	30.6	3.73	—	—	—	—
49.5	14.6	40.9	6.88	1.67	1.14	201	27.6	3.71	—	—	—	—
45	15.9	36.4	6.16	1.66	1.09	181	25.0	3.70	—	—	—	—
41	14.0	41.2	7.14	1.85	1.39	74.2	14.6	2.48	—	—	—	—
37	15.7	36.0	6.25	1.82	1.32	66.9	13.3	2.48	—	—	—	—
34	16.9	32.6	5.69	1.81	1.29	60.7	12.1	2.46	—	—	—	—
30.5	18.5	28.9	5.07	1.80	1.25	53.7	10.7	2.45	—	—	0.973	108
26.5	18.8	27.6	4.94	1.88	1.38	28.8	7.16	1.92	—	—	0.958	109
24	20.3	24.9	4.48	1.87	1.35	25.7	6.40	1.91	—	—	0.882	114
21.5	22.4	21.9	3.98	1.86	1.31	22.6	5.65	1.89	0.947	130	0.775	122
19	22.7	23.3	4.22	2.04	1.54	13.3	3.94	1.55	0.934	130	0.760	123
17	24.5	20.9	3.83	2.04	1.53	11.7	3.45	1.53	0.857	136	0.669	131
15	25.6	19.0	3.55	2.07	1.58	9.79	2.91	1.49	0.810	140	0.610	137
13	27.3	17.3	3.31	2.12	1.72	4.45	1.77	1.08	0.737	147	0.537	146
11	29.9	14.8	2.91	2.14	1.76	3.50	1.40	1.04	0.621	160	0.447	160
168	4.7	190	31.2	1.96	2.31	593	88.6	3.47	—	—	—	—
152.5	5.0	162	27.0	1.90	2.16	525	79.3	3.42	—	—	—	—
139.5	5.2	141	24.1	1.86	2.05	469	71.3	3.38	—	—	—	—
126	5.5	121	20.9	1.81	1.92	414	63.6	3.34	—	—	—	—
115	5.9	106	18.5	1.77	1.82	371	57.5	3.31	—	—	—	—
105	6.2	92.1	16.4	1.73	1.72	332	51.9	3.28	—	—	—	—
95	6.8	79.0	14.2	1.68	1.62	295	46.5	3.25	—	—	—	—
85	7.3	67.8	12.3	1.65	1.52	259	41.2	3.22	—	—	—	—
76	7.9	58.5	10.8	1.62	1.43	227	36.4	3.19	—	—	—	—
68	8.5	50.6	9.46	1.59	1.35	199	32.1	3.16	—	—	—	—
60	9.2	43.4	8.22	1.57	1.28	172	28.0	3.13	—	—	—	—
53	10.6	36.3	6.91	1.53	1.19	151	24.7	3.11	—	—	—	—
48	11.6	32.0	6.12	1.51	1.13	135	22.2	3.09	—	—	—	—
43.5	12.2	28.9	5.60	1.50	1.10	120	19.9	3.07	—	—	—	—
39.5	13.2	25.8	5.03	1.49	1.06	108	17.9	3.05	—	—	—	—
36	14.2	23.2	4.54	1.48	1.02	97.5	16.2	3.04	—	—	—	—
32.5	15.5	20.6	4.06	1.47	0.985	87.2	14.5	3.02	—	—	—	—

Where no value of C_c' or Q_s is shown, the Tee complies with Specification Sect. 1.9.1.2.

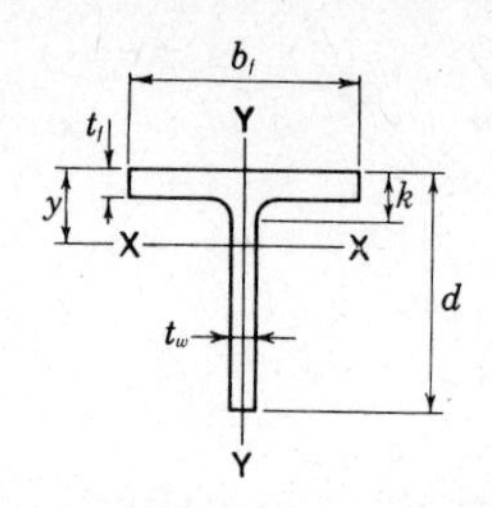

TABLE A-7. Properties of Structural Tees Cut from W Shapes (WT)

Designation	Area	Depth of Tee d		Stem Thickness t_w		Stem $\frac{t_w}{2}$	Area of Stem	Flange Width b_f		Flange Thickness t_f		Distance k
	In.²	In.		In.		In.	In.²	In.		In.		In.
WT 6x 29	8.52	6.095	6⅛	0.360	⅜	3/16	2.19	10.010	10	0.640	⅝	1⅜
x 26.5	7.78	6.030	6	0.345	⅜	3/16	2.08	9.995	10	0.575	9/16	1¼
WT 6x 25	7.34	6.095	6⅛	0.370	⅜	3/16	2.26	8.080	8⅛	0.640	⅝	1⅜
x 22.5	6.61	6.030	6	0.335	5/16	3/16	2.02	8.045	8	0.575	9/16	1¼
x 20	5.89	5.970	6	0.295	5/16	3/16	1.76	8.005	8	0.515	½	1¼
WT 6x 17.5	5.17	6.250	6¼	0.300	5/16	3/16	1.88	6.560	6½	0.520	½	1
x 15	4.40	6.170	6⅛	0.260	¼	⅛	1.60	6.520	6½	0.440	7/16	15/16
x 13	3.82	6.110	6⅛	0.230	¼	⅛	1.41	6.490	6½	0.380	⅜	⅞
WT 6x 11	3.24	6.155	6⅛	0.260	¼	⅛	1.60	4.030	4	0.425	7/16	⅞
x 9.5	2.79	6.080	6⅛	0.235	¼	⅛	1.43	4.005	4	0.350	⅜	13/16
x 8	2.36	5.995	6	0.220	¼	⅛	1.32	3.990	4	0.265	¼	¾
x 7	2.08	5.955	6	0.200	3/16	⅛	1.19	3.970	4	0.225	¼	11/16
WT 5x56	16.5	5.680	5⅝	0.755	¾	⅜	4.29	10.415	10⅜	1.250	1¼	1⅞
x50	14.7	5.550	5½	0.680	11/16	⅜	3.77	10.340	10⅜	1.120	1⅛	1¾
x44	12.9	5.420	5⅜	0.605	⅝	5/16	3.28	10.265	10¼	0.990	1	1⅝
x38.5	11.3	5.300	5¼	0.530	½	¼	2.81	10.190	10¼	0.870	⅞	1½
x34	9.99	5.200	5¼	0.470	½	¼	2.44	10.130	10⅛	0.770	¾	1⅜
x30	8.82	5.110	5⅛	0.420	7/16	¼	2.15	10.080	10⅛	0.680	11/16	1 5/16
x27	7.91	5.045	5	0.370	⅜	3/16	1.87	10.030	10	0.615	⅝	1¼
x24.5	7.21	4.990	5	0.340	5/16	3/16	1.70	10.000	10	0.560	9/16	1 3/16
WT 5x22.5	6.63	5.050	5	0.350	⅜	3/16	1.77	8.020	8	0.620	⅝	1¼
x19.5	5.73	4.960	5	0.315	5/16	3/16	1.56	7.985	8	0.530	½	1⅛
x16.5	4.85	4.865	4⅞	0.290	5/16	3/16	1.41	7.960	8	0.435	7/16	1 1/16
WT 5x15	4.42	5.235	5¼	0.300	5/16	3/16	1.57	5.810	5¾	0.510	½	15/16
x13	3.81	5.165	5⅛	0.260	¼	⅛	1.34	5.770	5¾	0.440	7/16	⅞
x11	3.24	5.085	5⅛	0.240	¼	⅛	1.22	5.750	5¾	0.360	⅜	¾
WT 5x 9.5	2.81	5.120	5⅛	0.250	¼	⅛	1.28	4.020	4	0.395	⅜	13/16
x 8.5	2.50	5.055	5	0.240	¼	⅛	1.21	4.010	4	0.330	5/16	¾
x 7.5	2.21	4.995	5	0.230	¼	⅛	1.15	4.000	4	0.270	¼	11/16
x 6	1.77	4.935	4⅞	0.190	3/16	⅛	0.938	3.960	4	0.210	3/16	⅝

TABLE A-7. (*Continued*)

Nominal Weight per Ft.	$\frac{d}{t_w}$	AXIS X-X				AXIS Y-Y			$C_c' = \sqrt{\frac{2\pi^2 E}{Q_s Q_a F_y}}$, $Q_a = 1.0$			
		I	S	r	y	I	S	r	F_y = 36 ksi		F_y = 50 ksi	
Lb.		In.^{4}	In.^{3}	In.	In.	In.^{4}	In.^{3}	In.	Q_s	C_c'	Q_s	C_c'
29	16.9	19.1	3.76	1.50	1.03	53.5	10.7	2.51	—	—	—	—
26.5	17.5	17.7	3.54	1.51	1.02	47.9	9.58	2.48	—	—	—	—
25	16.5	18.7	3.79	1.60	1.17	28.2	6.97	1.96	—	—	—	—
22.5	18.0	16.6	3.39	1.58	1.13	25.0	6.21	1.94	—	—	0.998	107
20	20.2	14.4	2.95	1.57	1.08	22.0	5.51	1.93	—	—	0.887	114
17.5	20.8	16.0	3.23	1.76	1.30	12.2	3.73	1.54	—	—	0.856	116
15	23.7	13.5	2.75	1.75	1.27	10.2	3.12	1.52	0.891	134	0.710	127
13	26.6	11.7	2.40	1.75	1.25	8.66	2.67	1.51	0.767	144	0.565	142
11	23.7	11.7	2.59	1.90	1.63	2.33	1.16	0.847	0.891	134	0.710	127
9.5	25.9	10.1	2.28	1.90	1.65	1.88	0.939	0.822	0.797	141	0.596	139
8	27.2	8.70	2.04	1.92	1.74	1.41	0.706	0.773	0.741	146	0.541	146
7	29.8	7.67	1.83	1.92	1.76	1.18	0.594	0.753	0.626	159	0.450	159
56	7.5	28.6	6.40	1.32	1.21	118	22.6	2.68	—	—	—	—
50	8.2	24.5	5.56	1.29	1.13	103	20.0	2.65	—	—	—	—
44	9.0	20.8	4.77	1.27	1.06	89.3	17.4	2.63	—	—	—	—
38.5	10.0	17.4	4.04	1.24	0.990	76.8	15.1	2.60	—	—	—	—
34	11.1	14.9	3.49	1.22	0.932	66.8	13.2	2.59	—	—	—	—
30	12.2	12.9	3.04	1.21	0.884	58.1	11.5	2.57	—	—	—	—
27	13.6	11.1	2.64	1.19	0.836	51.7	10.3	2.56	—	—	—	—
24.5	14.7	10.0	2.39	1.18	0.807	46.7	9.34	2.54	—	—	—	—
22.5	14.4	10.2	2.47	1.24	0.907	26.7	6.65	2.01	—	—	—	—
19.5	15.7	8.84	2.16	1.24	0.876	22.5	5.64	1.98	—	—	—	—
16.5	16.8	7.71	1.93	1.26	0.869	18.3	4.60	1.94	—	—	—	—
15	17.4	9.28	2.24	1.45	1.10	8.35	2.87	1.37	—	—	—	—
13	19.9	7.86	1.91	1.44	1.06	7.05	2.44	1.36	—	—	0.902	113
11	21.2	6.88	1.72	1.46	1.07	5.71	1.99	1.33	0.999	126	0.836	117
9.5	20.5	6.68	1.74	1.54	1.28	2.15	1.07	0.874	—	—	0.872	115
8.5	21.1	6.06	1.62	1.56	1.32	1.78	0.888	0.844	—	—	0.841	117
7.5	21.7	5.45	1.50	1.57	1.37	1.45	0.723	0.810	0.977	128	0.811	119
6	26.0	4.35	1.22	1.57	1.36	1.09	0.551	0.785	0.793	142	0.592	139

Where no value of C_c' or Q_s is shown, the Tee complies with Specification Sect. 1.9.1.2.

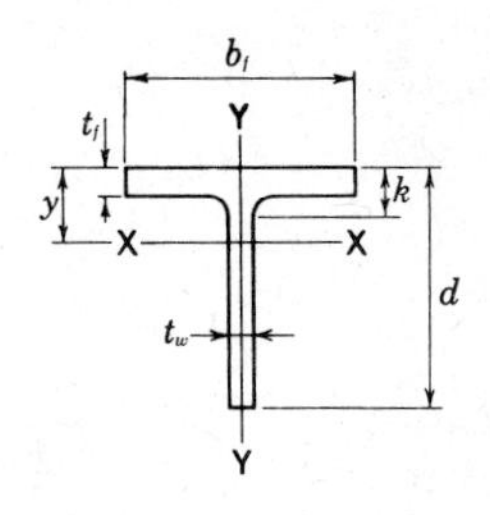

TABLE A-7. Properties of Structural Tees Cut from W Shapes (WT)

Designation	Area	Depth of Tee d		Stem Thickness t_w		Stem $\frac{t_w}{2}$	Area of Stem	Flange Width b_f		Flange Thickness t_f		Distance k
	In.^{2}	In.		In.		In.	In.^{2}	In.		In.		In.
WT 4 x33.5	9.84	4.500	4½	0.570	9/16	5/16	2.56	8.280	8¼	0.935	15/16	1 7/16
x29	8.55	4.375	4⅜	0.510	½	¼	2.23	8.220	8¼	0.810	13/16	1 5/16
x24	7.05	4.250	4¼	0.400	⅜	3/16	1.70	8.110	8⅛	0.685	11/16	1 3/16
x20	5.87	4.125	4⅛	0.360	⅜	3/16	1.48	8.070	8⅛	0.560	9/16	1 1/16
x17.5	5.14	4.060	4	0.310	5/16	3/16	1.26	8.020	8	0.495	½	1
x15.5	4.56	4.000	4	0.285	5/16	3/16	1.14	7.995	8	0.435	7/16	15/16
WT 4 x14	4.12	4.030	4	0.285	5/16	3/16	1.15	6.535	6½	0.465	7/16	15/16
x12	3.54	3.965	4	0.245	¼	⅛	0.971	6.495	6½	0.400	⅜	⅞
WT 4 x10.5	3.08	4.140	4⅛	0.250	¼	⅛	1.03	5.270	5¼	0.400	⅜	13/16
x 9	2.63	4.070	4⅛	0.230	¼	⅛	0.936	5.250	5¼	0.330	5/16	¾
WT 4 x 7.5	2.22	4.055	4	0.245	¼	⅛	0.993	4.015	4	0.315	5/16	¾
x 6.5	1.92	3.995	4	0.230	¼	⅛	0.919	4.000	4	0.255	¼	11/16
x 5	1.48	3.945	4	0.170	3/16	⅛	0.671	3.940	4	0.205	3/16	⅝
WT 3 x12.5	3.67	3.190	3¼	0.320	5/16	3/16	1.02	6.080	6⅛	0.455	7/16	13/16
x10	2.94	3.100	3⅛	0.260	¼	⅛	0.806	6.020	6	0.365	⅜	¾
x 7.5	2.21	2.995	3	0.230	¼	⅛	0.689	5.990	6	0.260	¼	⅝
WT 3 x 8	2.37	3.140	3⅛	0.260	¼	⅛	0.816	4.030	4	0.405	⅜	¾
x 6	1.78	3.015	3	0.230	¼	⅛	0.693	4.000	4	0.280	¼	⅝
x 4.5	1.34	2.950	3	0.170	3/16	⅛	0.502	3.940	4	0.215	3/16	9/16
WT 2.5x 9.5	2.77	2.575	2⅝	0.270	¼	⅛	0.695	5.030	5	0.430	7/16	13/16
x 8	2.34	2.505	2½	0.240	¼	⅛	0.601	5.000	5	0.360	⅜	¾
WT 2 x 6.5	1.91	2.080	2⅛	0.280	¼	⅛	0.582	4.060	4	0.345	⅜	11/16

TABLE A-7. (*Continued*)

Nominal Weight per Ft.	$\frac{d}{t_w}$	AXIS X-X				AXIS Y-Y			$C_c' = \sqrt{\frac{2\pi^2 E}{Q_s Q_a F_y}}$, $Q_a = 1.0$			
		I	S	r	y	I	S	r	F_y = 36 ksi		F_y = 50 ksi	
Lb.		In.^{4}	In.^{3}	In.	In.	In.^{4}	In.^{3}	In.	Q_s	C_c'	Q_s	C_c'
33.5	7.9	10.9	3.05	1.05	0.936	44.3	10.7	2.12	—	—	—	—
29	8.6	9.12	2.61	1.03	0.874	37.5	9.13	2.10	—	—	—	—
24	10.6	6.85	1.97	0.986	0.777	30.5	7.52	2.08	—	—	—	—
20	11.5	5.73	1.69	0.988	0.735	24.5	6.08	2.04	—	—	—	—
17.5	13.1	4.81	1.43	0.967	0.688	21.3	5.31	2.03	—	—	—	—
15.5	14.0	4.28	1.28	0.968	0.667	18.5	4.64	2.02	—	—	—	—
14	14.1	4.22	1.28	1.01	0.734	10.8	3.31	1.62	—	—	—	—
12	16.2	3.53	1.08	0.999	0.695	9.14	2.81	1.61	—	—	—	—
10.5	16.6	3.90	1.18	1.12	0.831	4.89	1.85	1.26	—	—	—	—
9	17.7	3.41	1.05	1.14	0.834	3.98	1.52	1.23	—	—	—	—
7.5	16.6	3.28	1.07	1.22	0.998	1.70	0.849	0.876	—	—	—	—
6.5	17.4	2.89	0.974	1.23	1.03	1.37	0.683	0.843	—	—	—	—
5	23.2	2.15	0.717	1.20	0.953	1.05	0.532	0.841	0.913	132	0.735	125
12.5	10.0	2.28	0.886	0.789	0.610	8.53	2.81	1.52	—	—	—	—
10	11.9	1.76	0.693	0.774	0.560	6.64	2.21	1.50	—	—	—	—
7.5	13.0	1.41	0.577	0.797	0.558	4.66	1.56	1.45	—	—	—	—
8	12.1	1.69	0.685	0.844	0.676	2.21	1.10	0.966	—	—	—	—
6	13.1	1.32	0.564	0.861	0.677	1.50	0.748	0.918	—	—	—	—
4.5	17.4	0.950	0.408	0.842	0.623	1.10	0.557	0.905	—	—	—	—
9.5	9.5	1.01	0.485	0.605	0.487	4.56	1.82	1.28	—	—	—	—
8	10.4	0.845	0.413	0.601	0.458	3.75	1.50	1.27	—	—	—	—
6.5	7.4	0.526	0.321	0.524	0.440	1.93	0.950	1.00	—	—	—	—

Where no value of C_c' or Q_s is shown, the Tee complies with Specification Sect. 1.9.1.2.

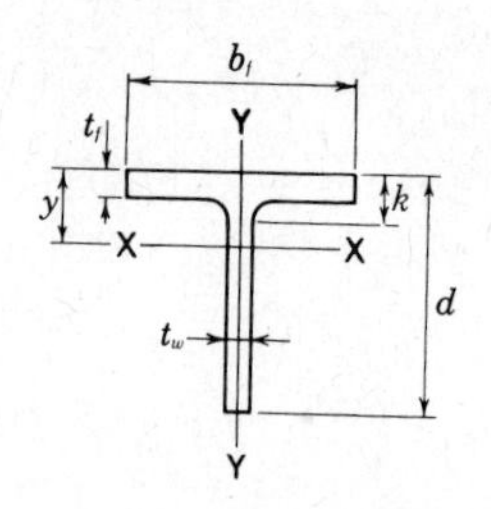

TABLE A-8. Properties of Structural Tees Cut from M Shapes (MT)

Designation	Area	Depth of Tee d		Stem			Area of Stem	Flange				Distance k	Grip	Max. Flge. Fastener
				Thickness t_w		$\frac{t_w}{2}$		Width b_f		Thickness t_f				
	In.^{2}	In.		In.		In.	In.^{2}	In.		In.		In.	In.	In.
MT 7 x 9	2.55	7.000	7	0.215	3/16	1/8	1.50	4.000	4	0.270	1/4	5/8	1/4	3/4
MT 6 x 5.9	1.73	6.000	6	0.177	3/16	1/8	1.06	3.065	3 1/8	0.225	1/4	9/16	1/4	—
MT 5 x 4.5	1.32	5.000	5	0.157	3/16	1/8	0.785	2.690	2 3/4	0.206	3/16	9/16	3/16	—
MT 4 x 3.25	0.958	4.000	4	0.135	1/8	1/16	0.540	2.281	2 1/4	0.189	3/16	1/2	3/16	—
MT 3 x 10	2.94	3.000	3	0.250	1/4	1/8	0.750	5.938	6	0.379	3/8	7/8	3/8	7/8
MT 3 x 2.2	0.646	3.000	3	0.114	1/8	1/16	0.342	1.844	1 7/8	0.171	3/16	7/16	3/16	—
MT 2.5 x 9.45	2.78	2.500	2 1/2	0.316	5/16	3/16	0.790	5.003	5	0.416	7/16	7/8	7/16	7/8
MT 2 x 6.5	1.90	2.000	2	0.254	1/4	1/8	0.508	3.940	4	0.371	3/8	13/16	3/8	3/4

TABLE A-8. (*Continued*)

Nominal Wt. per Ft.	$\frac{d}{t_w}$	AXIS X-X				AXIS Y-Y			$C_c' = \sqrt{\frac{2\pi^2 E}{Q_s Q_a F_y}}$, $Q_a = 1.0$			
		I	S	r	y	I	S	r	F_y = 36 ksi		F_y = 50 ksi	
Lb.		In.[4]	In.[3]	In.	In.	In.[4]	In.[3]	In.	Q_s	C_c'	Q_s	C_c'
9	32.6	13.1	2.69	2.27	2.12	1.32	0.660	0.719	0.523	174	0.376	174
5.9	33.9	6.60	1.60	1.95	1.89	0.490	0.320	0.532	0.483	181	0.348	181
4.5	31.8	3.46	0.997	1.62	1.53	0.305	0.227	0.480	0.549	170	0.396	170
3.25	29.6	1.57	0.556	1.28	1.17	0.172	0.150	0.423	0.634	158	0.457	158
10	12.0	1.54	0.624	0.724	0.531	5.80	1.95	1.40	—	—	—	—
2.2	26.3	0.577	0.267	0.945	0.836	0.083	0.090	0.358	0.780	143	0.578	141
9.45	7.9	1.05	0.527	0.615	0.511	3.93	1.57	1.19	—	—	—	—
6.5	7.9	0.431	0.271	0.476	0.410	1.68	0.853	0.939	—	—	—	—

Where no value of C'_c or Q_s is shown, the Tee complies with Specification Sect. 1.9.1.2.

TABLE A-9. Properties of Double Angles

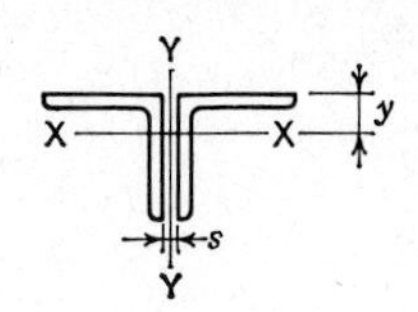

Two equal leg angles

Designation	Wt. per Ft. 2 Angles	Area of 2 Angles	AXIS X — X				AXIS Y — Y Radii of Gyration Back to Back of Angles, Inches			Q_s* Angles in Contact		Q_s* Angles Separated	
			I	S	r	y				F_y = 36 ksi	F_y = 50 ksi	F_y = 36 ksi	F_y = 50 ksi
	Lb.	In.^{2}	In.^{4}	In.^{3}	In.	In.	0	3/8	3/4				
L 8 x8 x1 1/8	113.8	33.5	195.0	35.1	2.42	2.41	3.42	3.55	3.69	—	—	—	—
1	102.0	30.0	177.0	31.6	2.44	2.37	3.40	3.53	3.67	—	—	—	—
7/8	90.0	26.5	159.0	28.0	2.45	2.32	3.38	3.51	3.64	—	—	—	—
3/4	77.8	22.9	139.0	24.4	2.47	2.28	3.36	3.49	3.62	—	—	—	—
5/8	65.4	19.2	118.0	20.6	2.49	2.23	3.34	3.47	3.60	—	—	.997	.935
1/2	52.8	15.5	97.3	16.7	2.50	2.19	3.32	3.45	3.58	.995	.921	.911	.834
L 6 x6 x1	74.8	22.0	70.9	17.1	1.80	1.86	2.59	2.73	2.87	—	—	—	—
7/8	66.2	19.5	63.8	15.3	1.81	1.82	2.57	2.70	2.85	—	—	—	—
3/4	57.4	16.9	56.3	13.3	1.83	1.78	2.55	2.68	2.82	—	—	—	—
5/8	48.4	14.2	48.3	11.3	1.84	1.73	2.53	2.66	2.80	—	—	—	—
1/2	39.2	11.5	39.8	9.23	1.86	1.68	2.51	2.64	2.78	—	—	—	.961
3/8	29.8	8.72	30.8	7.06	1.88	1.64	2.49	2.62	2.75	.995	.921	.911	.834
L 5 x5 x 7/8	54.4	16.0	35.5	10.3	1.49	1.57	2.16	2.30	2.45	—	—	—	—
3/4	47.2	13.9	31.5	9.06	1.51	1.52	2.14	2.28	2.42	—	—	—	—
1/2	32.4	9.50	22.5	6.31	1.54	1.43	2.10	2.24	2.38	—	—	—	—
3/8	24.6	7.22	17.5	4.84	1.56	1.39	2.09	2.22	2.35	—	—	.982	.919
5/16	20.6	6.05	14.8	4.08	1.57	1.37	2.08	2.21	2.34	.995	.921	.911	.834
L 4 x4 x 3/4	37.0	10.9	15.3	5.62	1.19	1.27	1.74	1.88	2.83	—	—	—	—
5/8	31.4	9.22	13.3	4.80	1.20	1.23	1.72	1.86	2.00	—	—	—	—
1/2	25.6	7.50	11.1	3.95	1.22	1.18	1.70	1.83	1.98	—	—	—	—
3/8	19.6	5.72	8.72	3.05	1.23	1.14	1.68	1.81	1.95	—	—	—	—
5/16	16.4	4.80	7.43	2.58	1.24	1.12	1.67	1.80	1.94	—	—	.997	.935
1/4	13.2	3.88	6.08	2.09	1.25	1.09	1.66	1.79	1.93	.995	.921	.911	.834
L 3 1/2x3 1/2x 3/8	17.0	4.97	5.73	2.30	1.07	1.01	1.48	1.61	1.75	—	—	—	—
5/16	14.4	4.18	4.90	1.95	1.08	.990	1.47	1.60	1.74	—	—	—	.986
1/4	11.6	3.38	4.02	1.59	1.09	.968	1.46	1.59	1.73	—	.982	.965	.897
L 3 x3 x 1/2	18.8	5.50	4.43	2.14	.898	.932	1.29	1.43	1.59	—	—	—	—
3/8	14.4	4.22	3.52	1.67	.913	.888	1.27	1.41	1.56	—	—	—	—
5/16	12.2	3.55	3.02	1.41	.922	.865	1.26	1.40	1.55	—	—	—	—
1/4	9.8	2.88	2.49	1.15	.930	.842	1.26	1.39	1.53	—	—	—	.961
3/16	7.42	2.18	1.92	.882	.939	.820	1.25	1.38	1.52	.995	.921	.911	.834
L 2 1/2x2 1/2x 3/8	11.8	3.47	1.97	1.13	.753	.762	1.07	1.21	1.36	—	—	—	—
5/16	10.0	2.93	1.70	.964	.761	.740	1.06	1.20	1.35	—	—	—	—
1/4	8.2	2.38	1.41	.789	.769	.717	1.05	1.19	1.34	—	—	—	—
3/16	6.14	1.80	1.09	.685	.778	.694	1.04	1.18	1.32	—	—	.982	.919
L 2 x2 x 3/8	9.4	2.72	.958	.702	.594	.636	.870	1.01	1.17	—	—	—	—
5/16	7.84	2.30	.832	.681	.601	.614	.859	1.00	1.16	—	—	—	—
1/4	6.38	1.88	.695	.494	.609	.592	.849	.989	1.14	—	—	—	—
3/16	4.88	1.43	.545	.381	.617	.569	.840	.977	1.13	—	—	—	—
1/8	3.30	.960	.380	.261	.626	.546	.831	.965	1.11	.995	.921	.911	.834

* Where no value of Q_s is shown, the angles comply with Specification Sect. 1.9.1.2 and may be considered fully effective.

For F_y = 36 ksi: $C'_c = 126.1/\sqrt{Q_s}$

For F_y = 50 ksi: $C'_c = 107.0/\sqrt{Q_s}$

TABLE A-9. (*Continued*)

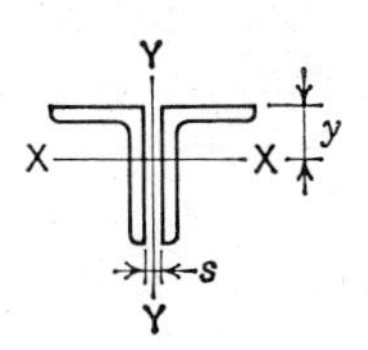

Two unequal leg angles

Long legs back to back

Designation	Wt. per Ft. 2 Angles	Area of 2 Angles	AXIS X — X				AXIS Y — Y: Radii of Gyration Back to Back of Angles, Inches			Q_s*			
			I	S	r	y				Angles in Contact		Angles Separated	
										F_y = 36 ksi	F_y = 50 ksi	F_y = 36 ksi	F_y = 50 ksi
	Lb.	In.^{2}	In.^{4}	In.^{3}	In.	In.	0	3/8	3/4				
L 8 x6 x1	88.4	26.0	161.0	30.2	2.49	2.65	2.39	2.52	2.66	—	—	—	—
3/4	67.6	19.9	126.0	23.3	2.53	2.56	2.35	2.48	2.62	—	—	—	—
1/2	46.0	13.5	88.6	16.0	2.56	2.47	2.32	2.44	2.57	—	—	.911	.834
L 8 x4 x1	74.8	22.0	139.0	28.1	2.52	3.05	1.47	1.61	1.75	—	—	—	—
3/4	57.4	16.9	109.0	21.8	2.55	2.95	1.42	1.55	1.69	—	—	—	—
1/2	39.2	11.5	77.0	15.0	2.59	2.86	1.38	1.51	1.64	—	—	.911	.834
L 7 x4 x 3/4	52.4	15.4	75.6	16.8	2.22	2.51	1.48	1.62	1.76	—	—	—	—
1/2	35.8	10.5	53.3	11.6	2.25	2.42	1.44	1.57	1.71	—	—	.965	.897
3/8	27.2	7.97	41.1	8.88	2.27	2.37	1.43	1.55	1.68	—	—	.839	.750
L 6 x4 x 3/4	47.2	13.9	49.0	12.5	1.88	2.08	1.55	1.69	1.83	—	—	—	—
5/8	40.0	11.7	42.1	10.6	1.90	2.03	1.53	1.67	1.81	—	—	—	—
1/2	32.4	9.50	34.8	8.67	1.91	1.99	1.51	1.64	1.78	—	—	—	.961
3/8	24.6	7.22	26.9	6.64	1.93	1.94	1.50	1.62	1.76	—	—	.911	.834
L 6 x3½x 3/8	23.4	6.84	25.7	6.49	1.94	2.04	1.26	1.39	1.53	—	—	.911	.834
5/16	19.6	5.74	21.8	5.47	1.95	2.01	1.26	1.38	1.51	—	—	.825	.733
L 5 x3½x 3/4	39.6	11.6	27.8	8.55	1.55	1.75	1.40	1.53	1.68	—	—	—	—
1/2	27.2	8.00	20.0	5.97	1.58	1.66	1.35	1.49	1.63	—	—	—	—
3/8	20.8	6.09	15.6	4.59	1.60	1.61	1.34	1.46	1.60	—	—	.982	.919
5/16	17.4	5.12	13.2	3.87	1.61	1.59	1.33	1.45	1.59	—	—	.911	.834
L 5 x3 x 1/2	25.6	7.50	18.9	5.82	1.59	1.75	1.12	1.25	1.40	—	—	—	—
3/8	19.6	5.72	14.7	4.47	1.61	1.70	1.10	1.23	1.37	—	—	.982	.919
5/16	16.4	4.80	12.5	3.77	1.61	1.68	1.09	1.22	1.36	—	—	.911	.834
1/4	13.2	3.88	10.2	3.06	1.62	1.66	1.08	1.21	1.34	—	—	.804	.708
L 4 x3½x 1/2	23.8	7.00	10.6	3.87	1.23	1.25	1.44	1.58	1.72	—	—	—	—
3/8	18.2	5.34	8.35	2.99	1.25	1.21	1.42	1.56	1.70	—	—	—	—
5/16	15.4	4.49	7.12	2.53	1.26	1.18	1.42	1.55	1.69	—	—	.997	.935
1/4	12.4	3.63	5.83	2.05	1.27	1.16	1.41	1.54	1.67	—	.982	.911	.834
L 4 x3 x 1/2	22.2	6.50	10.1	3.78	1.25	1.33	1.20	1.33	1.48	—	—	—	—
3/8	17.0	4.97	7.93	2.92	1.26	1.28	1.18	1.31	1.45	—	—	—	—
5/16	14.4	4.18	6.76	2.47	1.27	1.26	1.17	1.30	1.44	—	—	.997	.935
1/4	11.6	3.38	5.54	2.00	1.28	1.24	1.16	1.29	1.43	—	—	.911	.834

* Where no value of Q_s is shown, the angles comply with Specification Sect. 1.9.1.2 and may be considered fully effective.

For F_y = 36 ksi: $C'_c = 126.1/\sqrt{Q_s}$

For F_y = 50 ksi: $C'_c = 107.0/\sqrt{Q_s}$

Two unequal leg angles

Long legs back to back

L 3½x3 x ⅜	15.8	4.59	5.45	2.25	1.09	1.08	1.22	1.36	1.50	—	—	—	—
5/16	13.2	3.87	4.66	1.91	1.10	1.06	1.21	1.35	1.49	—	—	—	.986
¼	10.8	3.13	3.83	1.55	1.11	1.04	1.20	1.33	1.48	—	—	.965	.897
L 3½x2½x ⅜	14.4	4.22	5.12	2.19	1.10	1.16	.976	1.11	1.26	—	—	—	—
5/16	12.2	3.55	4.38	1.85	1.11	1.14	.966	1.10	1.25	—	—	—	.986
¼	9.8	2.88	3.60	1.51	1.12	1.11	.958	1.09	1.23	—	—	.965	.897
L 3 x2½x ⅜	13.2	3.84	3.31	1.62	.928	.956	1.02	1.16	1.31	—	—	—	—
¼	9.0	2.63	2.35	1.12	.945	.911	1.00	1.13	1.28	—	—	—	.961
3/16	6.77	1.99	1.81	.859	.954	.888	.993	1.12	1.27	—	—	.911	.834
L 3 x2 x ⅜	11.8	3.47	3.06	1.56	.940	1.04	.777	.917	1.07	—	—	—	—
5/16	10.0	2.93	2.63	1.33	.948	1.02	.767	.903	1.06	—	—	—	—
¼	8.2	2.38	2.17	1.08	.957	.993	.757	.891	1.04	—	—	—	.961
3/16	6.1	1.80	1.68	.830	.966	.970	.749	.879	1.03	—	—	.911	.834
L 2½x2 x ⅜	10.6	3.09	1.82	1.09	.768	.831	.819	.961	1.12	—	—	—	—
5/16	9.0	2.62	1.58	.932	.776	.809	.809	.948	1.10	—	—	—	—
¼	7.2	2.13	1.31	.763	.784	.787	.799	.935	1.09	—	—	—	—
3/16	5.5	1.62	1.02	.586	.793	.764	.790	.923	1.07	—	—	.982	.919

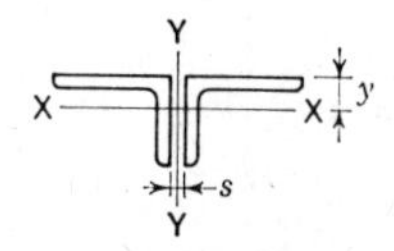

Two unequal leg angles

Short legs back to back

Designation	Wt. per Ft. 2 Angles	Area of 2 Angles	AXIS X — X				AXIS Y — Y Radii of Gyration Back to Back of Angles, Inches			Q_s* Angles in Contact		Q_s* Angles Separated	
			I	S	r	y				F_y =	F_y =	F_y =	F_y =
	Lb.	In.^{2}	In.^{4}	In.^{3}	In.	In.	0	⅜	¾	36 ksi	50 ksi	36 ksi	50 ksi
L 8 x6 x1	88.4	26.0	77.6	17.8	1.73	1.65	3.64	3.78	3.92	—	—	—	—
¾	67.6	19.9	61.4	13.8	1.76	1.56	3.60	3.74	3.88	—	—	—	—
½	46.0	13.5	43.4	9.58	1.79	1.47	3.56	3.69	3.83	.995	.921	.911	.834
L 8 x4 x1	74.8	22.0	23.3	7.88	1.03	1.05	3.95	4.10	4.25	—	—	—	—
¾	57.4	16.9	18.7	6.14	1.05	.953	3.90	4.05	4.19	—	—	—	—
½	39.2	11.5	13.5	4.29	1.08	.859	3.86	4.00	4.14	.995	.921	.911	.834
L 7 x4 x ¾	52.4	15.4	16.1	6.05	1.09	1.01	3.35	3.49	3.64	—	—	—	—
½	35.8	10.5	13.1	4.23	1.11	.917	3.30	3.44	3.59	—	.982	.965	.897
⅜	27.2	7.97	10.2	3.26	1.13	.870	3.28	3.42	3.56	.926	.838	.839	.750
L 6 x4 x ¾	47.2	13.9	17.4	5.94	1.12	1.08	2.80	2.94	3.09	—	—	—	—
⅝	40.0	11.7	15.0	5.07	1.13	1.03	2.78	2.92	3.06	—	—	—	—
½	32.4	9.50	12.5	4.16	1.15	.987	2.76	2.90	3.04	—	—	—	.961
⅜	24.6	7.22	9.81	3.21	1.17	.941	2.74	2.87	3.02	.995	.921	.911	.834
L 6 x3½x ⅜	23.4	6.84	6.68	2.46	.988	.787	2.81	2.95	3.09	.995	.921	.911	.834
5/16	19.6	5.74	5.70	2.08	.996	.763	2.80	2.94	3.08	.912	.822	.825	.733

* Where no value of Q_s is shown, the angles comply with Specification Sect. 1.9.1.2 and may be considered fully effective.

For F_y = 36 ksi: $C'_c = 126.1/\sqrt{Q_s}$

For F_y = 50 ksi: $C'_c = 107.0/\sqrt{Q_s}$

TABLE A-9. (*Continued*)

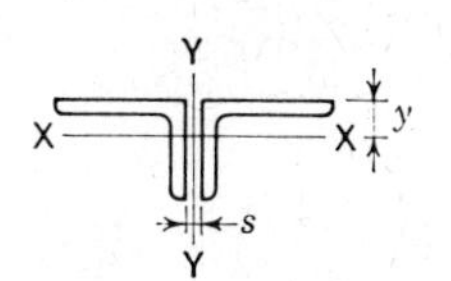

Two unequal leg angles

Short legs back to back

Designation	Wt. per Ft. 2 Angles	Area of 2 Angles	AXIS X – X				AXIS Y – Y			Q_s*			
							Radii of Gyration Back to Back of Angles, Inches			Angles in Contact		Angles Separated	
			I	S	r	y				F_y = 36 ksi	F_y = 50 ksi	F_y = 36 ksi	F_y = 50 ksi
	Lb.	In.^{2}	In.^{4}	In.^{3}	In.	In.	0	3/8	3/4				
L 5 x3½x 3/4	39.6	11.6	11.1	4.43	.977	.996	2.33	2.48	2.63	—	—	—	—
1/2	27.2	8.00	8.10	3.12	1.01	.906	2.29	2.43	2.57	—	—	—	—
3/8	20.8	6.09	6.37	2.41	1.02	.861	2.27	2.41	2.55	—	—	.982	.919
5/16	17.4	5.12	5.44	2.04	1.03	.838	2.26	2.39	2.54	.995	.921	.911	.834
L 5 x3 x 1/2	25.6	7.50	5.16	2.29	.829	.750	2.36	2.50	2.65	—	—	—	—
3/8	19.6	5.72	4.08	1.78	.845	.704	2.34	2.48	2.63	—	—	.982	.919
5/16	16.4	4.80	3.49	1.51	.853	.681	2.33	2.47	2.61	.995	.921	.911	.834
1/4	13.2	3.88	2.88	1.23	.861	.657	2.32	2.46	2.60	.891	.797	.804	.708
L 4 x3½x 1/2	23.8	7.00	7.58	3.03	1.04	1.00	1.76	1.89	2.04	—	—	—	—
3/8	18.2	5.34	5.97	2.35	1.06	.955	1.74	1.87	2.01	—	—	—	—
5/16	15.4	4.49	5.10	1.99	1.07	.932	1.73	1.86	2.00	—	—	.997	.935
1/4	12.4	3.63	4.19	1.62	1.07	.909	1.72	1.85	1.99	.995	.921	.911	.834
L 4 x3 x 1/2	22.2	6.50	4.85	2.23	.864	.827	1.82	1.96	2.11	—	—	—	—
3/8	17.0	4.97	3.84	1.73	.879	.782	1.80	1.94	2.08	—	—	—	—
5/16	14.4	4.18	3.29	1.47	.887	.759	1.79	1.93	2.07	—	—	.997	.935
1/4	11.6	3.38	2.71	1.20	.896	.736	1.78	1.92	2.06	.995	.921	.911	.834
L 3½x3 x 3/8	15.8	4.59	3.69	1.70	.897	.830	1.53	1.67	1.82	—	—	—	—
5/16	13.2	3.87	3.17	1.44	.905	.808	1.52	1.66	1.80	—	—	—	.986
1/4	10.8	3.13	2.61	1.18	.914	.785	1.52	1.65	1.79	—	.982	.965	.897
L 3½x2½x 3/8	14.4	4.22	2.18	1.18	.719	.660	1.60	1.74	1.89	—	—	—	—
5/16	12.2	3.55	1.88	1.01	.727	.637	1.59	1.73	1.88	—	—	—	.986
1/4	9.8	2.88	1.55	.824	.735	.614	1.58	1.72	1.86	—	.982	.965	.897
L 3 x2½x 3/8	13.2	3.84	2.08	1.16	.736	.706	1.33	1.47	1.62	—	—	—	—
1/4	9.0	2.63	1.49	.808	.753	.661	1.31	1.45	1.60	—	—	—	.961
3/16	6.77	1.99	1.15	.620	.761	.638	1.30	1.44	1.58	.995	.921	.911	.834
L 3 x2 x 3/8	11.8	3.47	1.09	.743	.559	.539	1.40	1.55	1.70	—	—	—	—
5/16	10.0	2.93	.941	.634	.567	.516	1.39	1.53	1.68	—	—	—	—
1/4	8.2	2.38	.784	.520	.574	.493	1.38	1.52	1.67	—	—	—	.961
3/16	6.1	1.80	.613	.401	.583	.470	1.37	1.51	1.66	.995	.921	.911	.834
L 2½x2 x 3/8	10.6	3.09	1.03	.725	.577	.581	1.13	1.28	1.43	—	—	—	—
5/16	9.0	2.62	.893	.620	.584	.559	1.12	1.26	1.42	—	—	—	—
1/4	7.2	2.13	.745	.509	.592	.537	1.11	1.25	1.40	—	—	—	—
3/16	5.5	1.62	.583	.392	.600	.514	1.10	1.24	1.39	—	—	.982	.919

* Where no value of Q_s is shown, the angles comply with Specification Sect. 1.9.1.2 and may be considered fully effective.

For F_y = 36 ksi: $C'_c = 126.1/\sqrt{Q_s}$

For F_y = 50 ksi: $C'_c = 107.0/\sqrt{Q_s}$

TABLE A-10. Properties of Round Steel Pipe

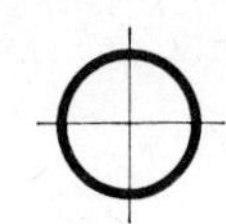

Dimensions				Weight per Foot Lbs. Plain Ends	Properties			
Nominal Diameter In.	Outside Diameter In.	Inside Diameter In.	Wall Thickness In.		A In.^{2}	I In.^{4}	S In.^{3}	r In.
Standard Weight								
½	.840	.622	.109	.85	.250	.017	.041	.261
¾	1.050	.824	.113	1.13	.333	.037	.071	.334
1	1.315	1.049	.133	1.68	.494	.087	.133	.421
1¼	1.660	1.380	.140	2.27	.669	.195	.235	.540
1½	1.900	1.610	.145	2.72	.799	.310	.326	.623
2	2.375	2.067	.154	3.65	1.07	.666	.561	.787
2½	2.875	2.469	.203	5.79	1.70	1.53	1.06	.947
3	3.500	3.068	.216	7.58	2.23	3.02	1.72	1.16
3½	4.000	3.548	.226	9.11	2.68	4.79	2.39	1.34
4	4.500	4.026	.237	10.79	3.17	7.23	3.21	1.51
5	5.563	5.047	.258	14.62	4.30	15.2	5.45	1.88
6	6.625	6.065	.280	18.97	5.58	28.1	8.50	2.25
8	8.625	7.981	.322	28.55	8.40	72.5	16.8	2.94
10	10.750	10.020	.365	40.48	11.9	161	29.9	3.67
12	12.750	12.000	.375	49.56	14.6	279	43.8	4.38
Extra Strong								
½	.840	.546	.147	1.09	.320	.020	.048	.250
¾	1.050	.742	.154	1.47	.433	.045	.085	.321
1	1.315	.957	.179	2.17	.639	.106	.161	.407
1¼	1.660	1.278	.191	3.00	.881	.242	.291	.524
1½	1.900	1.500	.200	3.63	1.07	.391	.412	.605
2	2.375	1.939	.218	5.02	1.48	.868	.731	.766
2½	2.875	2.323	.276	7.66	2.25	1.92	1.34	.924
3	3.500	2.900	.300	10.25	3.02	3.89	2.23	1.14
3½	4.000	3.364	.318	12.50	3.68	6.28	3.14	1.31
4	4.500	3.826	.337	14.98	4.41	9.61	4.27	1.48
5	5.563	4.813	.375	20.78	6.11	20.7	7.43	1.84
6	6.625	5.761	.432	28.57	8.40	40.5	12.2	2.19
8	8.625	7.625	.500	43.39	12.8	106	24.5	2.88
10	10.750	9.750	.500	54.74	16.1	212	39.4	3.63
12	12.750	11.750	.500	65.42	19.2	362	56.7	4.33
Double-Extra Strong								
2	2.375	1.503	.436	9.03	2.66	1.31	1.10	.703
2½	2.875	1.771	.552	13.69	4.03	2.87	2.00	.844
3	3.500	2.300	.600	18.58	5.47	5.99	3.42	1.05
4	4.500	3.152	.674	27.54	8.10	15.3	6.79	1.37
5	5.563	4.063	.750	38.55	11.3	33.6	12.1	1.72
6	6.625	4.897	.864	53.16	15.6	66.3	20.0	2.06
8	8.625	6.875	.875	72.42	21.3	162	37.6	2.76

The listed sections are available in conformance with ASTM Specification A53 Grade B or A501. Other sections are made to these specifications. Consult with pipe manufacturers or distributors for availability.

TABLE A-11. Properties of Rectangular Steel Tubing

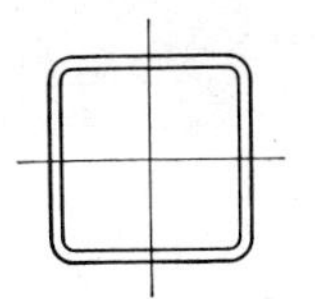

Square

DIMENSIONS				PROPERTIES**			
Nominal* Size	Wall Thickness		Weight per Foot	Area	I	S	r
In.	In.		Lb.	In.^{2}	In.^{4}	In.^{3}	In.
16 x 16	.5000	1/2	103.30	30.4	1200	150	6.29
	.3750	3/8	78.52	23.1	931	116	6.35
	.3125	5/16	65.87	19.4	789	98.6	6.38
14 x 14	.5000	1/2	89.68	26.4	791	113	5.48
	.3750	3/8	68.31	20.1	615	87.9	5.54
	.3125	5/16	57.36	16.9	522	74.6	5.57
12 x 12	.5000	1/2	76.07	22.4	485	80.9	4.66
	.3750	3/8	58.10	17.1	380	63.4	4.72
	.3125	5/16	48.86	14.4	324	54.0	4.75
	.2500	1/4	39.43	11.6	265	44.1	4.78
10 x 10	.6250	5/8	76.33	22.4	321	64.2	3.78
	.5000	1/2	62.46	18.4	271	54.2	3.84
	.3750	3/8	47.90	14.1	214	42.9	3.90
	.3125	5/16	40.35	11.9	183	36.7	3.93
	.2500	1/4	32.63	9.59	151	30.1	3.96
8 x 8	.6250	5/8	59.32	17.4	153	38.3	2.96
	.5000	1/2	48.85	14.4	131	32.9	3.03
	.3750	3/8	37.69	11.1	106	26.4	3.09
	.3125	5/16	31.84	9.36	90.9	22.7	3.12
	.2500	1/4	25.82	7.59	75.1	18.8	3.15
	.1875	3/16	19.63	5.77	58.2	14.6	3.18
7 x 7	.5000	1/2	42.05	12.4	84.6	24.2	2.62
	.3750	3/8	32.58	9.58	68.7	19.6	2.68
	.3125	5/16	27.59	8.11	59.5	17.0	2.71
	.2500	1/4	22.42	6.59	49.4	14.1	2.74
	.1875	3/16	17.08	5.02	38.5	11.0	2.77
6 x 6	.5000	1/2	35.24	10.4	50.5	16.8	2.21
	.3750	3/8	27.48	8.08	41.6	13.9	2.27
	.3125	5/16	23.34	6.86	36.3	12.1	2.30
	.2500	1/4	19.02	5.59	30.3	10.1	2.33
	.1875	3/16	14.53	4.27	23.8	7.93	2.36
5 x 5	.5000	1/2	28.43	8.36	27.0	10.8	1.80
	.3750	3/8	22.37	6.58	22.8	9.11	1.86
	.3125	5/16	19.08	5.61	20.1	8.02	1.89
	.2500	1/4	15.62	4.59	16.9	6.78	1.92
	.1875	3/16	11.97	3.52	13.4	5.36	1.95

* Outside dimensions across flat sides.

** Properties are based upon a nominal outside corner radius equal to two times the wall thickness.

TABLE A-11. (*Continued*)

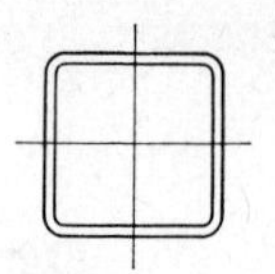

Square

DIMENSIONS				PROPERTIES**			
Nominal* Size	Wall Thickness		Weight per Foot	Area	I	S	r
In.	In.		Lb.	In.^{2}	In.^{4}	In.^{3}	In.
4 x 4	.5000	1/2	21.63	6.36	12.3	6.13	1.39
	.3750	3/8	17.27	5.08	10.7	5.35	1.45
	.3125	5/16	14.83	4.36	9.58	4.79	1.48
	.2500	1/4	12.21	3.59	8.22	4.11	1.51
	.1875	3/16	9.42	2.77	6.59	3.30	1.54
3.5 x 3.5	.3125	5/16	12.70	3.73	6.09	3.48	1.28
	.2500	1/4	10.51	3.09	5.29	3.02	1.31
	.1875	3/16	8.15	2.39	4.29	2.45	1.34
3 x 3	.3125	5/16	10.58	3.11	3.58	2.39	1.07
	.2500	1/4	8.81	2.59	3.16	2.10	1.10
	.1875	3/16	6.87	2.02	2.60	1.73	1.13
2.5 x 2.5	.2500	1/4	7.11	2.09	1.69	1.35	.899
	.1875	3/16	5.59	1.64	1.42	1.14	.930
2 x 2	.2500	1/4	5.41	1.59	.766	.766	.694
	.1875	3/16	4.32	1.27	.668	.668	.726

* Outside dimensions across flat sides.
** Properties are based upon a nominal outside corner radius equal to two times the wall thickness.

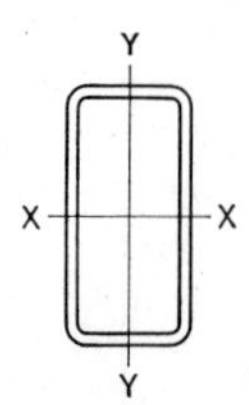

Rectangular

DIMENSIONS				PROPERTIES**						
					X-X AXIS			Y-Y AXIS		
Nominal* Size	Wall Thickness		Weight per Foot	Area	I_x	S_x	r_x	I_y	S_y	r_y
In.	In.		Lb.	In.^{2}	In.^{4}	In.^{3}	In.	In.^{4}	In.^{3}	In.
20 x 12	.5000	1/2	103.30	30.4	1650	165	7.37	750	125.0	4.97
	.3750	3/8	78.52	23.1	1280	128	7.44	583	97.2	5.03
	.3125	5/16	65.87	19.4	1080	108	7.47	495	82.5	5.06
20 x 8	.5000	1/2	89.68	26.4	1270	127	6.94	300	75.1	3.38
	.3750	3/8	68.31	20.1	988	98.8	7.02	236	59.1	3.43
	.3125	5/16	57.36	16.9	838	83.8	7.05	202	50.4	3.46
20 x 4	.5000	1/2	76.07	22.4	889	88.9	6.31	61.6	30.8	1.66
	.3750	3/8	58.10	17.1	699	69.9	6.40	50.3	25.1	1.72
	.3125	5/16	48.86	14.4	596	59.6	6.44	43.7	21.8	1.74

* Outside dimensions across flat sides.
** Properties are based upon a nominal outside corner radius equal to two times the wall thickness.

TABLE A-11. (*Continued*)

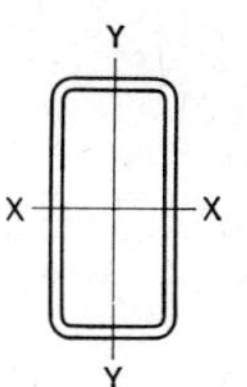

Rectangular

DIMENSIONS				PROPERTIES**						
					X-X AXIS			Y-Y AXIS		
Nominal* Size	Wall Thickness		Weight per Foot	Area	I_x	S_x	r_x	I_y	S_y	r_y
In.	In.		In.	In.^{2}	In.^{4}	In.^{3}	In.	In.^{4}	In.^{3}	In.
18 x 6	.5000	1/2	76.07	22.4	818	90.9	6.05	141	47.2	2.52
	.3750	3/8	58.10	17.1	641	71.3	6.13	113	37.6	2.57
	.3125	5/16	48.86	14.4	546	60.7	6.17	97.0	32.3	2.60
16 x 12	.5000	1/2	89.68	26.4	962	120	6.04	618	103.0	4.84
	.3750	3/8	68.31	20.1	748	93.5	6.11	482	80.3	4.90
	.3125	5/16	57.36	16.9	635	79.4	6.14	409	68.2	4.93
16 x 8	.5000	1/2	76.07	22.4	722	90.2	5.68	244	61.0	3.30
	.3750	3/8	58.10	17.1	565	70.6	5.75	193	48.2	3.36
	.3125	5/16	48.86	14.4	481	60.1	5.79	165	41.2	3.39
16 x 4	.5000	1/2	62.46	18.4	481	60.2	5.12	49.3	24.6	1.64
	.3750	3/8	47.90	14.1	382	47.8	5.21	40.4	20.2	1.69
	.3125	5/16	40.35	11.9	327	40.9	5.25	35.1	17.6	1.72
14 x 10	.5000	1/2	76.07	22.4	608	86.9	5.22	361	72.3	4.02
	.3750	3/8	58.10	17.1	476	68.0	5.28	284	56.8	4.08
	.3125	5/16	48.86	14.4	405	57.9	5.31	242	48.4	4.11
14 x 6	.5000	1/2	62.46	18.4	426	60.8	4.82	111	37.1	2.46
	.3750	3/8	47.90	14.1	337	48.1	4.89	89.1	29.7	2.52
	.3125	5/16	40.35	11.9	288	41.2	4.93	76.7	25.6	2.54
	.2500	1/4	32.63	9.59	237	33.8	4.97	63.4	21.1	2.57
14 x 4	.5000	1/2	55.66	16.4	335	47.8	4.52	43.1	21.5	1.62
	.3750	3/8	42.79	12.6	267	38.2	4.61	35.4	17.7	1.68
	.3125	5/16	36.10	10.6	230	32.8	4.65	30.9	15.4	1.71
	.2500	1/4	29.23	8.59	189	27.0	4.69	25.8	12.9	1.73
12 x 8	.6250	5/8	76.33	22.4	418	69.7	4.32	221	55.3	3.14
	.5000	1/2	62.46	18.4	353	58.9	4.39	188	46.9	3.20
	.3750	3/8	47.90	14.1	279	46.5	4.45	149	37.3	3.26
	.3125	5/16	40.35	11.9	239	39.8	4.49	128	32.0	3.28
	.2500	1/4	32.63	9.59	196	32.6	4.52	105	26.3	3.31
12 x 6	.5000	1/2	55.66	16.4	287	47.8	4.19	96.0	32.0	2.42
	.3750	3/8	42.79	12.6	228	38.1	4.26	77.2	25.7	2.48
	.3125	5/16	36.10	10.6	196	32.6	4.30	66.6	22.2	2.51
	.2500	1/4	29.23	8.59	161	26.9	4.33	55.2	18.4	2.53
	.1875	3/16	22.18	6.52	124	20.7	4.37	42.8	14.3	2.56
12 x 4	.5000	1/2	48.85	14.4	221	36.8	3.92	36.9	18.5	1.60
	.3750	3/8	37.69	11.1	178	29.6	4.01	30.5	15.2	1.66
	.3125	5/16	31.84	9.36	153	25.5	4.05	26.6	13.3	1.69
	.2500	1/4	25.82	7.59	127	21.1	4.09	22.3	11.1	1.71
	.1875	3/16	19.63	5.77	98.2	16.4	4.13	17.5	8.75	1.74
12 x 2	.2500	1/4	22.42	6.59	92.2	15.4	3.74	4.62	4.62	.837
	.1875	3/16	17.08	5.02	72.0	12.0	3.79	3.76	3.76	.865

* Outside dimensions across flat sides.

** Properties are based upon a nominal outside corner radius equal to two times the wall thickness.

TABLE A-11. *(Continued)*

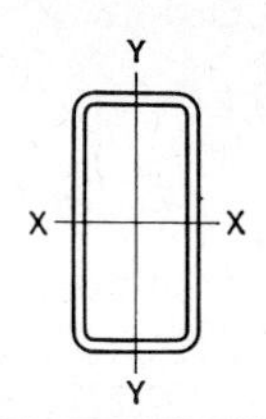

Rectangular

DIMENSIONS				PROPERTIES**						
				X-X AXIS			Y-Y AXIS			
Nominal* Size	Wall Thickness		Weight per Foot	Area	I_x	S_x	r_x	I_y	S_y	r_y
In.	In.		Lb.	In.^{2}	In.^{4}	In.^{3}	In.	In.^{4}	In.^{3}	In.
10 x 6	.5000	1/2	48.85	14.4	181	36.2	3.55	80.8	26.9	2.37
	.3750	3/8	37.69	11.1	145	29.0	3.62	65.4	21.8	2.43
	.3125	5/16	31.84	9.36	125	25.0	3.65	56.5	18.8	2.46
	.2500	1/4	25.82	7.59	103	20.6	3.69	46.9	15.6	2.49
	.1875	3/16	19.63	5.77	79.8	16.0	3.72	36.5	12.2	2.51
10 x 4	.5000	1/2	42.05	12.4	136	27.1	3.31	30.8	15.4	1.58
	.3750	3/8	32.58	9.58	110	22.0	3.39	25.5	12.8	1.63
	.3125	5/16	27.59	8.11	95.0	19.1	3.43	22.4	11.2	1.66
	.2500	1/4	22.42	6.59	79.3	15.9	3.47	18.8	9.39	1.69
	.1875	3/16	17.08	5.02	61.7	12.3	3.51	14.8	7.39	1.72
10 x 2	.3750	3/8	27.48	8.08	75.4	15.1	3.06	4.85	4.85	.775
	.3125	5/16	23.34	6.86	66.1	13.2	3.10	4.42	4.42	.802
	.2500	1/4	19.02	5.59	55.5	11.1	3.15	3.85	3.85	.830
	.1875	3/16	14.53	4.27	43.7	8.74	3.20	3.14	3.14	.858
8 x 6	.5000	1/2	42.05	12.4	103	25.8	2.89	65.7	21.9	2.31
	.3750	3/8	32.58	9.58	83.7	20.9	2.96	53.5	17.8	2.36
	.3125	5/16	27.59	8.11	72.4	18.1	2.99	46.4	15.5	2.39
	.2500	1/4	22.42	6.59	60.1	15.0	3.02	38.6	12.9	2.42
	.1875	3/16	17.08	5.02	46.8	11.7	3.05	30.1	10.0	2.45
8 x 4	.5000	1/2	35.24	10.4	75.1	18.8	2.69	24.6	12.3	1.54
	.3750	3/8	27.48	8.08	61.9	15.5	2.77	20.6	10.3	1.60
	.3125	5/16	23.34	6.86	53.9	13.5	2.80	18.1	9.05	1.62
	.2500	1/4	19.02	5.59	45.1	11.3	2.84	15.3	7.63	1.65
	.1875	3/16	14.53	4.27	35.3	8.83	2.88	12.0	6.02	1.68
8 x 3	.3750	3/8	24.93	7.33	51.0	12.7	2.64	10.4	6.92	1.19
	.3125	5/16	21.21	6.23	44.7	11.2	2.68	9.25	6.16	1.22
	.2500	1/4	17.32	5.09	37.6	9.40	2.72	7.90	5.26	1.25
	.1875	3/16	13.25	3.89	29.6	7.40	2.76	6.31	4.21	1.27
8 x 2	.3750	3/8	22.37	6.58	40.1	10.0	2.47	3.85	3.85	.765
	.3125	5/16	19.08	5.61	35.5	8.87	2.51	3.52	3.52	.792
	.2500	1/4	15.62	4.59	30.1	7.52	2.56	3.08	3.08	.819
	.1875	3/16	11.97	3.52	23.9	5.97	2.60	2.52	2.52	.847
7 x 5	.5000	1/2	35.24	10.4	63.5	18.1	2.48	37.2	14.9	1.90
	.3750	3/8	27.48	8.08	52.2	14.9	2.54	30.8	12.3	1.95
	.3125	5/16	23.34	6.86	45.5	13.0	2.58	26.9	10.8	1.98
	.2500	1/4	19.02	5.59	38.0	10.9	2.61	22.6	9.04	2.01
	.1875	3/16	14.53	4.27	29.8	8.50	2.64	17.7	7.10	2.04
7 x 4	.3750	3/8	24.93	7.33	44.0	12.6	2.45	18.1	9.06	1.57
	.3125	5/16	21.21	6.23	38.5	11.0	2.49	16.0	7.98	1.60
	.2500	1/4	17.32	5.09	32.3	9.23	2.52	13.5	6.75	1.63
	.1875	3/16	13.25	3.89	25.4	7.26	2.55	10.7	5.34	1.66

* Outside dimensions across flat sides.

** Properties are based upon a nominal outside corner radius equal to two times the wall thickness.

TABLE A-11. (*Continued*)

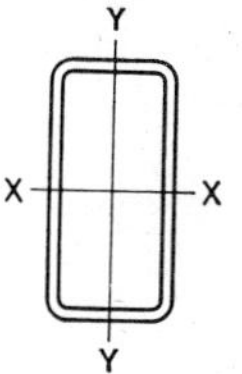

Rectangular

DIMENSIONS				PROPERTIES**						
					X-X AXIS			Y-Y AXIS		
Nominal* Size	Wall Thickness		Weight per Foot	Area	I_x	S_x	r_x	I_y	S_y	r_y
In.	In.		In.	In.^{2}	In.^{4}	In.^{3}	In.	In.^{4}	In.^{3}	In.
7 x 3	.3750	3/8	22.37	6.58	35.7	10.2	2.33	9.08	6.05	1.18
	.3125	5/16	19.08	5.61	31.5	9.00	2.37	8.11	5.41	1.20
	.2500	1/4	15.62	4.59	26.6	7.61	2.41	6.95	4.63	1.23
	.1875	3/16	11.97	3.52	21.1	6.02	2.45	5.57	3.71	1.26
6 x 4	.5000	1/2	28.43	8.36	35.3	11.8	2.06	18.4	9.21	1.48
	.3750	3/8	22.37	6.58	29.7	9.90	2.13	15.6	7.82	1.54
	.3125	5/16	19.08	5.61	26.2	8.72	2.16	13.8	6.92	1.57
	.2500	1/4	15.62	4.59	22.1	7.36	2.19	11.7	5.87	1.60
	.1875	3/16	11.97	3.52	17.4	5.81	2.23	9.32	4.66	1.63
6 x 3	.3750	3/8	19.82	5.83	23.8	7.92	2.02	7.78	5.19	1.16
	.3125	5/16	16.96	4.98	21.1	7.03	2.06	6.98	4.65	1.18
	.2500	1/4	13.91	4.09	17.9	5.98	2.09	6.00	4.00	1.21
	.1875	3/16	10.70	3.14	14.3	4.76	2.13	4.83	3.22	1.24
6 x 2	.3750	3/8	17.27	5.08	17.8	5.94	1.87	2.84	2.84	.748
	.3125	5/16	14.83	4.36	16.0	5.34	1.92	2.62	2.62	.775
	.2500	1/4	12.21	3.59	13.8	4.60	1.96	2.31	2.31	.802
	.1875	3/16	9.42	2.77	11.1	3.70	2.00	1.90	1.90	.829
5 x 4	.3750	3/8	19.82	5.83	18.7	7.50	1.79	13.2	6.58	1.50
	.3125	5/16	16.96	4.98	16.6	6.65	1.83	11.7	5.85	1.53
	.2500	1/4	13.91	4.09	14.1	5.65	1.86	9.98	4.99	1.56
	.1875	3/16	10.70	3.14	11.2	4.49	1.89	7.96	3.98	1.59
5 x 3	.5000	1/2	21.63	6.36	16.9	6.75	1.63	7.33	4.88	1.07
	.3750	3/8	17.27	5.08	14.7	5.89	1.70	6.48	4.32	1.13
	.3125	5/16	14.83	4.36	13.2	5.27	1.74	5.85	3.90	1.16
	.2500	1/4	12.21	3.59	11.3	4.52	1.77	5.05	3.37	1.19
	.1875	3/16	9.42	2.77	9.06	3.62	1.81	4.08	2.72	1.21
5 x 2	.3125	5/16	12.70	3.73	9.74	3.90	1.62	2.16	2.16	.76
	.2500	1/4	10.51	3.09	8.48	3.39	1.66	1.92	1.92	.78
	.1875	3/16	8.15	2.39	6.89	2.75	1.70	1.60	1.60	.81
4 x 3	.3125	5/16	12.70	3.73	7.45	3.72	1.41	4.71	3.14	1.12
	.2500	1/4	10.51	3.09	6.45	3.23	1.45	4.10	2.74	1.15
	.1875	3/16	8.15	2.39	5.23	2.62	1.48	3.34	2.23	1.18
4 x 2	.3125	5/16	10.58	3.11	5.32	2.66	1.31	1.71	1.71	.743
	.2500	1/4	8.81	2.59	4.69	2.35	1.35	1.54	1.54	.770
	.1875	3/16	6.87	2.02	3.87	1.93	1.38	1.29	1.29	.798
3 x 2	.2500	1/4	7.11	2.09	2.21	1.47	1.03	1.15	1.15	.742
	.1875	3/16	5.59	1.64	1.86	1.24	1.06	.977	.977	.771

* Outside dimensions across flat sides.

** Properties are based upon a nominal outside corner radius equal to two times the wall thickness.

B

Section Modulus and Moment Resistance for Selected Rolled Structural Shapes

The following tables list the rolled structural shapes that are commonly used as beams. Shapes are listed in sequence in accordance with the value of the section modulus about their principal axes (S_x). Moment values are the total resisting moment of the section with a steel yield strength of 36 ksi [250 MPa] and are based on a maximum bending stress of 24 ksi [165 MPa] for compact sections, 22 ksi [150 MPa] for noncompact sections.

Beams whose designations appear in boldface type are *least weight members*. These members have a section modulus value that is higher then other members with greater weight. When bending resistance alone is the critical design factor, these members constitute highly efficient shapes.

To assist in the identification of concern for buckling the limiting lateral unsupported length limits of L_c and L_u are given for

TABLE B-1. Section Modulus and Moment of Resistance for Selected Rolled Structural Shapes

S_x	Shape	F_y = 36 ksi		
		L_c	L_u	M_R
In.^{3}		Ft.	Ft.	Kip-ft.
1110	**W 36x300**	**17.6**	**35.3**	**2220**
1030	**W 36x280**	**17.5**	**33.1**	**2060**
953	**W 36x260**	**17.5**	**30.5**	**1910**
895	**W 36x245**	**17.4**	**28.6**	**1790**
837	**W 36x230**	**17.4**	**26.8**	**1670**
829	W 33x241	16.7	30.1	1660
757	**W 33x221**	**16.7**	**27.6**	**1510**
719	**W 36x210**	**12.9**	**20.9**	**1440**
684	**W 33x201**	**16.6**	**24.9**	**1370**
664	**W 36x194**	**12.8**	**19.4**	**1330**
663	W 30x211	15.9	29.7	1330
623	**W 36x182**	**12.7**	**18.2**	**1250**
598	W 30x191	15.9	26.9	1200
580	**W 36x170**	**12.7**	**17.0**	**1160**
542	**W 36x160**	**12.7**	**15.7**	**1080**
539	W 30x173	15.8	24.2	1080
504	**W 36x150**	**12.6**	**14.6**	**1010**
502	W 27x178	14.9	27.9	1000
487	W 33x152	12.2	16.9	974
455	W 27x161	14.8	25.4	910
448	**W 33x141**	**12.2**	**15.4**	**896**
439	**W 36x135**	**12.3**	**13.0**	**878**
414	W 24x162	13.7	29.3	828
411	W 27x146	14.7	23.0	822
406	**W 33x130**	**12.1**	**13.8**	**812**
380	W 30x132	11.1	16.1	760
371	W 24x146	13.6	26.3	742
359	**W 33x118**	**12.0**	**12.6**	**718**
355	W 30x124	11.1	15.0	710
329	**W 30x116**	**11.1**	**13.8**	**658**
329	W 24x131	13.6	23.4	658
329	W 21x147	13.2	30.3	658
299	**W 30x108**	**11.1**	**12.3**	**598**
299	W 27x114	10.6	15.9	598
295	W 21x132	13.1	27.2	590
291	W 24x117	13.5	20.8	582
273	W 21x122	13.1	25.4	546
269	**W 30x 99**	**10.9**	**11.4**	**538**
267	W 27x102	10.6	14.2	534
258	W 24x104	13.5	18.4	516
249	W 21x111	13.0	23.3	498
243	**W 27x 94**	**10.5**	**12.8**	**486**
231	W 18x119	11.9	29.1	462
227	W 21x101	13.0	21.3	454
222	**W 24x 94**	**9.6**	**15.1**	**444**
213	**W 27x 84**	**10.5**	**11.0**	**426**
204	W 18x106	11.8	26.0	408
196	**W 24x 84**	**9.5**	**13.3**	**392**
192	W 21x 93	8.9	16.8	384
190	W 14x120	15.5	44.1	380
188	W 18x 97	11.8	24.1	376
176	**W 24x 76**	**9.5**	**11.8**	**352**
175	W 16x100	11.0	28.1	350
173	W 14x109	15.4	40.6	346
171	W 21x 83	8.8	15.1	342
166	W 18x 86	11.7	21.5	332
157	W14x 99	15.4	37.0	314
155	W 16x 89	10.9	25.0	310
154	**W 24x 68**	**9.5**	**10.2**	**308**
151	W 21x 73	8.8	13.4	302
146	W 18x 76	11.6	19.1	292
143	W 14x 90	15.3	34.0	286
140	**W 21x 68**	**8.7**	**12.4**	**280**
134	W 16x 77	10.9	21.9	268
131	**W 24x 62**	**7.4**	**8.1**	**262**
127	**W 21x 62**	**8.7**	**11.2**	**254**
127	W 18x 71	8.1	15.5	254
123	W 14x 82	10.7	28.1	246
118	W 12x 87	12.8	36.2	236
117	W 18x 65	8.0	14.4	234
117	W 16x 67	10.8	19.3	234
114	**W 24x 55**	**7.0**	**7.5**	**228**
112	W 14x 74	10.6	25.9	224
111	W 21x 57	6.9	9.4	222
108	W 18x 60	8.0	13.3	216
107	W 12x 79	12.8	33.3	214
103	W 14x 68	10.6	23.9	206
98.3	**W 18x 55**	**7.9**	**12.1**	**197**
97.4	W 12x 72	12.7	30.5	195

TABLE B-1. (*Continued*)

S_x	Shape	F_y = 36 ksi		
		L_c	L_u	M_R
In.^{3}		Ft.	Ft.	Kip-ft.
94.5	**W 21x50**	**6.9**	**7.8**	**189**
92.2	W 16x57	7.5	14.3	184
92.2	W 14x61	10.6	21.5	184
88.9	**W 18x50**	**7.9**	**11.0**	**178**
87.9	W 12x65	12.7	27.7	176
81.6	**W 21x44**	**6.6**	**7.0**	**163**
81.0	W 16x50	7.5	12.7	162
78.8	W 18x46	6.4	9.4	158
78.0	W 12x58	10.6	24.4	156
77.8	W 14x53	8.5	17.7	156
72.7	W 16x45	7.4	11.4	145
70.6	W 12x53	10.6	22.0	141
70.3	W 14x48	8.5	16.0	141
68.4	**W 18x40**	**6.3**	**8.2**	**137**
66.7	W 10x60	10.6	31.1	133
64.7	**W 16x40**	**7.4**	**10.2**	**129**
64.7	W 12x50	8.5	19.6	129
62.7	W 14x43	8.4	14.4	125
60.0	W 10x54	10.6	28.2	120
58.1	W 12x45	8.5	17.7	116
57.6	**W 18x35**	**6.3**	**6.7**	**115**
56.5	W 16x36	7.4	8.8	113
54.6	W 14x38	7.1	11.5	109
54.6	W 10x49	10.6	26.0	109
51.9	W 12x40	8.4	16.0	104
49.1	W 10x45	8.5	22.8	98
48.6	**W 14x34**	**7.1**	**10.2**	**97**
47.2	**W 16x31**	**5.8**	**7.1**	**94**
45.6	W 12x35	6.9	12.6	91
42.1	W 10x39	8.4	19.8	84
42.0	**W 14x30**	**7.1**	**8.7**	**84**
38.6	**W 12x30**	**6.9**	**10.8**	**77**
38.4	**W 16x26**	**5.6**	**6.0**	**77**
35.3	**W 14x26**	**5.3**	**7.0**	**71**
35.0	W 10x33	8.4	16.5	70
33.4	**W 12x26**	**6.9**	**9.4**	**67**
32.4	W 10x30	6.1	13.1	65
31.2	W 8x35	8.5	22.6	62

S_x	Shape	F_y = 36 ksi		
		L_c	L_u	M_R
In.^{3}		Ft.	Ft.	Kip-ft.
29.0	**W 14x22**	**5.3**	**5.6**	**58**
27.9	W 10x26	6.1	11.4	56
27.5	W 8x31	8.4	20.1	55
25.4	**W 12x22**	**4.3**	**6.4**	**51**
24.3	W 8x28	6.9	17.5	49
23.2	**W 10x22**	**6.1**	**9.4**	**46**
21.3	**W 12x19**	**4.2**	**5.3**	**43**
21.1	**M 14x18**	**3.6**	**4.0**	**42**
20.9	W 8x24	6.9	15.2	42
18.8	W 10x19	4.2	7.2	38
18.2	W 8x21	5.6	11.8	36
17.1	**W 12x16**	**4.1**	**4.3**	**34**
16.7	W 6x25	6.4	20.0	33
16.2	W 10x17	4.2	6.1	32
15.2	W 8x18	5.5	9.9	30
14.9	**W 12x14**	**3.5**	**4.2**	**30**
13.8	W 10x15	4.2	5.0	28
13.4	W 6x20	6.4	16.4	27
13.0	M 6x20	6.3	17.4	26
12.0	**M 12x11.8**	**2.7**	**3.0**	**24**
11.8	W 8x15	4.2	7.2	24
10.9	W 10x12	3.9	4.3	22
10.2	W 6x16	4.3	12.0	20
10.2	W 5x19	5.3	19.5	20
9.91	W 8x13	4.2	5.9	20
9.72	W 6x15	6.3	12.0	19
9.63	M 5x18.9	5.3	19.3	19
8.51	W 5x16	5.3	16.7	17
7.81	**W 8x10**	**4.2**	**4.7**	**16**
7.76	**M 10x 9**	**2.6**	**2.7**	**16**
7.31	W 6x12	4.2	8.6	15
5.56	**W 6x 9**	**4.2**	**6.7**	**11**
5.46	W 4x13	4.3	15.6	11
5.24	M 4x13	4.2	16.9	10
4.62	**M 8x 6.5**	**2.4**	**2.5**	**9**
2.40	**M 6x 4.4**	**1.9**	**2.4**	**5**

each shape. If the actual unsupported lengths exceed these values, the graphs in Appendix C should be used rather than the tables in this section.

These data have been reproduced from the *Manual of Steel Construction,* 8th ed. (Ref. 1) with the permission of the publishers, the American Institute of Steel Construction.

C

Allowable Bending Moment for Beams with Various Unbraced Lengths

These charts may be used to determine the allowable bending moment for shapes used as beams. Graphs apply to beams of steel with a yield stress of 36 ksi [250 MPa]. Dashed lines on the graphs indicate that the shape is heavier than one that has a higher bending resistance. To find the most efficient shapes proceed on the chart upward and toward the right until the first solid line graph is encountered.

These charts have been reproduced and adapted from the *Manual of Steel Construction,* 8th ed. (Ref. 1), with permission of the publishers, American Institute of Steel Construction.

Example. Find the lightest shape that may be used to resist a moment of 325 kip-ft [441 kN-m] if the unbraced length is 10 ft [3.05 m].

Solution: Find the graph page that includes the value for the given moment. Proceed across the horizontal line for the moment of 325 kip-ft until you come to the vertical line for an unbraced length of 10 ft. The first shape whose graph is above and to the right of this point is a W 16 × 96. However, the line for this shape is dashed to indicate that there is a lighter shape with greater moment of resistance. Proceed upward and to the right and you will encounter the solid line for the shape W 24 × 76, which is the lightest beam for this situation.

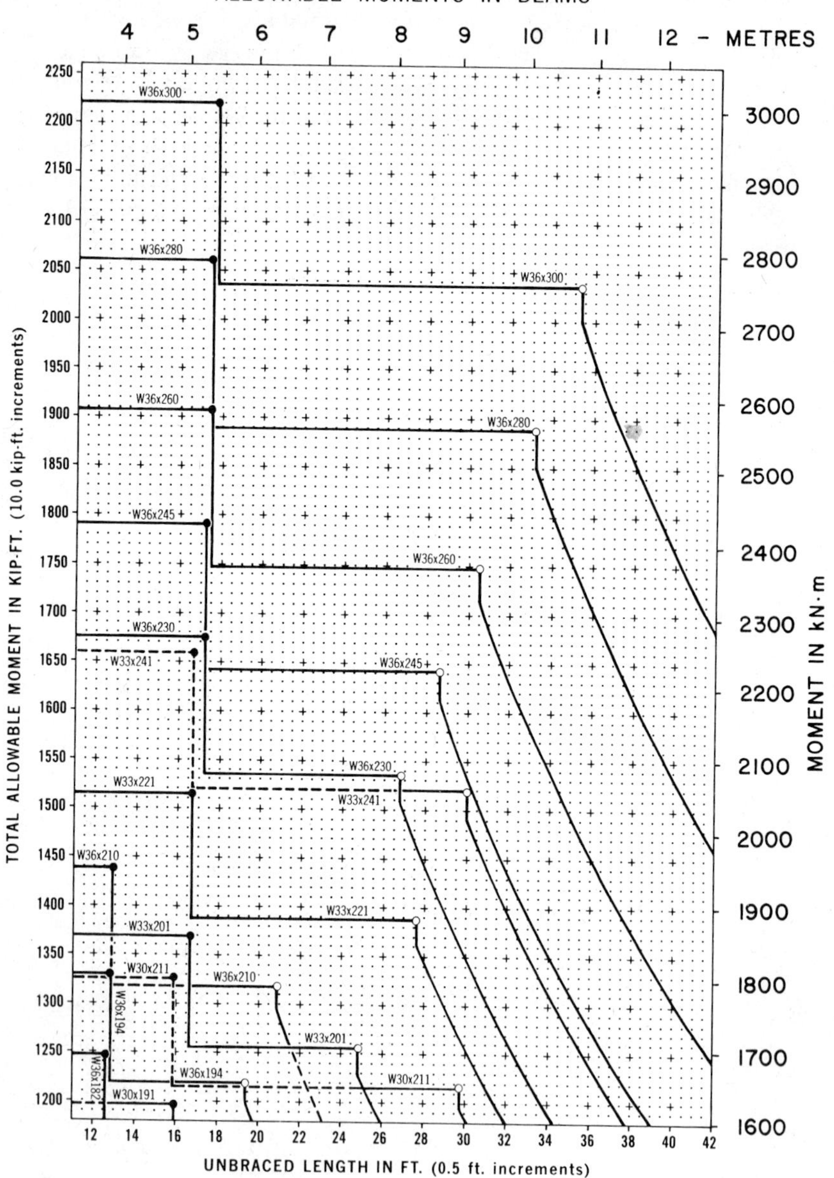

ALLOWABLE MOMENTS IN BEAMS
4
5
6
7
8
9
10
11
12 - METRES
TOTAL ALLOWABLE MOMENT IN KIP-FT. (10.0 kip-ft. increments)
MOMENT IN kN·m
UNBRACED LENGTH IN FT. (0.5 ft. increments)
W36x300
W36x280
W36x260
W36x245
W36x230
W33x241
W33x221
W36x210
W33x201
W30x211
W36x194
W36x182
W30x191
2250
2200
2150
2100
2050
2000
1950
1900
1850
1800
1750
1700
1650
1600
1550
1500
1450
1400
1350
1300
1250
1200
3000
2900
2800
2700
2600
2500
2400
2300
2200
2100
2000
1900
1800
1700
1600
12
14
16
18
20
22
24
26
28
30
32
34
36
38
40
42

ALLOWABLE MOMENTS IN BEAMS

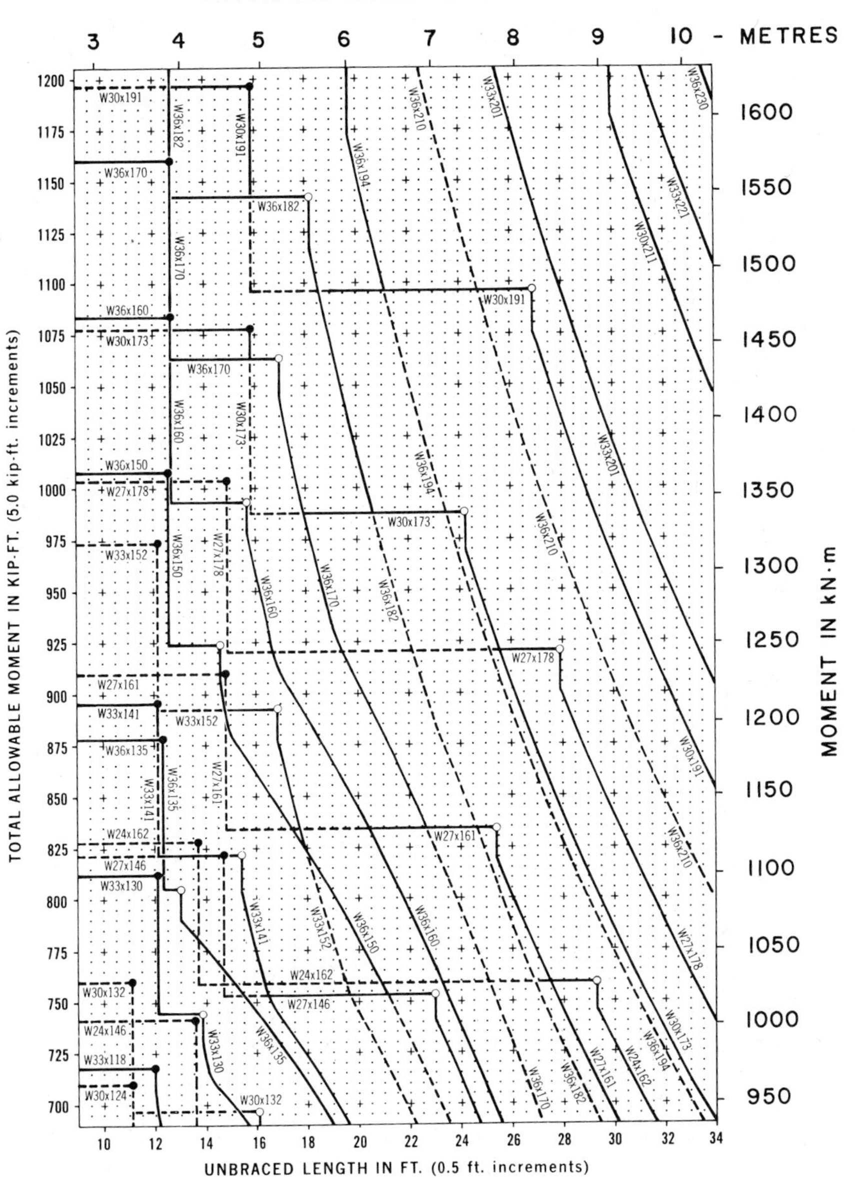

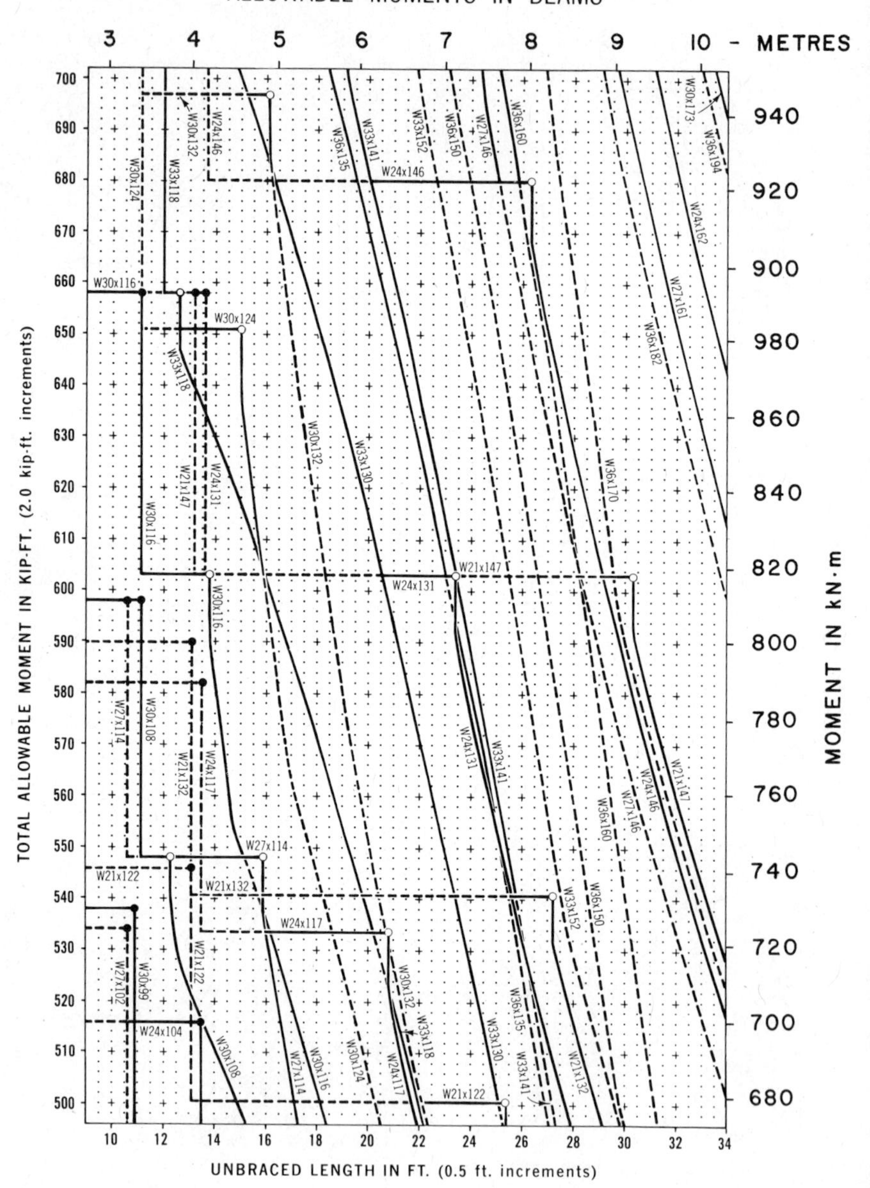
ALLOWABLE MOMENTS IN BEAMS
METRES
TOTAL ALLOWABLE MOMENT IN KIP-FT. (2.0 kip-ft. increments)
MOMENT IN kN·m
UNBRACED LENGTH IN FT. (0.5 ft. increments)

ALLOWABLE MOMENTS IN BEAMS

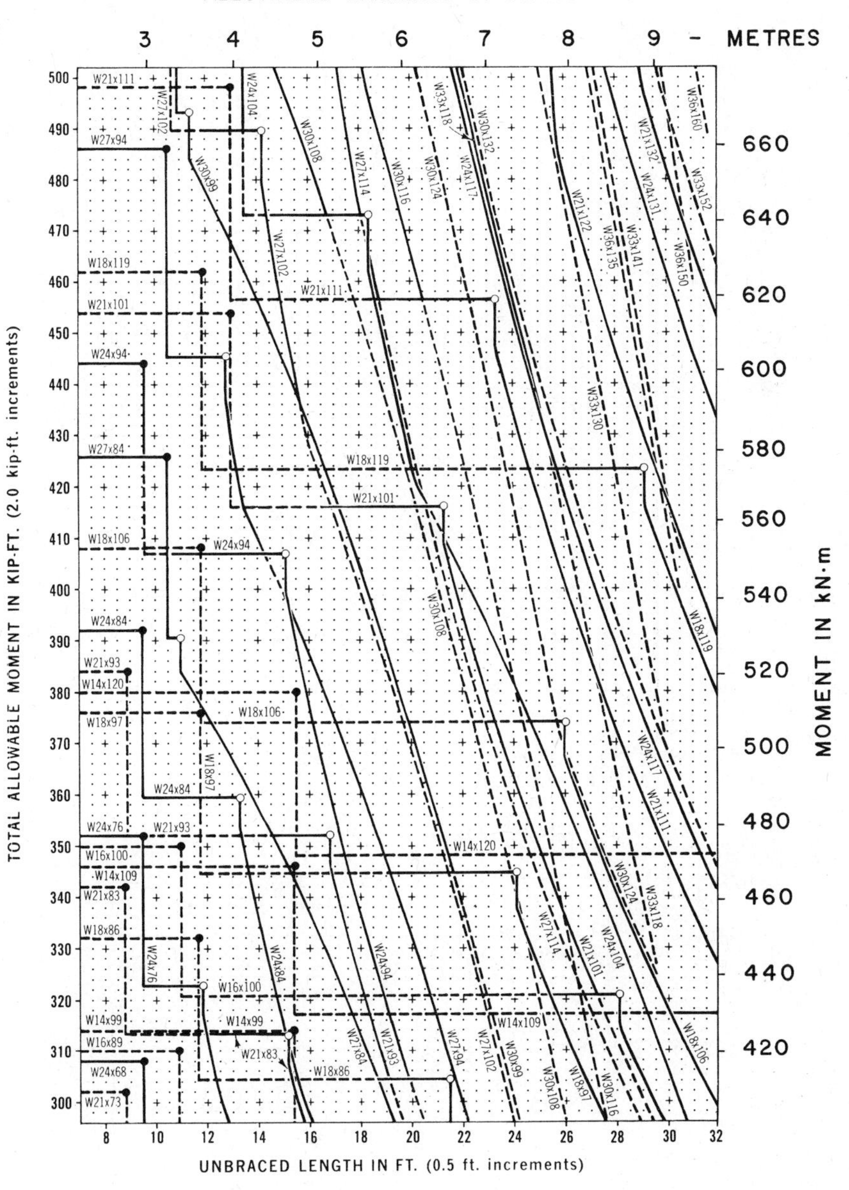

ALLOWABLE MOMENTS IN BEAMS

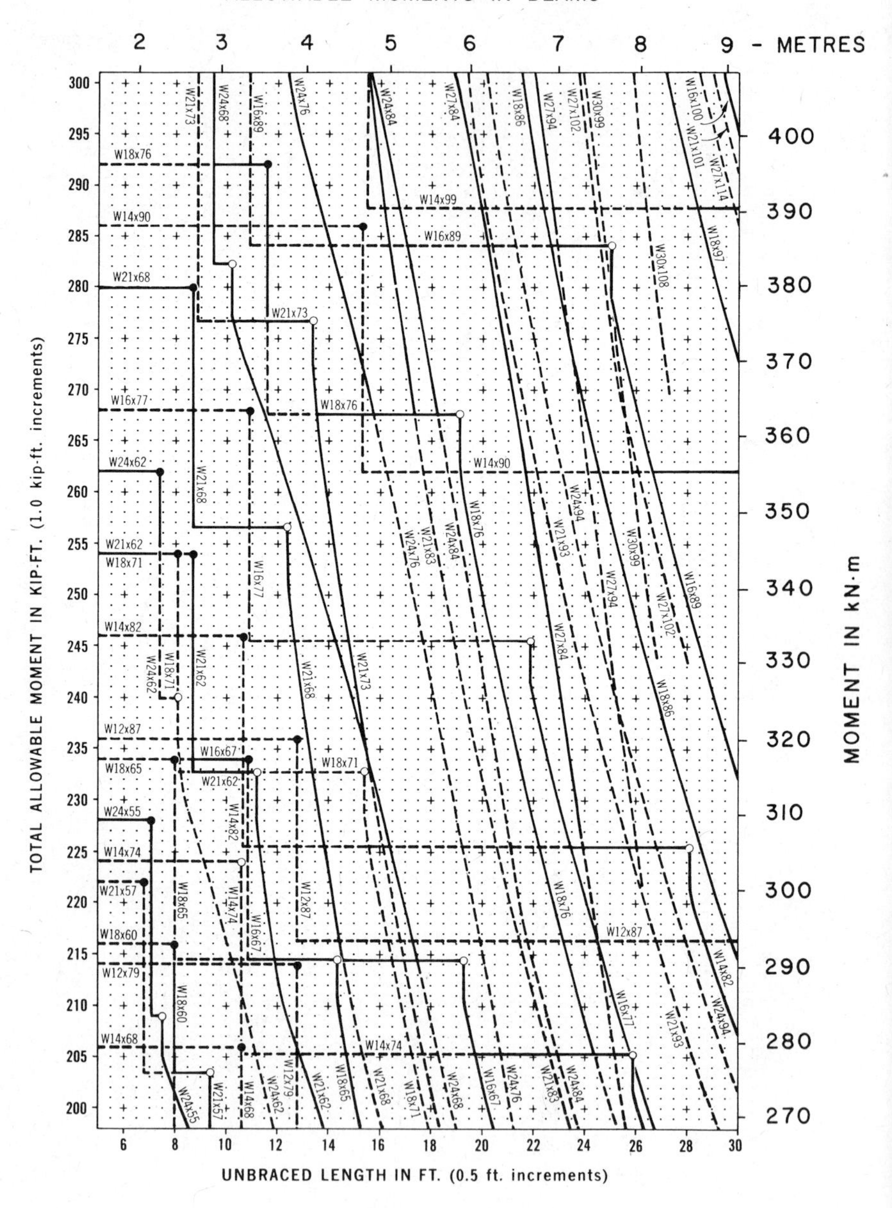

ALLOWABLE MOMENTS IN BEAMS

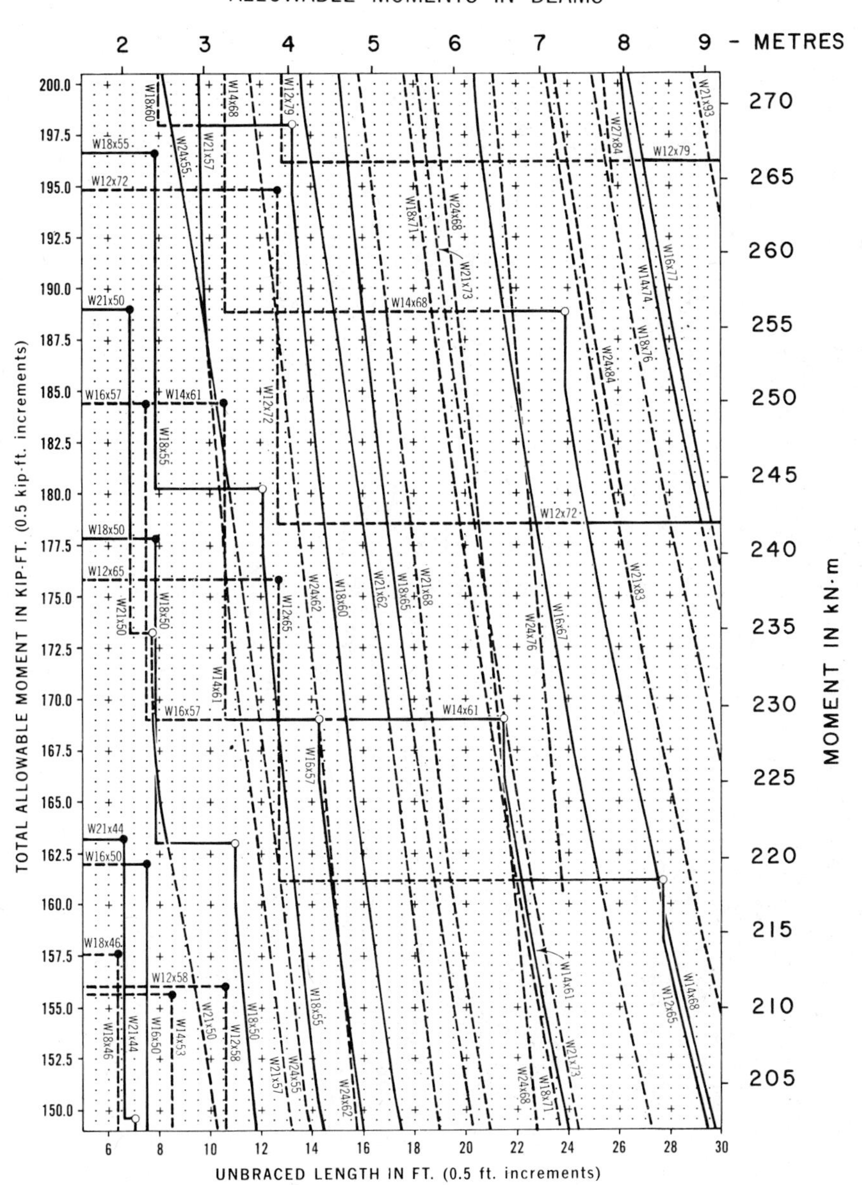

ALLOWABLE MOMENTS IN BEAMS

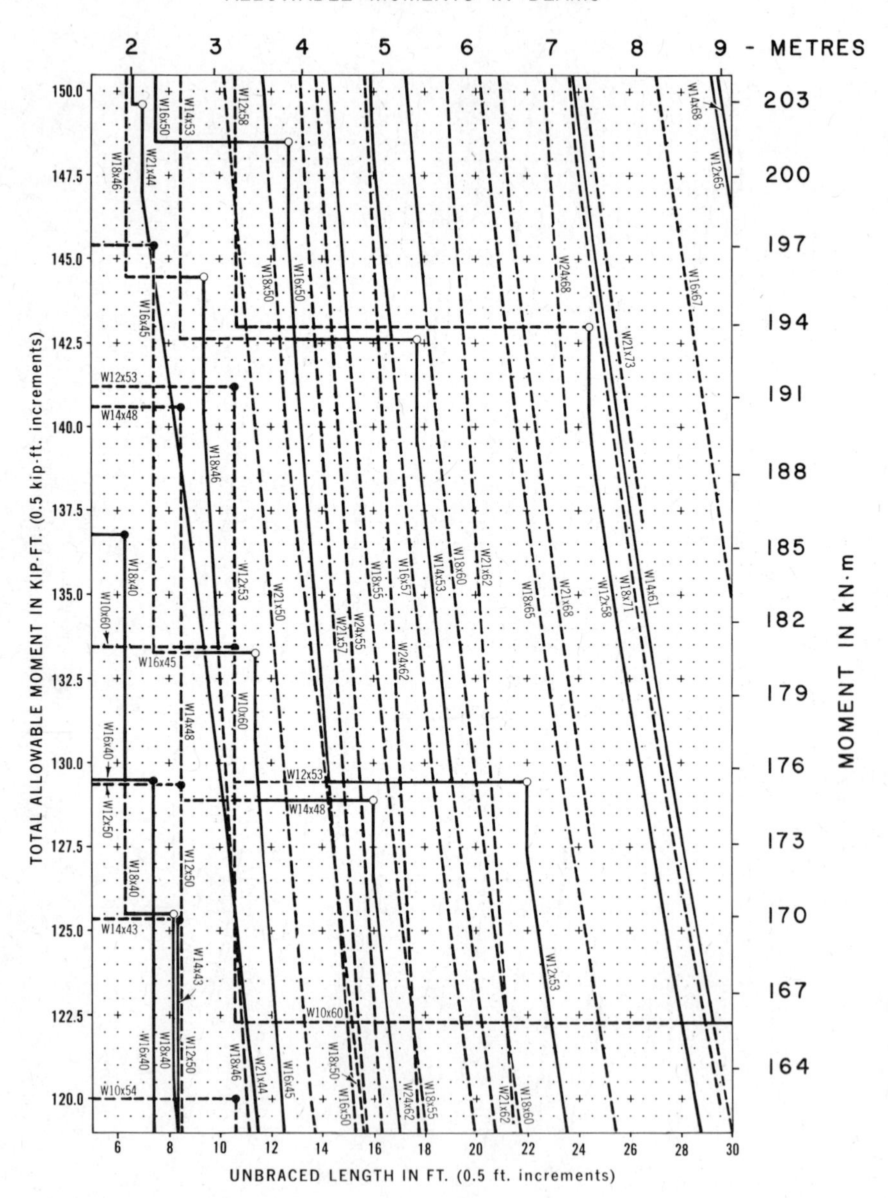

ALLOWABLE MOMENTS IN BEAMS

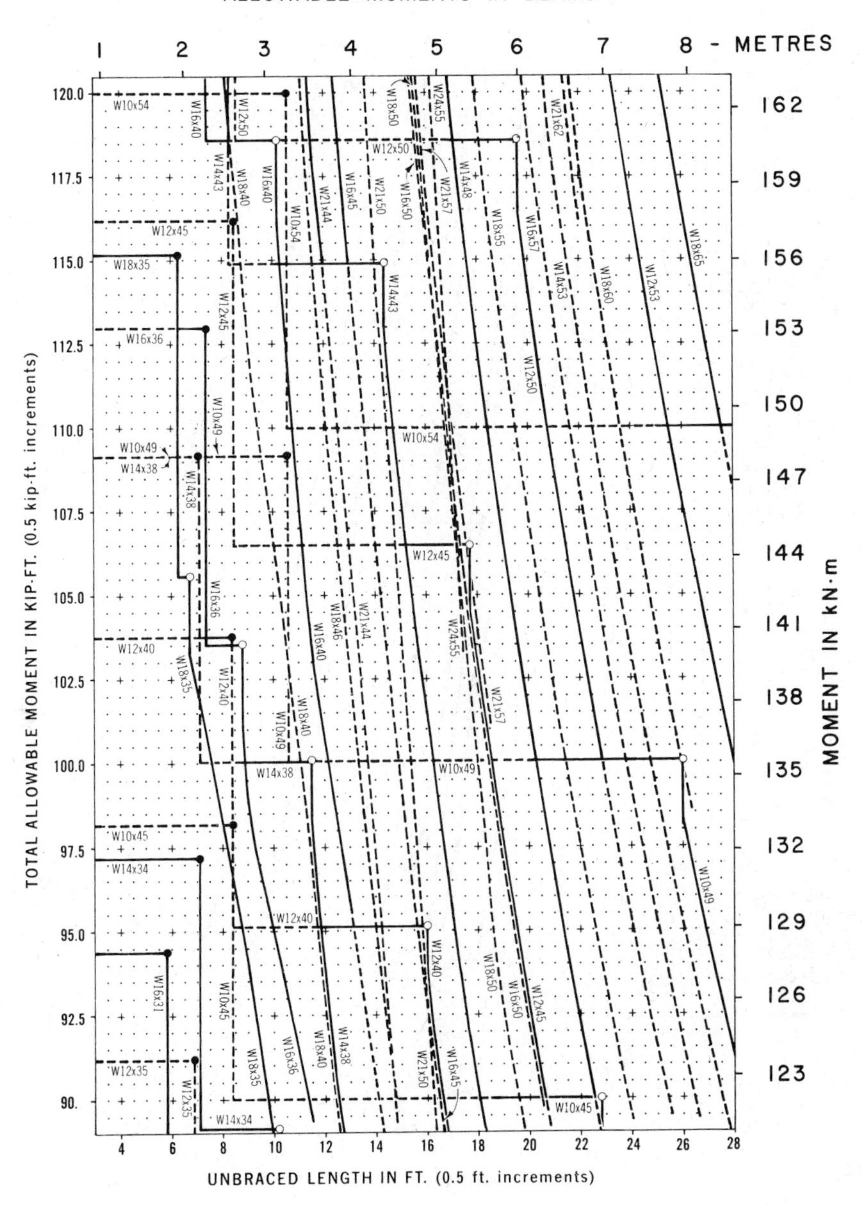

ALLOWABLE MOMENTS IN BEAMS

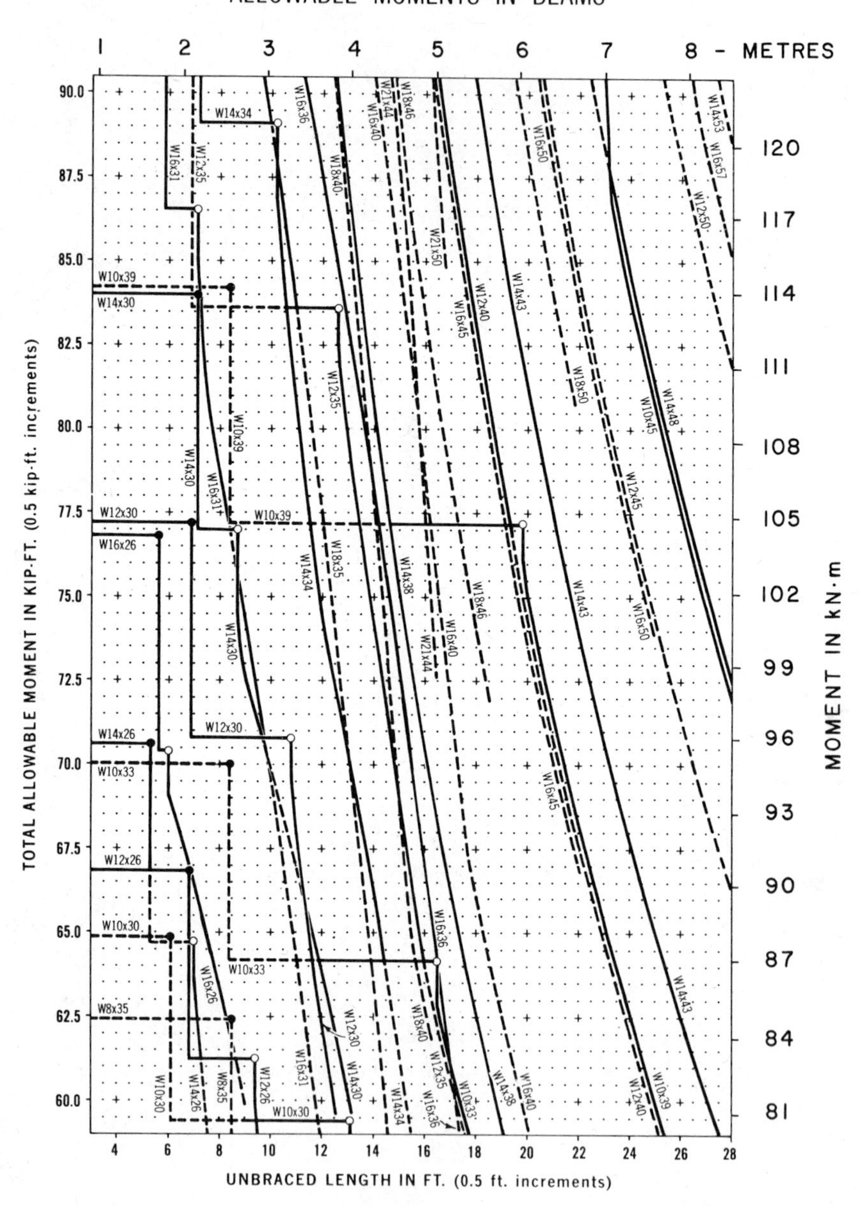

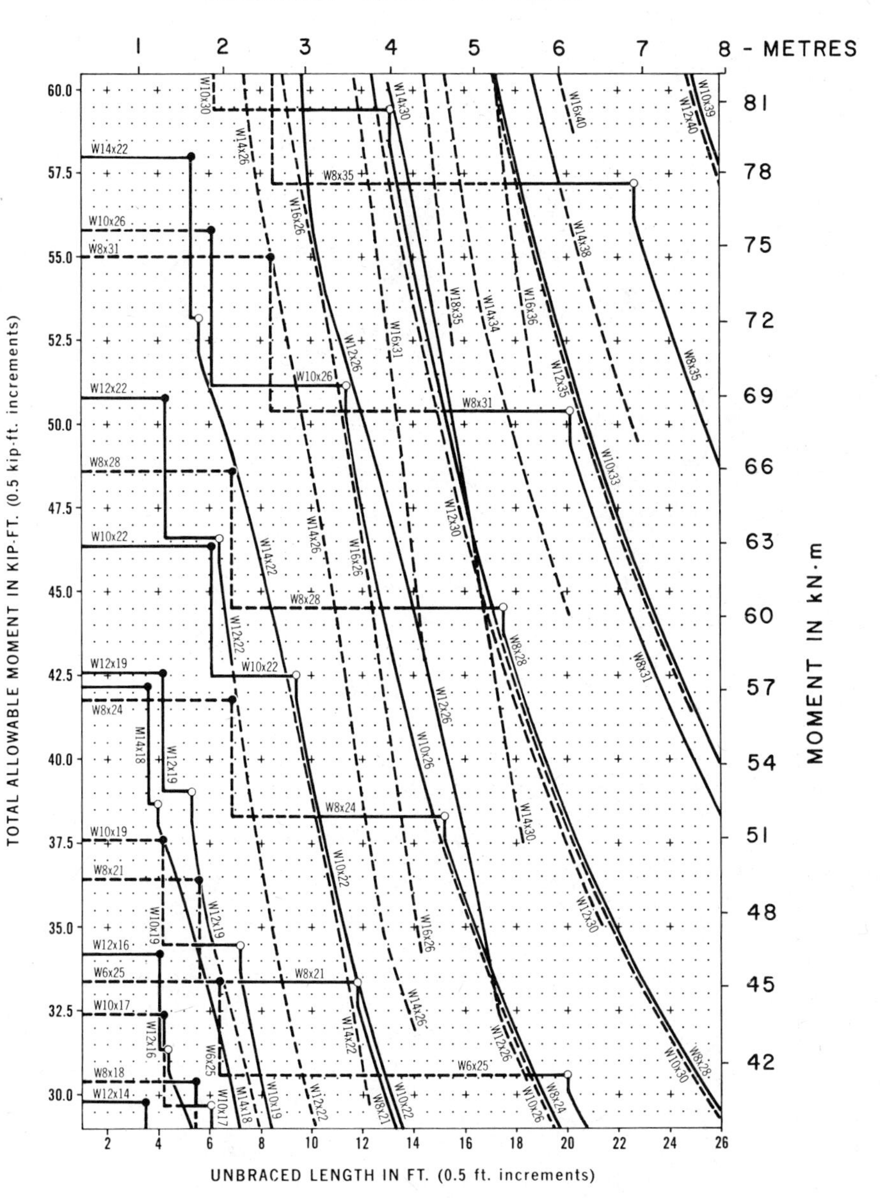
ALLOWABLE MOMENTS IN BEAMS
METRES
TOTAL ALLOWABLE MOMENT IN KIP-FT. (0.5 kip-ft. increments)
UNBRACED LENGTH IN FT. (0.5 ft. increments)
MOMENT IN kN·m

ALLOWABLE MOMENTS IN BEAMS

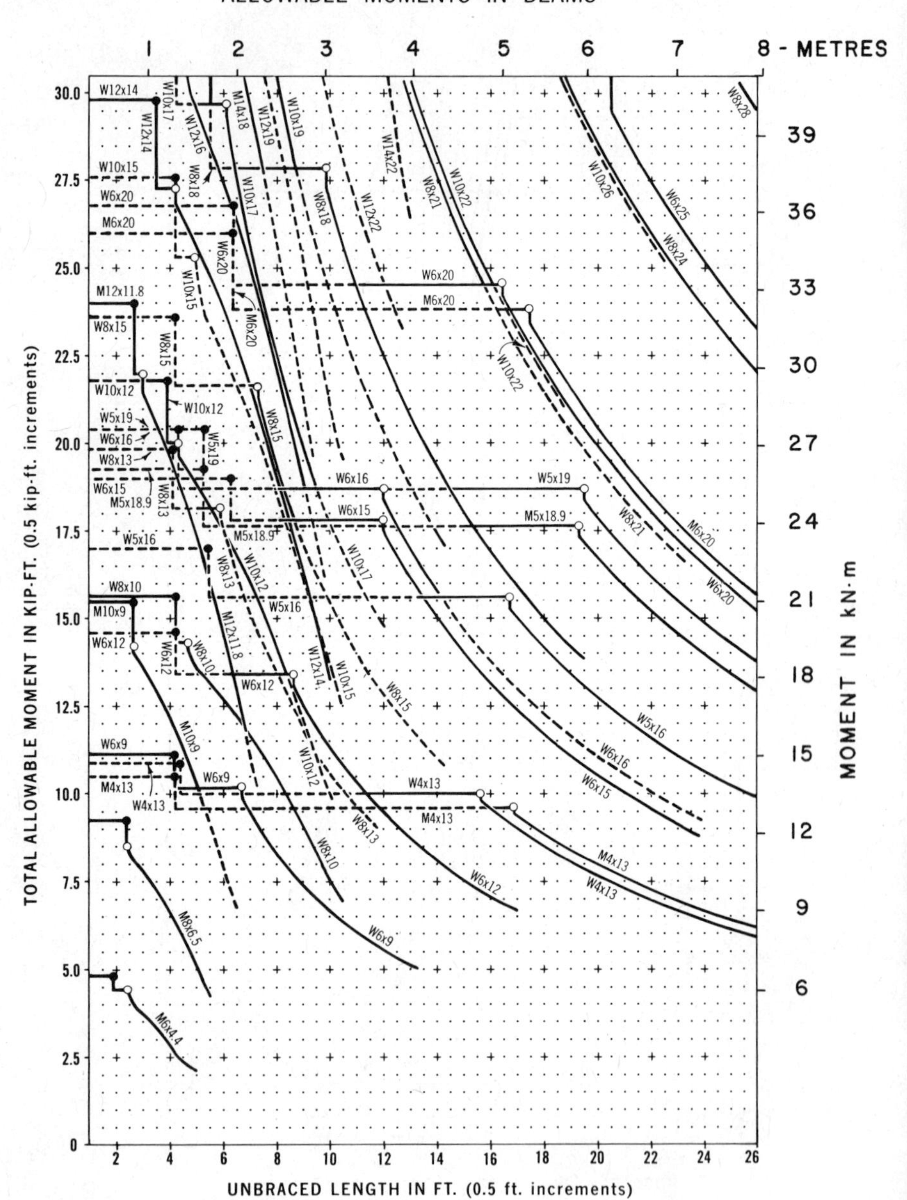

D

Load-Span Values for Beams

The following tables give values for the total allowable uniformly distributed load for simple span beams. Table values are based on a maximum bending stress of 24 ksi [165 MPa]; this assumes a yield stress of 36 ksi [250 MPa], moment due to loading in the plane of the *y*-axis of the section, and adequate conditions of lateral support. Smaller shapes and shapes more commonly used as columns are omitted. Beams are listed in order of the value of their section modulus with respect to the *x*-axis (S_x).

Table values may be used for beams when the distance between points of lateral support does not exceed the L_c value for the shape. These values are given in the table. If the distance between points of lateral support exceeds L_c, the charts in Appendix C should be used instead of these tables.

Table values are not given for spans in excess of 27 times the beam depth, which is the approximate feasible limit based on reasonable limitation of deflection. Deflections in excess of $\frac{1}{360}$ of the span will result for table loads to the right of the heavy vertical lines. Actual values of deflections in inches at the center of the span may be obtained by dividing the deflection factor given for the span by the depth of the beam in inches.

Examples of the use of these tables for common design situations appear in Art. 9-9.

TABLE D-1. Load-Span Values for Beams[a]

Shape	Span (ft) / Deflection factor[b] / $L_c{}^c$ (ft)	8	10	12	14	16	18	20	22	24	26	28	30
		1.59	2.48	3.58	4.87	6.36	8.05	9.93	12.0	14.3	16.8	19.5	22.3
M 8 × 6.5	2.4	9.24	7.39	6.16	5.28	4.62	4.11						
M 10 × 9	2.6	15.5	12.4	10.3	8.87	7.76	6.90	6.21	5.64				
W 8 × 10	4.2	15.6	12.5	10.4	8.92	7.81	6.94						
W 8 × 13	4.2	19.8	15.9	13.2	11.3	9.91	8.81						
W 10 × 12	3.9	21.8	17.4	14.5	12.5	10.9	9.69	8.72	7.93				
W 8 × 15	4.2	23.6	18.9	15.7	13.5	11.8	10.5						
M 12 × 11.8	2.7	24.0	19.2	16.0	13.7	12.0	10.7	9.60	8.73	8.00	7.38	6.86	
W 10 × 15	4.2	27.6	22.1	18.4	15.8	13.8	12.3	11.0	10.0				
W 12 × 14	3.5	29.8	23.8	19.9	17.0	14.9	13.2	11.9	10.8	9.93	9.17	8.51	
W 8 × 18	5.5	30.4	24.3	20.3	17.4	15.2	13.5						
W 10 × 17	4.2	32.4	25.9	21.6	18.5	16.2	14.4	13.0	11.8				
W 12 × 16	4.1	34.2	27.4	22.8	19.5	17.1	15.2	13.7	12.4	11.4	10.5	9.77	
W 8 × 21	5.6	36.4	29.1	24.3	20.8	18.2	16.2						
W 10 × 19	4.2	37.6	30.1	25.1	21.5	18.8	16.7	15.0	13.7				
W 8 × 24	6.9	41.8	33.4	27.9	23.9	20.9	18.6						
M 14 × 18	3.6	42.2	33.8	28.1	24.1	21.1	18.7	16.9	15.3	14.1	13.0	12.0	11.2
W 12 × 19	4.2	42.6	34.1	28.4	24.3	21.3	18.9	17.0	15.5	14.2	13.1	12.2	
W 10 × 22	6.1	46.4	37.1	30.9	26.5	23.2	20.6	18.5	16.9				
W 8 × 28	6.9	48.6	38.9	32.4	27.8	24.3	21.6						

TABLE D-1. *(Continued)*

Shape	Span (ft)	12	14	16	18	20	22	24	26	28	30	32	34
	$L_c{}^c$ (ft) / Deflection factor[b]	3.58	4.87	6.36	8.05	9.93	12.0	14.3	16.8	19.5	22.3	25.4	28.7
W 12 × 22	4.3	33.9	29.0	25.4	22.6	20.3	18.5	16.9	15.6	14.5			
W 10 × 26	6.1	37.2	31.9	27.9	24.8	22.3	20.3						
W 14 × 22	5.3	38.7	33.1	29.0	25.8	23.2	21.1	19.3	17.8	16.6	15.5	14.5	
W 10 × 30	6.1	43.2	37.0	32.4	28.8	25.9	23.6						
W 12 × 26	6.9	44.5	38.2	33.4	29.7	26.7	24.3	22.3	20.5	19.1			
W 10 × 33	8.4	46.7	40.0	35.0	31.0	28.0	25.4						
W 14 × 26	5.3	47.1	40.3	35.3	31.4	28.2	25.7	23.5	21.7	20.2	18.8	17.6	
W 16 × 26	5.6	51.2	43.9	38.4	34.1	30.7	27.9	25.6	23.6	21.9	20.5	19.2	18.1
W 12 × 30	6.9	51.5	44.1	38.6	34.3	30.9	28.1	25.7	23.8	22.0			
W 14 × 30	7.1	56.0	48.0	42.0	37.3	33.6	30.5	28.0	25.8	24.0	22.4	21.0	
W 10 × 39	8.4	56.1	48.1	42.1	37.4	33.7	30.6						
W 12 × 35	6.9	60.8	52.1	45.6	40.5	36.5	33.2	30.4	28.1	26.0			
W 16 × 31	5.8	62.9	53.9	47.2	41.9	37.8	34.3	31.5	29.0	27.0	25.2	23.6	22.2
W 14 × 34	7.1	64.8	55.5	48.6	43.2	38.9	35.3	32.4	29.9	27.8	25.9	24.3	
W 10 × 45	8.5	65.5	56.1	49.1	43.6	39.3	35.7						

TABLE D-1. (*Continued*)

Shape	Span (ft)	16	18	20	22	24	26	28	30	32	34	36	38
	Deflection factor[b] / L_c^c (ft)	6.36	8.05	9.93	12.0	14.3	16.8	19.5	22.3	25.4	28.7	32.2	35.9
W 12 × 40	8.4	51.9	46.1	41.5	37.7	34.6	31.9	29.6					
W 14 × 38	7.1	54.6	48.5	43.7	39.7	36.4	33.6	31.2	29.1	27.3			
W 16 × 36	7.4	56.5	50.2	45.2	41.1	37.7	34.8	32.3	30.1	28.2	26.6	25.1	
W 18 × 35	6.3	57.8	51.4	46.2	42.0	38.5	35.6	33.0	30.8	28.9	27.2	25.7	24.3
W 12 × 45	8.5	58.1	51.6	46.5	42.2	38.7	35.7	33.2					
W 14 × 43	8.4	62.7	55.7	50.1	45.6	41.8	38.6	35.8	33.4	31.3			
W 12 × 50	8.5	64.7	57.5	51.7	47.0	43.1	39.8	37.0					
W 16 × 40	7.4	64.7	57.5	51.7	47.0	43.1	39.8	37.0	34.5	32.3	30.4	28.7	
W 18 × 40	6.3	68.4	60.8	54.7	49.7	45.6	42.1	39.1	36.5	34.2	32.2	30.4	28.8
W 14 × 48	8.5	70.3	62.5	56.2	51.1	46.9	43.3	40.2	37.5	35.1			
W 12 × 53	10.6	70.6	62.7	56.5	51.3	47.1	43.4	40.3					
W 16 × 45	7.4	72.7	64.6	58.2	52.9	48.5	44.7	41.5	38.8	36.3	34.2	32.3	
W 14 × 53	8.5	77.8	69.1	62.2	56.6	51.9	47.9	44.4	41.5	38.9			
W 18 × 46	6.4	78.8	70.0	63.0	57.3	52.5	48.5	45.0	42.0	39.4	37.1	35.0	33.2
W 16 × 50	7.5	81.0	72.0	64.8	58.9	54.0	49.8	46.3	43.2	40.5	38.1	36.0	

TABLE D-1. *(Continued)*

Shape	Span (ft)	16	18	20	22	24	27	30	33	36	39	42	45
	Deflection factor[b] / $L_c{}^c$ (ft)	6.36	8.05	9.93	12.0	14.3	18.1	22.3	27.0	32.2	37.8	43.8	50.3
W 21 × 44	6.6	81.6	72.5	65.3	59.3	54.4	48.3	43.5	39.6	36.3	33.5	31.1	29.0
W 18 × 50	7.9	88.9	79.0	71.1	64.6	59.3	52.7	47.4	43.1	39.5	36.5		
W 14 × 61	10.6	92.2	81.9	73.8	67.0	61.5	54.6	49.2	44.7				
W 16 × 57	7.5	92.2	81.9	73.8	67.0	61.5	54.6	49.2	44.7	41.0			
W 21 × 50	6.9	94.5	84.0	75.6	68.7	63.0	56.0	50.4	45.8	42.0	38.8	36.0	33.6
W 18 × 55	7.9	98.3	87.4	78.6	71.5	65.5	58.2	52.4	47.7	43.7	40.3		
W 18 × 60	8.0	108	96.0	86.4	78.5	72.0	64.0	57.6	52.4	48.0	44.3		
W 21 × 57	6.9	111	98.7	88.6	80.7	74.0	65.8	59.2	53.8	49.3	45.5	42.3	39.5
W 24 × 55	7.0	114	101	91.2	82.9	76.0	67.5	60.8	55.3	50.7	46.8	43.4	40.5
W 16 × 67	10.8	117	104	93.6	85.1	78.0	69.3	62.4	56.7	52.0			
W 18 × 65	8.0	117	104	93.6	85.1	78.0	69.3	62.4	56.7	52.0	48.0		
W 18 × 71	8.1	127	113	102	92.4	84.7	72.2	67.7	61.5	56.4	52.1		
W 21 × 62	8.7	127	113	102	92.4	84.7	72.2	67.7	61.5	56.4	52.1	48.4	45.1
W 24 × 62	7.4	131	116	105	95.3	87.3	77.6	69.9	63.5	58.2	53.7	49.9	46.6
W 16 × 77	10.9	134	119	107	97.4	89.3	79.4	71.5	65.0	59.5			
W 21 × 68	8.7	140	124	112	102	93.3	83.0	74.7	67.9	62.2	57.4	53.3	49.8
W 18 × 76	11.6	146	130	117	106	97.3	86.5	77.9	70.8	64.9	59.9		
W 21 × 73	8.8	151	134	121	110	101	89.5	80.5	73.2	67.1	61.9	57.5	53.7
W 24 × 68	9.5	154	137	123	112	103	91.2	82.1	74.7	68.4	63.2	58.7	54.7
W 18 × 86	11.7	166	147	133	121	111	98.4	88.5	80.5	73.8	68.1		
W 21 × 83	8.8	171	152	137	124	114	101	91.2	82.9	76.0	70.1	65.1	60.8

TABLE D-1. *(Continued)*

Shape	Span (ft)	24	27	30	33	36	39	42	45	48	52	56	60
	Deflection factor[b] / $L_c{}^c$ (ft)	14.3	18.1	22.3	27.0	32.2	37.8	43.8	50.3	57.2	67.1	77.9	89.4
W 24 × 76	9.5	117	104	93.9	▌85.3	78.2	72.2	67.0	62.6	58.7			
W 21 × 93	8.9	128	114	▌102	93.1	85.3	78.8	73.1	68.3				
W 24 × 84	9.5	131	116	104	▌95.0	87.1	80.4	74.7	69.7	65.3			
W 27 × 84	10.5	142	126	114	103	94.7	87.4	81.1	75.7	71.0	65.5	60.8	
W 24 × 94	9.6	148	131	118	▌108	98.7	91.1	84.6	78.9	74.0			
W 21 × 101	13.0	151	134	▌121	110	101	93.1	86.5	80.7				
W 27 × 94	10.5	162	144	130	113	108	▌99.7	92.6	86.4	81.0	74.8	69.4	
W 24 × 104	13.5	172	153	138	▌125	115	106	98.3	91.7	86.0			
W 27 × 102	10.6	178	158	142	129	119	▌109	102	94.9	89.0	82.1	76.3	
W 30 × 99	10.9	179	159	143	130	120	110	▌102	95.6	89.7	82.8	76.9	71.7
W 24 × 117	13.5	194	172	155	▌141	129	119	111	103	97.0			
W 27 × 114	10.6	199	177	159	145	133	▌123	114	106	99.7	92.0	85.4	
W 30 × 108	11.1	199	177	159	145	133	123	▌114	106	99.7	92.0	85.4	79.7
W 30 × 116	11.1	219	195	175	159	146	135	▌125	117	110	101	94.0	87.7
W 30 × 124	11.1	237	210	189	172	158	146	▌135	126	118	109	101	94.7

TABLE D-1. *(Continued)*

Shape	Span (ft)	30	33	36	39	42	45	48	52	56	60	65	70
	Deflection factor[b] / L_c[c] (ft)	22.3	27.0	32.2	37.8	43.8	50.3	57.2	67.1	77.9	89.4	105	122
W 33 × 118	12.0	191	174	159	147	137	128	120	110	103	95.7	88.4	
W 30 × 132	11.1	203	184	169	156	145	135	127	117	109	101		
W 33 × 130	12.1	216	197	180	166	155	144	135	125	116	108	99.9	
W 27 × 146	14.7	219	199	183	169	156	146	137	126	117			
W 36 × 135	12.3	234	213	195	180	167	156	146	135	125	117	108	100
W 33 × 141	12.2	239	217	199	184	171	159	149	138	128	119	110	
W 33 × 152	12.2	260	236	216	200	185	173	162	150	139	130	120	
W 36 × 150	12.6	269	244	224	207	192	179	168	155	144	134	124	115
W 30 × 173	15.8	287	261	239	221	205	192	180	166	154	144		
W 36 × 160	12.7	289	263	241	222	206	193	181	167	155	144	133	124
W 36 × 170	12.7	309	281	258	238	221	206	193	178	166	155	143	132
W 30 × 191	15.9	319	290	268	245	228	213	199	184	171	159		
W 36 × 182	12.7	332	302	277	256	237	221	208	192	178	166	153	142
W 36 × 194	12.8	354	322	295	272	253	236	221	204	190	177	163	152
W 33 × 201	16.6	365	332	304	281	260	243	228	210	195	182	168	
W 36 × 210	12.9	383	349	319	295	274	256	240	221	205	192	177	164
W 33 × 221	16.7	404	367	336	310	288	269	252	233	216	202	186	
W 33 × 241	16.7	442	402	368	340	316	295	276	255	237	221	204	
W 36 × 230	17.4	446	406	372	343	319	298	279	257	239	223	206	191
W 36 × 245	17.4	477	434	398	367	341	318	298	275	256	239	220	204
W 36 × 260	17.5	508	462	423	391	363	339	318	293	272	254	234	218
W 36 × 280	17.5	549	499	458	422	392	366	343	317	294	275	253	235
W 36 × 300	17.6	592	538	493	455	423	395	370	341	317	296	273	254

[a] Total allowable uniformly distributed load in kips for simple span beams with yield stress of 36 ksi [250 MPa]. Loads to the right of the heavy vertical lines will cause deflections in excess of $\frac{1}{360}$ of the span.

[b] Maximum deflection in inches at the center of the span may be obtained by dividing the factor given by the depth of the beam in inches.

[c] Maximum permitted distance between points of lateral support. For greater distances use the charts in Appendix C to obtain the required beam size.

Answers to Selected Problems

The answers given here are for those problems marked with an asterisk (*) in the text.

Chapter 3

3.3.A 12,160 psi [83 MPa]
3.3.C 1.94, or 2 in. [49.8, or 50 mm]
3.5.B 10,938 psi [75.4 MPa]
3.6.B 29,400,000 psi [210 GPa]
3.8.A Stress is 28,294 psi [196 MPa], rod not sufficient
3.8.E 0.004 in. [0.1 mm]

Chapter 5

5.6.C R_1 = 2620 lb [11.589 kN], R_2 = 2980 lb [13.051 kN]
5.6.E R_1 = 4103 lb [18.128 kN], R_2 = 7997 lb [35.276 kN]
5.9.D Maximum shear = 14 kips [61.7 kN]; zero shear at 6 ft [1.8 m] from left end
5.9.E Maximum shear = 3333 lb [14.60 kN]; zero shear at left support and at 5.33 ft [1.60 m] from right support

5.12.A Maximum moment = 48 kip-ft [64.08 kN-m]

5.19.C Maximum shear = 7290 lb [32.4 kN], maximum moment = 42,100 ft-lb [56.1 kN-m], inflection at 1.23 ft [0.37 m] to left of right support

Chapter 6

6.4.A c' = 1.65625 in. [41.375 mm] to bottom, c = 3.34375 in. [83.625 mm] to top

6.9.A I = 1120 in.^{4} [466 × 10^6 mm^4], S = 97.4 in.^{3} [1596 × 10^3 mm^3] (bottom), S = 166 in.^{3} [2721 × 10^3 mm^3] (top)

6.9.C I = 9.99 in.^{4} [4.041 × 10^6 mm^4], S = 2.99 in.^{3} [48.32 × 10^3 mm^3] (top), r = 1.58 in. [39.4 mm]

Chapter 7

7.2.A Yes, maximum f_b = 22 ksi [152 MPa]

7.2.C (a) W = 8.8 kips [39 kN], (b) W = 4.6 kips [20 kN]

7.4.B W 21 × 44

7.4.C W 14 × 30

Chapter 8

8.4.A 0.653 in. [17.4 mm]

Chapter 9

9.3.A Beam is OK; W 18 × 76 is lightest permitted

9.3.B W 30 × 99 is lightest but deflects too much; required beam is W 30 × 108

9.5.A (a) 80.4 kips [358 kN]

9.7.A Pl 1 × 8 × 15 in. [25 × 200 × 385 mm]

9.9.A (a) M 10 × 9, (b) W 8 × 10

9.9.E (a) W 18 × 35, deflection = 0.98 in.; (b) W 14 × 48, deflection = 1.034 in.

9.11.A W 10 × 22 is lightest
9.11.B W 16 × 40 is lightest

Chapter 10

10.9.A 24 H 8
10.9.C 20 H 6 is lightest, but total load deflection is too large; use 22 H 6

Chapter 11

11.3.A $L/r = 95$
11.6.A 430 kips [1912 kN]
11.6.C 3612 kips [16065 kN]
11.7.A W 8 × 31, allowable load is 149 kips
11.9.B TS 4 × 4 × $\frac{1}{4}$
11.10.A 78 kips [347 kN]
11.13.D W 14 × 61
11.13.H 137 kips [609 kN]
11.13.J (approximately) *PL* 1.25 × 15 × 16 in. [32 × 380 × 410 mm]

Chapter 12

12.8.A 6 bolts, outer plate $\frac{7}{8}$ in. [20 mm], inner plate 1 in. [24 mm]
12.10.A (a) 8 bolt with $\frac{3}{4}$-in. bolts, 6 bolt with $\frac{7}{8}$-in. bolts, 5 bolt with 1-in. bolts; (b) bearing, 284 kips [1263 kN]; shear, 336 kips [1494 kN]; 189 kips [841 kN]

Chapter 13

13.5.A $L_1 = 11$ in. [265 mm], $L_2 = 5$ in. [110 m]

Index